Aggregation and Fusion
of Imperfect Information

Studies in Fuzziness and Soft Computing

Editor-in-chief
Prof. Janusz Kacprzyk
Systems Research Institute
Polish Academy of Sciences
u. Newelska 6
01-447 Warsaw, Poland
E-mail: kacprzyk @ ibspan.waw.pl

Vol. 1. J. Kacprzyk and M. Fedrizzi (Eds.)
Fuzzy Regression Analysis, 1992
ISBN 3-7908-0591-2 (ISBN 83-85262-07-5)

Vol. 2. A.M. Delgado et al. (Eds.)
Fuzzy Optimization, 1994
ISBN 3-7908-0749-4

Vol. 3. A. Geyer-Schulz
Fuzzy Rule-Based Expert Systems and Genetic Machine Learning, 2nd ed. 1996
ISBN 3-7908-0964-0

Vol. 4. T. Onisawa and J. Kacprzyk (Eds.)
Reliability and Safety Analyses under Fuzziness, 1995
ISBN 3-7908-0837-7

Vol. 5. P. Bosc and J. Kacprzyk (Eds.)
Fuzziness in Database Management Systems, 1995
ISBN 3-7908-0858-X

Vol. 6. E.S. Lee and Q. Zhu
Fuzzy and Evidence Reasoning, 1995
ISBN 3-7908-0880-6

Vol. 7. B.A. Juliano and W. Bandler
Tracing Chains-of-Thought, 1996
ISBN 3-7908-0922-5

Vol. 8. F. Herrera and J.L. Verdegay (Eds.)
Genetic Algorithms and Soft Computing, 1996
ISBN 3-7908-0956-X

Vol. 9. M. Sato et al.
Fuzzy Clustering Models and Applications, 1997
ISBN 3-7908-1026-6

Vol. 10. L.C. Jain (Ed.)
Soft Computing Techniques in Knowledge-based Intelligent Engineering Systems, 1997
ISBN 3-7908-1035-5

Vol. 11. W. Mielczarski (Ed.)
Fuzzy Logic Techniques in Power Systems, 1998
ISBN 3-7908-1044-4

Bernadette Bouchon-Meunier (Ed.)

Aggregation and Fusion of Imperfect Information

With 60 Figures
and 3 Tables

Physica-Verlag
A Springer-Verlag Company

Dr. Bernadette Bouchon-Meunier
Centre National de la Recherche Scientifique
LIP6 – Laboratoire d'Informatique de Paris 6
Université Pierre et Marie Curie
Case 169
4, Place Jussieu
F-75252 Paris Cedex 05, France

ISBN 3-7908-1048-7 Physica-Verlag Heidelberg New York

Library of Congress Cataloging-in-Publication Data
Die Deutsche Bibliothek – CIP-Einheitsaufnahme
Aggregation and fusion of imperfect information: with 3 tables / Bernadette Bouchon-Meunier (Ed.). – Heidelberg: Physica-Verl., 1998
 (Studies in fuzziness and soft computing; Vol. 12)
 ISBN 3-7908-1048-7

This work is subject to copyright. All rights are reserved, whether the whole or part of the material is concerned, specifically the rights of translation, reprinting, reuse of illustrations, recitation, broadcasting, reproduction on microfilm or in any other way, and storage in data banks. Duplication of this publication or parts thereof is permitted only under the provisions of the German Copyright Law of September 9, 1965, in its current version, and permission for use must always be obtained from Physica-Verlag. Violations are liable for prosecution under the German Copyright Law.

© Physica-Verlag Heidelberg 1998
Printed in Germany

The use of general descriptive names, registered names, trademarks, etc. in this publication does not imply, even in the absence of a specific statement, that such names are exempt from the relevant protective laws and regulations and therefore free for general use.

Hardcover Design: Erich Kirchner, Heidelberg

SPIN 10640242   88/2202-5 4 3 2 1 0 – Printed on acid-free paper

Preface

Aggregation and fusion of information are major problems for all kinds of knowledge based systems, from image processing to decision making, from pattern recognition to automatic learning. The two words are often used for the same general purpose: how to use simultaneously pieces of information provided by several sources in order to come to a conclusion or a decision. Nevertheless, there exist two general approaches to this general scheme, depending on the problem to be dealt with.

The first one corresponds to the *aggregation of preferences* given by several individuals of a group or the aggregation of criteria to satisfy in order to make a decision. In both cases, a consensus is looked for, which means that a solution of the problem must satisfy as much as possible individuals or criteria.

The second approach concerns the *fusion of evidence* provided by several sources, exemplified by cameras or sensors, regarding a unique object. In this framework, complementary information is used to construct a global description of the object from partial views and to eliminate inconsistencies from the elementary descriptions.

In many cases, the available information is imperfect which means that some of its elements are *uncertain* and/or *imprecise*, and some of them are missing (information is then *incomplete*). This imperfection can be caused by the relative accuracy or physical limitations of devices, such as sensors, or the reliability of agents providing information, by the difficulty for an individual to express crisp preferences or to precisely describe an object. It can also be a component of the object itself, with limits which are not well defined (the limits of a forest, for instance), properties (color, size, ...) which can change from one situation to another, similarities between different objects which are difficult to analyse. Finally, imperfection comes also from the environment, with objects hidden by something, lure pervading images, etc,

Several methodologies are useful to manage such imperfect information. Among the most important ones are *probability theory, evidence theory, fuzzy set theory and possibility theory*. In this book, all these methodologies are represented, with more developments on fuzzy set based approaches because of their importance in works on aggregation and fusion of information, due to their ability to manage several kinds of imperfections and to the diversity of available tools.

The book contains both theoretical and applied studies of aggregation methods in all the sections. The first one is essentially formal, the two next ones are oriented towards decision making and control, and fusion of complementary information provided by various sources, respectively.

The first section is devoted to the presentation of properties of *aggregation operators for fusion under fuzziness*, which can be used for several kinds of purpose. The first paper, by S. Ovchinnikov, addresses an important concept of *robustness* of aggregation procedures. The second one, by R. Mesiar and M. Komorníková, discusses the *compensatory behavior* of some aggregation operators based on triangular norms and conorms. The third paper, by J. Fodor and T. Calvo, studies properties *of associativity and idempotency* of a class of aggregation operators based on triangular norms and conorms.

The second section is more dedicated to the potential utilization of aggregation methods in *decision making and control*, even though it contains theoretical studies of the proposed tools. The first paper, by M. Grabisch, is a comprehensive study of *fuzzy integral* with respect to fuzzy measures, including the key problems of interpretation and identification of fuzzy measures. Then, A. Kelman and R.R. Yager present methods to manage *priorities between criteria*, on the basis of a possibilistic approach and OWA operators, with details about the parameter learning process. The next paper, by V. Cutello and J. Montero, studies aggregation operators for rational fuzzy preferences through *fuzzy rationality measures*. While all the previous works deal with fuzzy set based techniques, M.T. Lamata's paper presents the general framework of aggregation of evidence in *Dempster-Shafer belief theory*. The last paper of this section, by J. Kacprzyk, develops a multistage fuzzy control model with a *fuzzy linguistic quantifier based aggregation* of stage scores.

In the third section of the book, *heterogenous information* is supposed to be provided by several sources on a given object or phenomenon and a phase of fusion is necessary to obtain a unified view of this object or phenomenon. First, S. Benferhat, D. Dubois and H. Prade are interested in using *possibilistic logic* to represent and manage uncertainty in knowledge bases and they propose solutions to take into account priorities of expert opinions, reliability of sources, conflict between sources, etc. in a semantic approach as well as its syntactic counterpart. A *probabilistic point of view* is preferred by S. Moral and J. del Sagrado to represent expert uncertainties and they present general methods of probability consensus, employing the *Dempster-Shafer theory of evidence* to measure expert reliability.

The other papers of this section are devoted to important fields of applications. The case of fusion of *image information* is addressed by I. Bloch and H. Maître, who give a review of techniques available in the *three main approaches* to manage uncertainty at all the basic levels of image processing, exploring also the field of spatial information in images. The next paper, by G. Mauris, E. Benoit and L. Foulloy, presents an artificial intelligent approach to the aggregation of information provided by *multisensor systems*, using a *fuzzy set based* knowledge representation, and it shows an application of the proposed methods to the perception of environment by robot sensors. Multisensor analysis is also the topic of A. Appriou's paper who focuses on *target classification and tracking* procedures using *evidence theory* to aggregate observations delivered by a set of disparate and possibly delocalized sensors. The last paper, by M. Sato and Y. Sato, tackles *clustering* problems and proposes a model based on *fuzzy similarities*.

Finally, we express our thanks to all the authors for their cooperation.

Paris, March 1997   B. Bouchon-Meunier

Table of contents

Preface v
    B. Bouchon-Meunier

1. AGGREGATION OPERATORS FOR FUSION UNDER FUZZINESS

On robust aggregation procedures 3
    S. Ovchinnikov

Triangular norm-based aggregation of evidence under fuzziness 11
    R. Mesiar and M. Komorniková

Aggregation functions defined by t-norms and t-conorms 36
    J. Fodor and T. Calvo

2. AGGREGATION IN DECISION MAKING AND CONTROL

Fuzzy integral as a flexible and interpretable tool of aggregation 51
    M. Grabisch

Using priorities in aggregation connectives 73
    A. Kelman and R.R. Yager

Aggregation operators for fuzzy rationality measures 98
    V. Cutello and J. Montero

Aggregation in decision making with belief structures 106
    M.T. Lamata

Multistage fuzzy control with a soft aggregation of stage scores 118
    J. Kacprzyk

3. FUSION OF COMPLEMENTARY INFORMATION

From semantic to syntactic approaches to information combination in possibilistic logic 141
    S. Benferhat, D. Dubois and H. Prade

Aggregation of imprecise probabilities 162
    S. Moral and J. del Sagrado

Fusion of image information under imprecision 189
    I. Bloch and H. Maître

*Fuzzy linguistic methods for the aggregation of complementary
sensor information* 214
    G. Mauris, E. Benoit and L. Foulloy

Uncertain data aggregation in classification and tracking processes 231
    A. Appriou

*A generalized fuzzy clustering model based on aggregation operators
and its applications* 261
    M. Sato and Y. Sato

AGGREGATION OPERATORS FOR FUSION UNDER FUZZINESS

On Robust Aggregation Procedures

Sergei Ovchinnikov

Mathematics Department, San Francisco State University
San Francisco, CA 94132

Abstract. A notion of a robust aggregation procedure is introduced and studied in connection with invariance properties of aggregation procedures. We prove that a robust aggregation procedure is invariant if measurements are in the ordinal scale. We also show that any robust mean on the set of all fuzzy sets on a finite universe is a pointwise order statistic.

1. Introduction

In a rather informal way, the aggregation problem is that of aggregating n-tuples of objects all belonging to a given set X, into a single object from the same set.

We begin with a special case when X is the set $\mathbb{R}$ of all real numbers, although our main results are valid in much more general settings. Thus we assume that our objects are represented by real numbers; in other words, we deal with *measurements* of real objects rather than with objects themselves.

In this setting, an aggregation procedure is simply a function

$$y = M(x_1, x_2, \ldots, x_n)$$

which assigns a real number y to any n-tuple $(x_1, x_2, \ldots, x_n)$ of real numbers. Naturally, we should impose certain conditions on M to justify the name "aggregation procedure". First we assume that M is a *symmetric* function of its arguments. This property is also known in social sciences as *anonymity* condition. We also assume that M satisfies

$$\min\{x_1, x_2, \ldots, x_n\} \leq M(x_1, x_2, \ldots, x_n) \leq \max\{x_1, x_2, \ldots, x_n\}$$

for all n-tuples $(x_1, x_2, \ldots, x_n)$. In social sciences this is the *Pareto* property. Here we prefer to say (following Cauchy [1]) that M is a *mean*. We shall also assume that M is a continuous function. Yager's OWA operators [8] are good examples of aggregation procedures satisfying the above conditions.

Since real numbers $x_1, x_2, \ldots, x_n$ are measurements of certain objects, we should specify a *scale* in which these measurements were performed. Moreover, we want the aggregation function M to represent a *meaningful* relation with respect to the given scale. The notion of *mearingfulness* is formalized in the representational theory of measurement [3] as the *invariance* property:

For any *admissible* transformation $f : \mathbb{R} \to \mathbb{R}$,

$$f(y) = M(f(x_1), f(x_2), \ldots, f(x_n))$$

is equivalent to

$$y = M(x_1, x_2, \ldots, x_n).$$

There are plenty of means in mathematics and statistics. One of the most popular symmetric means is the *arithmetic mean*

$$y = \frac{x_1 + x_2 + \ldots + x_n}{n}.$$

It is well known [3] that this mean is a meaningful relation in the framework of interval scales, but it is a not a meaningful relation with respect to ordinal scales. The following theorem [4, 6, 7] provides a complete description of all meaningful means in ordinal scales. By definition, the *p'th order statistic* $x^{(p)}$ is obtained by arranging the n-tuple $(x_1, x_2, \ldots, x_n)$ in the increasing order

$$x^{(1)} \leq x^{(2)} \leq \ldots \leq x^{(n)}.$$

Theorem. Suppose M is a continuous symmetric mean which is invariant with respect to all increasing bijections of $\mathbb{R}$ onto itself. Then M is an order statistic.

In what follows we assume that all measurements are in the ordinal scale. We shall denote by $A(\mathbb{R})$ the set of all increasing bijections of $\mathbb{R}$ onto itself (*admissible transformations* [3]). This set is a group with respect to the operation of function composition. The identity function is the identity element of $A(\mathbb{R})$, and the inverse function f^{-1} is the inverse of $f \in A(\mathbb{R})$.

In this paper we introduce and study the *robustness* property of means which is, roughly speaking, a restricted version of the invariance property. Namely, suppose that, in the definition of invariance, we can apply only transformations that are "very close" to the identity transformation. In other words, we can change our scales only "a little bit". To formalize this idea we need to introduce a topology on $A(\mathbb{R})$ which makes it a *topological group*.

The paper is organized as follows. In Section 2 we first introduce formal definitions of invariance and robustness properties in a rather general setting. Then we show that in the particular case of ordinal measurements, these two properties are equivalent and discuss some generalizations of this result. Section 3 presents mathematical foundations for our models.

2. Invariance and Robustness

Let X be a set on which we want to define a mean M. We want this mean to be a continuous function satisfying the Pareto condition. Therefore we assume that X is a *lattice* and a *topological space* with respect to the lattice topology. We develop this framework bearing in mind the following two special cases:

1. $X = \mathbb{R}$. Aggregation in the ordinal scale.

2. $X = [0,1]^m$. Aggregation of fuzzy sets on a finite universe.

In both cases, the lattice and topological structures are defined naturally.

Definition 2.1. A *mean* M on X is a symmetric continuous function with values in X satisfying

$$\bigwedge_i x_i \leq M(x_1, x_2, \ldots, x_n) \leq \bigvee_i x_i$$

for all n-tuples $(x_1, x_2, \ldots, x_n)$ with elements from X.

To formalize the notion of invariance, we introduce the group $A(X)$ of all admissible transformations of X. The elements of this group are bijections $f : X \to X$ that preserve the lattice structure on X.

Definition 2.2. A mean M on X is *invariant* if

$$M(f(x_1), f(x_2), \ldots, f(x_n)) = f(M(x_1, x_2, \ldots, x_n))$$

for all n-tuples $(x_1, x_2, \ldots, x_n)$ and all transformations $f \in A(X)$.

The invariance condition greatly restricts the set of means on X. The following theorem was established in [7] for a wide class of topological lattices X including $X = \mathbb{R}$ and $X = [0,1]^m$.

Theorem 2.1. An invariant mean on X is an order statistic.

The set $A(X)$ of all admissible transformations is a huge set. For instance, in the case $X = \mathbb{R}$ it consists of all strictly increasing functions mapping $\mathbb{R}$ onto itself. What happens if we consider only "small" admissible transformations from $A(X)$ which are "near" the identity transformation? In mathematics, the notion of "nearness" is formalized by introducing a topological structure on the set under consideration. In our case it is possible to introduce a topology on $A(X)$ which makes this set a *topological group*. In the case of $A(\mathbb{R})$, we present a detailed study of topological properties in Section 3. Here we just assume that $A(X)$ is endowed with topology consistent with the topological structure of X. Then transformations that are "close" to the identity transformation belong to some *neighborhood* V of the identity. These observations motivate the following definition.

Definition 2.3. A mean M on X is V-*robust* if

$$M(f(x_1), f(x_2), \ldots, f(x_n)) = f(M(x_1, x_2, \ldots, x_n))$$

for all n-tuples $(x_1, x_2, \ldots, x_n)$ and all transformations $f \in V$ where V is a neighborhood of the identity in $A(X)$.

Obviously, any invariant mean is robust. It turns out that in the case $X = \mathbb{R}$ the converse is also true. Namely, we have the following theorem.

Theorem 2.2. Let V be a neighborhood of the identity in $A(\mathbb{R})$. Any V-robust mean on $\mathbb{R}$ is invariant.

Proof. It is shown in Section 3 (Theorem 3.6) that $A(\mathbb{R})$ is generated by V. It means that for a given $f \in A(\mathbb{R})$ there are transformations $f_1, f_2, \ldots, f_k \in V$ such that

$$f(x) = (f_1 f_2 \cdots f_k)(x)$$

for all $x \in A(\mathbb{R})$. Since

$$M(f_i(x_1), f_i(x_2), \ldots, f_i(x_n)) = f_i(M(x_1, x_2, \ldots, x_n))$$

for all i, $1 \leq i \leq k$, we conclude that

$$M(f(x_1), f(x_2), \ldots, f(x_n)) = f(M(x_1, x_2, \ldots, x_n)). \qquad \square$$

Note, that in this proof we do not use the fact that M is a mean. Actually, the result is true for any function on $\mathbb{R}$. Namely, any function on $\mathbb{R}$ is V-robust if and only if it is invariant.

The situation is already quite different in the case $X = [0,1]^m$ if $m > 1$. It can be shown that in this case the group $A(X)$ is not connected and therefore is not generated by some neighborhoods of the identity. Actually, the class of V-robust means depends on the neighborhood V and is much larger than the class of invariant means. Although it is possible to describe all V-robust means in this case, we consider here only the most interesting case when V is a "sufficiently small" neighborhood of the identity.

We shall use the following model for $X = [0,1]^m$. Let U be a finite set containing m elements. Then X can be viewed as a set of all fuzzy sets with the universe of discourse U; an element $x \in X$ is a fuzzy set with the membership function $x(u)$, $u \in U$. Consider the set of all admissible transformations of X of the form

$$F(x)(u) = f_u(x(u))$$

where, for each $u \in U$, f_u is an increasing bijection of the unit interval onto itself. Thus each f_u belongs to the set $A([0,1])$ of all admissible transformations of the unit interval in the ordinal scale. One can say that these are "pointwise" admissible transformations. It is easy to see that such transformations preserve lattice structure on X and form a group $A^o(X)$ which is a proper subgroup of the group $A(X)$ of all admissible transformations. We have the following theorem.

Theorem 2.3. Let V be a neighborhood of the identity of $A(X)$ such that $V \subseteq A^o(X)$. Then any V-robust mean is $A^o(X)$-robust.

Proof. The group $A^o(X)$ is a direct product of m copies of groups $A([0, 1])$, i.e.,

$$A^o(X) = (A([0, 1]))^m.$$

Using methods developed in Section 3, one can prove that $A([0, 1])$ is a connected topological group. Thus $A^o(X)$ is also connected and therefore is generated by V. Now we can apply the same argument as in the proof of Theorem 2.2 to complete the proof of Theorem 2.3. □

The following result follows immediately from the proof of Theorem 6.1 in [7].

Theorem 2.4. Let

$$y(u) = M(x_1(u), x_2(u), \ldots, x_n(u))$$

be a V-robust mean on $X = [0, 1]^m$, where $V \subseteq A^o(X)$ is a neighborhood of the identity. There exists function $p: U \to \{1, 2, \ldots, m\}$ such that, for any given $u \in U$, $y(u)$ is $p(u)$'th order statistic.

It is proven in [7] that any *invariant* mean on X is a p'th order statistic for some p which does not depend on $u \in U$. Thus, in the case $X = [0, 1]^m$, the class of V-robust means is much wider than the class of invariant means.

3. Topologies on $A(\mathbb{R})$

We consider $\mathbb{R}$ as a topological space with respect to the usual topology. For the general theory of topological function spaces the reader is referred to [2, 5].

For each pair of sets $A \subseteq \mathbb{R}$ and $B \subseteq \mathbb{R}$, we denote (A, B) the set of all functions $f \in A(\mathbb{R})$ such that $f(A) \subseteq B$.

Definition 3.1. The pointwise topology (*p*-topology) $\mathfrak{T}_p$ on $A(\mathbb{R})$ is that having as subbasis all sets $(\{x\}, V)$ where $x \in \mathbb{R}$ and V belongs to the subbasis of the topology on $\mathbb{R}$ consisting of open rays.

Definition 3.2. A topology $\mathfrak{T}$ on $A(\mathbb{R})$ is *admissible* if the *evaluation* mapping $\mathcal{E}: A(\mathbb{R}) \times \mathbb{R} \to \mathbb{R}$ defined by

$$\mathcal{E}(f, x) = f(x)$$

for all $f \in A(\mathbb{R})$ and $x \in \mathbb{R}$ is continuous.

Theorem 3.1. $\mathfrak{T}_p$ is the smallest admissible topology on $A(\mathbb{R})$.

Proof. (a) First we show that $\mathfrak{T}_p$ is admissible. It suffices to show that the inverse image W of any open ray in $\mathbb{R}$ is an open set in $A(\mathbb{R}) \times \mathbb{R}$. Consider an open ray (a, ∞) (the case of open rays in the form $(-\infty, a)$ is treated similarly). Then

$$W = \mathcal{E}^{-1}((a, \infty)) = \{(f, x) : f(x) > a\}$$

Let $(f_0, x_0) \in W$. Then $f_0(x_0) > a$ or, equivalently, $f_0^{-1}(a) < x_0$. There exists b such that $f_0^{-1}(a) < b < x_0$ or, equivalently, $a < f_0(b) < f_0(x_0)$. Then

$$x_0 \in V = (b, \infty) \text{ and } f_0 \in U = \{f : f(b) > a\}.$$

$U \times V$ is an open neighborhood of (f_0, x_0). Suppose $(f, x) \in U \times V$. Then

$$f(x) > f(b) > a$$

implying $U \times V \subseteq W$. Therefore W is an open set.

(b) We prove now that $\mathfrak{T}_p$ is the smallest admissible topology on $A(\mathbb{R})$. Let $\mathfrak{T}$ be an admissible topology. For any given $x \in \mathbb{R}$, the mapping $\mathcal{E}_x : A(\mathbb{R}) \to \mathbb{R}$ defined by $f \mapsto f(x)$ is continuous, since $\mathcal{E}$ is a continuous mapping. Let U be an open set in $\mathbb{R}$. Then

$$\mathcal{E}_x^{-1}(U) = \{f : f(x) \in U\}$$

These sets are open in $\mathfrak{T}$ and form a subbasis for $\mathfrak{T}_p$. Thus $\mathfrak{T}_p \subseteq \mathfrak{T}$. □

Since any topology containing an admissible topology is admissible, we can reformulate the previous theorem as follows.

Theorem 3.2. A topology $\mathfrak{T}$ on $A(\mathbb{R})$ is admissible if and only if $\mathfrak{T} \supseteq \mathfrak{T}_p$.

We now prove that group operations are continuous in p-topology on $A(\mathbb{R})$.

Theorem 3.3. $A(\mathbb{R})$ endowed with p-topology is a topological group.

Proof. We prove first that $f \mapsto f^{-1}$ is a continuous mapping of $A(\mathbb{R})$ onto itself. Let

$$(\{a\}, (b, \infty)) = \{f : f(a) > b\}$$

be an element of subbasis for $\mathfrak{T}_p$. The inverse image of this set is given by

$$\{f : f^{-1}(a) > b\} = \{f : f(b) < a\} = (\{b\}, (-\infty, a))$$

which is an element of the same subbasis. Similarly, the inverse image of $(\{a\}, (-\infty, b))$ is $(\{b\}, (a, \infty))$. Thus $f \mapsto f^{-1}$ is continuous.

Now we prove that the binary group operation in $A(\mathbb{R})$ is continuous. Let $W = (\{a\}, (b, \infty))$ be an element of the subbasis in $A(X)$ (the case of elements in the form $(\{a\}, (-\infty, b))$ is treated similarly) and $h_0 = f_0 g_0$ be an element of W. Then $f_0(g_0(a)) > b$. To prove continuity of the composition operation in $A(\mathbb{R})$ it suffices to find open neighborhoods U and V of f_0 and g_0, respectively, such that, for any $f \in U$ and $g \in V$, $fg \in W$. There is $c \in \mathbb{R}$ such that

$$b < c < f_0(g_0(a)).$$

Then

$$f_0^{-1}(b) < f_0^{-1}(c) < g_0(a).$$

Let $V = (\{a\}, (d, \infty))$ and $U = (\{d\}, (b, \infty))$, where $d = f_0^{-1}(c)$. We have $f_0 \in U$, since $f_0(d) = c > b$, and $g_0 \in V$, since $g_0(a) > d$. Thus U and V are neighborhoods of f_0 and g_0, respectively. Let $f \in U$ and $g \in V$. Then

$$f(g(a)) > f(d) > b$$

implying $fg \in W$. □

We proved that p-topology is the smallest admissible topology on $A(\mathbb{R})$ and $A(\mathbb{R})$ is a topological group with respect to $\mathfrak{T}_p$. Our next result shows that many other topologies that play an important role in the theory of transformation groups coincide with p-topology in the case of groups $A(\mathbb{R})$. First we introduce the following definition [5].

Definition 3.3. Let S be the set of all closed subsets of $\mathbb{R}$. The S-topology $\mathfrak{T}_s$ on $A(\mathbb{R})$ is defined by its subbasis which consists of sets (B, U) were $B \in S$ and U is an open ray in $\mathbb{R}$.

$\mathfrak{T}_s$ is the largest set-open topology which contains the compact-open and pointwise topologies.

The following theorem shows that S-topology coincides with the pointwise topology in the case of group $A(\mathbb{R})$.

Theorem 3.4. $\mathfrak{T}_s = \mathfrak{T}_p$ on $A(\mathbb{R})$.

Proof. Since $\mathfrak{T}_s \supseteq \mathfrak{T}_p$, it suffices to prove that $\mathfrak{T}_s \subseteq \mathfrak{T}_p$. Let $W = (B, (a, \infty))$ be a nonempty element of the subbasis for $\mathfrak{T}_s$ (the case of elements in the form $(B, (-\infty, a))$ is treated similarly). For a given $f \in W$, we have $f(x) > a$ for all $x \in B$. Thus $x > f^{-1}(a)$ for all $x \in B$ and B is bounded below. Let $b = \inf B$. Since B is a closed set, $b \in B$. Since all functions in $A(\mathbb{R})$ are strictly increasing, we have

$$(B, (a, \infty)) = (\{b\}, (a, \infty)).$$

Thus W is open in p-topology and $\mathfrak{T}_s \subseteq \mathfrak{T}_p$. □

Our results show that the pointwise topology is the only set-open topology on $A(\mathbb{R})$ and $A(\mathbb{R})$ is a topological group with respect to this topology. Our next result shows that $A(\mathbb{R})$ is a connected topological group. Actually we prove even more.

Theorem 3.5. $A(\mathbb{R})$ is a path connected topological group.

Proof. For given $f, g \in A(\mathbb{R})$ we define a function $[0, 1] \to A(\mathbb{R})$ by

$$h_t(x) = (1 - t)f(x) + t g(x)$$

for all $t \in [0, 1]$ and $x \in \mathbb{R}$. To prove that this function is continuous, it suffices to prove that the inverse image of any element of the subbasis in $A(\mathbb{R})$ is an open subset in $[0, 1]$.

Let $U = (\{a\}, (b, \infty))$. The inverse image of U is the set
$$\{t \in [0,1] : h_t(a) > b\} = \{t \in [0,1] : (1-t)f(a) + tg(a) > b\},$$
which is obviously open in $[0, 1]$. The case when $U = (\{a\}, (-\infty, b))$ is treated similarly. □

Since any connected topological group is generated by any neighborhood of the identity element, we have the following theorem.

Theorem 3.6. $A(\mathbb{R})$ is generated by any neighborhood of the identity element.

Therefore, for a given neighborhood V of the identity element in $A(\mathbb{R})$ and any function $f \in A(\mathbb{R})$, f can be represented as a finite composition
$$f = f_1 f_2 \cdots f_k$$
of functions from V.

References

1. A.L. Cauchy: *Cours d'analyse de l'École Royale Polytechnique, Ire partie, Analyse algébrique*, Paris (1821)
2. J. Dugundji: *Topology*, Allyn and Bacon, Boston (1965)
3. R.D. Luce, D.H. Krantz, P. Suppes, A. Tversky: *Foundations of Measurement*, Academic Press, New York (1990)
4. J.-L. Marichal, M. Roubens: Characterization of some stable aggregation functions. In: Proceedings of the International Conference on Industrial Engineering and Product Management, Mons 1993, 187-196
5. R.A. McCoy, I. Ntantu: *Topological Properties of Spaces of Continuous Functions*, Lecture Notes in Mathematics, v. 1315, Springer-Verlag (1988)
6. A. Orlov: The connection between mean quantities and admissible transformations. Mathematical Notes 30, 774-778 (1981)
7. S. Ovchinnikov: Means on ordered sets. Mathematical Social Sciences 32, 39-56 (1996)
8. R.R. Yager: Decision making under Dempster-Shafer uncertainties. International Journal of General Systems 20, 233-245 (1992)

Triangular Norm-Based Aggregation of Evidence Under Fuzziness

Radko Mesiar, Magda Komorníková

Department of Mathematics, Faculty of Civil Engineering, Slovak Technical University
SK-81368 Bratislava, Slovakia
and
ÚTIA AV ČR, P. O. Box 18, 18208 Praha 8, Czech Republic

Abstract. Several types of aggregation operators on fuzzy sets are discussed, all of them based on triangular norms and triangular conorms. The main stress is given to the operators allowing to compensate a low input value by another high input value, i. e., to the compensatory operators. We discuss only the pointwise aggregation and the crisp output can be obtained afterwards by any defuzzification procedure. The properties of exponential compensatory operators, convex-linear compensatory operators, convex non-linear compensatory operators, uni-norms and generated aggregative operators are investigated.

Keywords: aggregation, compensation, triangular norm.

1. Introduction

Applications of many intelligent systems are based on the aggregation of incoming data. Recall for example multi-criteria decision making, expert systems, fuzzy controllers, pattern recognition, information retrieval, etc. Especially the recently emerging technologies of neural (fuzzy neural) networks and fuzzy logic require an extensive use of the aggregation process.

The vagueness of incoming data is commonly expressed by means of fuzzy set theory. Recall that the fundamental fuzzy set theory operations of intersection and union are built up pointwisely, i. e., they are based upon aggregations of values in the unit interval. These operations are implemented by multivalued logic conjunctive and disjunctive operators which are characterized in terms of triangular norms (t-norms, in short) and triangular conorms (t-conorms), respectively [9, 13, 14]. However as noticed by Zimmermann and Zysno [23]: "*The interpretation of the intersection of fuzzy sets, computed by applying any t-norm based operator implies that there is no compensation between low and high degrees of membership. If, on the other hand, a decision is defined to be the union of fuzzy sets, represented by some t-conorm, full compensation is assumed. Managerial decisions hardly ever represent any of these extremes.*" To avoid such inaccuracies, several operators on the unit interval based on triangular norm and triangular conorms were suggested.

We recall some of these aggregation operators. Moreover, we propose some new types of aggregation operators based on an additive generator. Further, we investigate and compare some properties of introduced operators.

For the sake of completeness, note that as a result of aggregation procedure of evidence under fuzziness often a crisp output is required. However, applying the introduced aggregation operators to the fuzzy input data (pointwiselly), a fuzzy output will be reached only. Thus a defuzzification procedures should complete our procedures. We do not focus on defuzzification procedure here, so we only recall for interested readers, that basic information on defuzzification procedures can be found e. g. in [16, 17].

2. Triangular norms and conorms

The concept of a triangular norm appeared firstly in the Menger's paper [10] when generalizing the triangular inequality of a metric. The nowadays notion of a triangular norm T and its dual operation triangular t-conorm S is due to Schweizer and Sklar [13].

However, both these operations can be understood also as generalizations of the Boolean logic connectives (and consequently as generalizations of Cantorian set-theoretical operations) from the two-valued set {0, 1 } to the whole unit interval [0,1]. Recall that both Boolean conjunction and disjunction are associative, commutative, non-decreasing binary operations on {0, 1} with neutral elements 1 and 0, respectively. An immediate extension to [0, 1] leads to a tnorm and a t-conorm, respectively. A pointwise extension of t-norms and tconorms leads to the fuzzy intersection and the fuzzy union, respectively.

Note that the strongest t-norm T_M is just the minimum, $T_M(x, y) = min(x, y)$. Each t-norm is determined on the border of the unit square, $T(x, 1) = T(1, x) = x$ and $T(x, 0) = T(0, x) = 0$ for all $x \in [0, 1]$. Hence two t-norms may differ only on the open square $]0,1[^2$ (when looking on T as to a function of two variables). Putting $T(x, y) = 0$ on $]0,1[^2$ we get the weakest t-norm T_W. For each t-norm T it holds

$T_W \leq T \leq T_M$.

In general, the structure of a t-norm cannot be characterized. Recall that even a non-measurable (with respect to Borel subsets) t-norm T can be constructed [5]. However, requiring the continuity of T the following characterization holds [9].

Theorem. *T is a continuous triangular norm if and only if there is a disjoint system $\{]\alpha_k, \beta_k[\}_{k \in K}$ of open subintervals of the unit interval [0, 1] (where the index set K may be also an empty set) and a system $\{f_k\}_{k \in K}$ of continuous strictly decreasing mappings $f_k : [\alpha_k, \beta_k] \to [0, 1]$, $f_k(\beta_k) = 0$, $k \in K$, so that*

$$T(x, y) = \begin{cases} f_k^{-1} (min (f_k(\alpha_k), f_k(x) + f_k(y))) & \text{if } (x, y) \in [\alpha_k, \beta_k]^2 \\ min (x, y) & \text{otherwise} \end{cases}.$$

Note that K empty implies $T = T_M$. Further, let $K = \{1\}$, $\alpha_1 = 0$, $\beta_1 = 1$. Then $f_1(x) = 1 - x$ leads to the Lukasiewicz t-norm $T_L(x, y) = max(0, x + y - 1)$. Arbitrary bounded f_1 leads to a nilpotent t-norm T isomorphic with T_L.

Further, let $f_1(x) = -log(x)$. Then the corresponding t-norm is the usual product, $T_P(x, y) = x.y$. Arbitrary unbounded f_1 leads to a strict t-norm T isomorphic with T_P.

Note that the function f_1 (when $K = \{1\}$ and $\alpha_1 = 0$, $\beta_1 = 1$) is called an additive generator of the corresponding t-norm T and it is unique up to a positive multiplicative constant. Recall that even non-continuous functions f_1 can be used for construction of a t-norm [6]. So, e. g., for

$$f_1(x) = \begin{cases} 2 - x & \text{if } x \in [0, 1[\\ 0 & \text{if } x = 1 \end{cases},$$

we get $T = T_W$ the weakest t-norm, $T_W(x, y) = f_1^{-1}(min(f_1(0), f_1(x) + f_1(y)))$, see also Ling [9].

For more details about t-norms see e. g. [14].

Triangular conorms can be introduced and investigated independently of the concept of triangular norms. However, if the duality operator **D** is applied to a t-conorm **S**,

D(S) (x, y) =1- **S**(1- x,1- y),

the resulting operation **D(S)** is a t-norm **T**. Similarly, **D(T)** is a t-conorm. Recall that the duality operator **D** is involutive, **D(D(T)) = T** and **D(D(S)) = S**. This allows to derive the properties of t-conorms from the corresponding properties of t-norms. So, e. g., if a t-norm **T** has an additive generator f, the corresponding dual t-conorm **S = D(T)** has an additive generator

$g: [0, 1] \to [0, \infty]$, $g(x) = f(1 - x)$, and $S(x, y) = g^{-1}(min(g(1), g(x) + g(y)))$.

When looking on t-norms and t-conorms from the aggregation point of view, recall the objections of Zimmermann and Zysno [23]. Namely, for each x, y $\in$ [0, 1] it is

$T(x, y) \leq min(x, y)$ and $max(x, y) \leq S(x, y)$.

The idempotency property $T(x, x) = x$ and $S(x, x) = x$ is fulfilled only by the limit t-norm T_M and its dual S_M (= max).

For arbitrary generated t-norm T (t-conorm S) it is

$T(x, x) < x$

and

$\lim_{n \to \infty} T(\underbrace{x, ..., x}_{n\text{-times}}) = 0$ for $x < 1$

($S(x, x) > x$ and $\lim_{n \to \infty} S(\underbrace{x, ..., x}_{n\text{-times}}) = 1$ for $x > 0$).

Example 1. For $A = \{x_1, ..., x_n\}$ denote by $T(A) = T(x_1, ..., x_n)$, which is defined correctly due to the associativity of **T**. Similarly, we define $S(A)$. Let $A = \{0.1, 0.1\}$, $B = \{0.9, 0.9\}$, $C = \{0.1, 0.9\}$, $D = \{\underbrace{0.1, ..., 0.1}_{9\text{-times}}, 0.9\}$, $E = \{0.1, \underbrace{0.9, ..., 0.9}_{9\text{-times}}\}$.

Then

	A	B	C	D	E
T_L	0.000	0.800	0.000	0.000	0.000
T_P	0.010	0.810	0.090	$9*10^{-10}$	0.039
T_M	0.100	0.900	0.100	0.100	0.100
S_M	0.100	0.900	0.900	0.900	0.900
S_P	0.190	0.990	0.910	0.961	$1-9*10^{-10}$
S_L	0.200	1.000	1.000	1.000	1.000

Note that $S_P(x, y) = x + y - x.y$ is the dual t-conorm to T_P (additive generator is $g(x) = -log(1-x)$) and $S_L(x, y) = min(1, x + y)$ is the dual t-conorm to T_L (additive generator is $g(x) = x$). Further note that the inputs A, B, C, D, E will be used throughout this paper in another examples, too.

3. Exponential compensatory operators

The lack of compensation by t-norms, as well as full compensation by t-conorms, can be understood as limit cases of an aggregative operator $A: [0, 1]^n \to [0, 1]$. In any case, the commutativity and the non-decreasigness of A should be required. The other t-norm (t-conorm) axioms, namely the associativity and the existence of a neutral element, may be violated, in general. Recall, e. g., the common arithmetical mean.

First aggregation operator based on triangular norms and conorms was the γ-operator Γ_γ suggested by Zimmermann and Zysno [23]

$$\Gamma_\gamma(x, y) = (x \cdot y)^{1-\gamma}(x + y - x \cdot y)^\gamma, \quad \gamma \in [0, 1].$$

Here the parameter γ indicates the degree of compensation. Gamma operators are a special class of the *exponential compensatory operators* [15] based on the product t-norm T_P and the probabilistic sum S_P (which is a dual t-conorm to T_P). The general form of an exponential compensatory operator E based on a (continuous) t-norm T and t-conorm S (not necessarily dual to T) is the next:

$$E_{\gamma, T, S}(x, y) = (T(x, y))^{1-\gamma}(S(x, y))^\gamma.$$

The exponential compensatory operators are defined on the unit interval $[0, 1]$, similarly as all following operators. For all $\gamma \in [0, 1]$ we have

$$T \leq E_{\gamma, T, S} \leq S, \qquad T = E_{0, T, S} \quad \text{and} \quad S = E_{1, T, S}.$$

The exponential compensatory operators are (up to the trivial cases when γ is either 0 or 1) non-associative and the n-ary operators have to be defined separately,

$$E_{\gamma, T, S}(x_1, \ldots, x_n) = (T(x_1, \ldots, x_n))^{1-\gamma}(S(x_1, \ldots, x_n))^\gamma.$$

In the case of γ-operators, we have

$$\Gamma_\gamma(x_1, \ldots, x_n) = \left(\prod_{i=1}^n x_i\right)^{1-\gamma}\left(1 - \prod_{i=1}^n (1 - x_i)\right)^\gamma.$$

Among several practical applications of the exponential compensatory operators recall, e. g., the control of an automatic model car [1].

The continuity of an exponential compensatory operator $E_{\gamma, T, S}$ is equivalent to the continuity of the underlying t-norm T and t-conorm S. The dual operator $D(E_{\gamma, T, S})$ is not an exponential compensatory operator (up to trivial cases $\gamma \in \{0, 1\}$) and it can be suggested as a new type of aggregative operators. The idempotency property $E_{\gamma, T, S}(x, x) = x$ is fulfilled (for each $\gamma \in [0, 1]$) for the limit t-norm T_M and t-conorm S_M only, i. e., when

$$E_\gamma(x_1, \ldots, x_n) = (min\ x_i)^{1-\gamma} (max\ x_i)^\gamma.$$

However, for some special γ, also some other pairs of T and S lead to the idempotency property. Take, e. g., $\gamma = 0.5$ and $T = T_0^H$ (Hamacher's product)

$$T(x, y) = \frac{x \cdot y}{x + y - x \cdot y} \quad \text{for } (x, y) \in \ ]0, 1[^2$$

and $S = S_P$.

Then for $V = E_{0.5, T, S}$ we have

$$V(x, y) = \sqrt{x \cdot y}$$

and

$$V(x_1, \ldots, x_n) = \sqrt{\frac{1 - \prod(1 - x_i)}{1 + \sum \frac{1 - x_i}{x_i}}}.$$

Now, it is evident that

$V(x, x) = \sqrt{x^2}$ for all $x \in [0, 1]$. However, $V(x, x, x) < x$ for all $x \in \ ]0, 1[$.

Note that the dual operator $D(V)$ fulfills the idempotency property, too. Further note that the exponential compensatory operators have the conjunctive property $E_{\gamma, T, S}(0, x_1, \ldots, x_n) = 0$ whenever $\gamma < 1$. Vice versa, the dual operators have the disjunctive property

$DE_{\gamma, T, S}(1, x_1, \ldots, x_n) = 1$ whenever $\gamma < 1$.

These facts immediately prove that there is no self-dual exponential compensatory operator.

Example 2.

	A	B	C	D	E
$\Gamma_{0.2}$	0.018	0.843	0.143	0.000	0.074
$\Gamma_{0.5}$	0.044	0.895	0.286	0.000	0.197
$\Gamma_{0.8}$	0.105	0.951	0.573	0.015	0.522
V	0.100	0.900	0.300	0.108	0.302
D(V)	0.100	0.900	0.700	0.698	0.892

4. Convex-linear compensatory operators

Another class of non-associative t-norm based compensatory operators is formed by the *convex-linear compensatory* operators [15],

$$L_{\gamma, T, S}(x_1, \ldots, x_n) = (1 - \gamma) \cdot T(x_1, \ldots, x_n) + \gamma \cdot S(x_1, \ldots, x_n).$$

Again we have

$T \leq L_{\gamma, T, S} \leq S$, $T = L_{0, T, S}$ and $S = L_{1, T, S}$.

Recall e. g. the applications of L_{γ, T_M, S_L} (Luhandjula's operators combining the minimum t-norm T_M and the Lukasiewicz t-conorm S_L) in linear programming with multiple objectives.

The dual $D(L_{\gamma, T, S})$ is again a convex-linear compensatory operator, namely $D(L_{\gamma, T, S}) = L_{1 - \gamma, T^*, S^*}$, where T^* and S^* are the dual t-norm and the dual t-conorm to the t-conorm S and the t-norm T, respectively. A convex-linear compensatory operator $L_{\gamma, T, S}$ is self-dual, $L_{\gamma, T, S} = D(L_{\gamma, T, S})$ if and only if $\gamma = 0.5$ and (T, S) form a dual pair of a t-norm and t-conorm.

The idempotency property holds for L_{γ, T_M, S_M} with arbitrary $\gamma \in [0, 1]$. The partial idempotency property holds also for $L_{0.5, T, S}$ where T and S are the solutions of Frank's functional equation:

$$T(x, y) + S(x, y) = x + y,$$

e. g. T_L and S_L, T_P and S_P, etc.

Example 3.

	A	B	C	D	E
$L_{0.2, T_M, S_L}$	0.120	0.920	0.280	0.280	0.280
$L_{0.5, T_M, S_L}$	0.150	0.950	0.550	0.550	0.550
$L_{0.8, T_M, S_L}$	0.180	0.980	0.820	0.820	0.820
$L_{0.5, T_P, S_P}$	0.100	0.900	0.500	0.481	0.520
$L_{0.5, T_L, S_L}$	0.100	0.900	0.500	0.500	0.500

5. Convex non-linear compensatory operators

Exponential compensatory operators as well as convex linear compensatory operators have some limitations with respect to the compensation. For a given input $x \in [0, 1[$ we have

$$0 \le E_{\gamma, T, S}(x, y) \le x^{1-\gamma} \qquad \text{(if } \gamma < 1)$$

and

$$\gamma x \le L_{\gamma, T, S}(x, y) \le (1 - \gamma) x + \gamma = 1 - (1 - \gamma) \cdot (1 - x).$$

The above limitation can be avoided by replacing the fixed coefficient γ (which is independent on the input) by some input dependent coefficient γ^*. First traces of this approach can be found in [12]. Recall that in the case of the convex linear compensatory operators this approach was suggested and investigated by Yager and Rybalov [20, 21] and it is based on the fuzzy systems modeling techniques. Omitting the details from [21], for a given input $A = (x_1, ..., x_n)$, we want to combine a t-norm **T** decreasing the low values and t-conorm **S** increasing the high values. The convex coefficient γ^* describe to "highness" of elements of A and is proportional to $T_1(A) = T_1(x_1, ..., x_n)$, where T_1 is some t-norm without zero-divizors. Similarly, the "lowness" of elements of A is proportional to

$T_1(A') = T_1(1 - x_1, ..., 1 - x_n)$.

Then

$$\gamma^* = \frac{T_1(A)}{T_1(A') + T_1(A)}.$$

The resulting aggregation operator $L_{T,S}^{(T_1)}$ is no more a linear convex combination of **T** and **S**, therefore it will be called a convex non-linear compensatory operator. $L_{T,S}^{(T_1)}$ is defined by

$$L_{T,S}^{(T_1)}(A) = (1 - \gamma^*) T(A) + \gamma^* S(A) = \frac{T_1(A') T(A) + T_1(A) S(A)}{T_1(A') + T_1(A)}, \text{ up to the case when}$$

$\{0,1\} \subset A$.

Now, starting from some $x \in]0,1[$, we have $L_{T,S}^{(T_1)}(x, 0) = 0$ and $L_{T,S}^{(T_1)}(x, 1) = 1$.

If $T = T_1$ and $S = D(T)$ is the dual t-conorm, we denote $L_{T,S}^{(T_1)}$ by L_T. Then

$$L_T(A) = \frac{T(A)}{T(A) + T(A')}, \text{ i. e.}$$

$$L_T(x_1, ..., x_n) = \frac{T(x_1, ..., x_n)}{T(x_1, ..., x_n) + T(1 - x_1, ..., 1 - x_n)},$$

up to the case when $\{0, 1\} \subset A$.

For the product t-norm T_P, we have

$$L_{T_P}(x_1, ..., x_n) = \frac{\prod_{i=1}^{n} x_i}{\prod_{i=1}^{n} x_i + \prod_{i=1}^{n}(1-x_i)},$$

i. e. L_{T_P} is just the self-dual associative compensatory operator $C^{(1)}$ described in Example 5.

For the strongest t-norm T_M, we have

$$L_{T_M}(x_1, ..., x_n) = \frac{\min(x_1, ..., x_n)}{\min(x_1, ..., x_n) + 1 - \max(x_1, ..., x_n)}.$$

Note that as far as the Lukasiewicz t-norm T_L has zero divizors, L_{T_L} is not defined.

For arbitrary t-norm T without zero divizors, the operator L_T is self-dual, $L_T = D(L_T)$.) The idempotency property is fulfilled for L_{T_M}.

Another class of convex non-linear compensatory operators based on fuzzy systems modeling methods forces the t-norm T aggregation when at least one input is low, i. e., when the "lowness" of A is proportional to $S_1(A')$, S_1 being the dual t-conorm to T_1. Now, the coefficient γ^* is given by

$$\gamma^* = \frac{T_1(A)}{T_1(A) + S_1(A')} = T_1(A)$$

and

$$L_{T,S}^{[T_1]} = (1 - T_1(A))\, T(A) + T_1(A)\, S(A).$$

As T_1 we usualy put $T_1 = T_M$ or $T_1 = T_P$. For more details see [21]. Again the idempotency of $L_{T,S}^{[T_1]}$ is ensured in the case $T = T_M$ and $S = S_M$. On the other hand, these operators are never self-dual, as far as

$$D(L^{[T_1]}_{T,S})(0, x) = 1 - L^{[T_1]}_{T,S}(1, 1-x) = 1 - (x(1-x) + 1 - x) = x^2$$

and

$L^{[T_1]}_{T,S}(0, x) = 0$. Further, $L^{[T_1]}_{T,S}(x, 1) = 2x - x^2 = L^{[T_1]}_{T,S}(x, \underbrace{1, ..., 1}_{n\text{-times}})$, i. e., there are some limitations in the upwards compensation.

In a similar way, we can modify the exponential compensatory operators,

$$E^{(T_1)}_{T,S}(A) = T(A)^{(1-\gamma^*)} S(A)^{\gamma^*}, \text{ up to the case when } \{0, 1\} \subset A, \text{ where}$$

$$\gamma^* = \frac{T_1(A)}{T_1(A) + T_1(A')}.$$

Again, for each $x \in]0, 1[$ we get $E^{(T_1)}_{T,S}(x, 0) = 0$ and $E^{(T_1)}_{T,S}(x, 1) = 1$. Further,

$$E^{[T_1]}_{T,S}(A) = (T(A))^{1-T_1(A)} S(A)^{T_1(A)}.$$

Here $E^{[T_1]}_{T,S}(x, 0) = 0$ and $E^{[T_1]}_{T,S}(x, 1) = x^{1-x}$ for each $x \in]0, 1[$.

Example 4.

	A	B	C	D	E
L_{T_P}	0.012	0.988	0.500	2.32×10^{-8}	1.000
L_{T_M}	0.100	0.900	0.500	0.500	0.500
L^*	0.028	0.972	0.500	0.481	0.519
E^*	0.013	0.970	0.286	2.9×10^{-5}	0.197
L^{**}	0.028	0.972	0.172	0.096	0.135
E^{**}	0.013	0.970	0.113	7.19×10^{-9}	0.054

where

$$L^* = L^{(T_M)}_{T_P,S_P}, \text{ i. e.,}$$

$$L^*(A) = \frac{1-max(A)}{min(A)+1-max(A)} T_P(A) + \frac{min(A)}{min(A)+1-max(A)} S_P(A)$$

and

$$E^* = E^{(T_M)}_{T_P,S_P}. \text{ Further, } L^{**} = L^{[T_M]}_{T_P,S_P} \text{ and } E^{**} = E^{[T_M]}_{T_P,S_P}.$$

6. Uni-norms

Recently, Yager and Rybalov [19] introduced the concept of uni-norms, see also [4]. An operator U: $[0, 1]^2 \to [0, 1]$ is a uni-norm if it is associative, commutative, non-decreasing and has a neutral element e ∈]0,1 [. Recall that the limit cases of neutral element e lead to the t-norms (when e = 1) and to the t-conorms (when e = 0). One can easily check that each uni-norm U acts on the square $[0, e]^2$ as a t-norm, on the square $[e, 1]^2$ as a t-conorm, and on the remainder is between the minimum and the maximum. Hence the aggregation of two low values cannot exceed their minimum and the aggregation of two high values leads at least to their maximum. For arbitrary t-norm T, t-conorm S and element e ∈]0, 1 [, two types of uni-norms can be defined via ordinal sum-like construction:

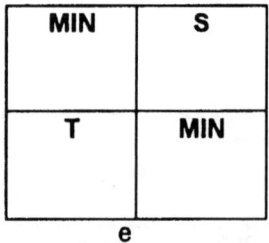

conjunctive uni-norm U_C

Fig. 1

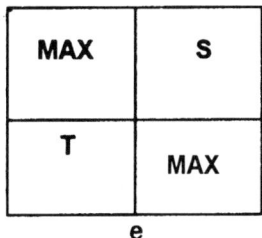

disjunctive uni-norm U_D

Fig. 2

Both U_C and U_D coincide with **T** and **S** on the half-open squares $[0, e[^2$ and $]e,1]^2$, respectively. On the remainder, U_C = MIN while U_D = MAX. Recall that the restriction of U_C and U_D to the corners of the unit square act as a conjunction and disjunction, respectively.

Another special type of a uni-norm is the so called associative compensatory operator **C** introduced by Klement et al. [7] (see also Dombi [2]). Operator **C** is generated by an additive generator $f: [0, 1] \to [-\infty, +\infty]$, f is continuous increasing bijection, so that

$$C(x, y) = f^{-1}(f(x) + f(y))$$

up to the case when $(x, y) \in \{(0, 1), (1, 0)\}$. Note that $C(0, 1) = C(1, 0) \in \{0, 1\}$ (0 corresponds to the conjunctive type, while 1 corresponds to the disjunctive type). Operator **C** is continuous up to points (0, 1) and (1, 0) and cancellative on the open square $]0,1[^2$. Note that the non-continuity in (0,1) and (1, 0) expresses the difficulties when aggregating the contradictory informations. The ordinal sum-like structure characterizes also the associative compensatory operators.

Namely, let **T** be a strict t-norm with additive generator g: $[0, 1] \to [0, \infty]$ (a decreasing bijection), let **S** be a strict t-conorm with additive generator

h: $[0,1] \to [0, \infty)$ (an increasing bijection) and let $e \in]0,1[$. Put

$$f(x) = \begin{cases} -g\left(\dfrac{x}{e}\right) & \text{if } x \le e \\ h\left(\dfrac{x - e}{1 - e}\right) & \text{if } x > e \end{cases}.$$

Then f generates an associative compensatory operator **C** such that **C** coincides on $[0,e]^2$ with the t-norm T_e and on $[e,1]^2$ with the t-conorm S_e. Here $T_e \approx (<0, e, T>)$ and $S_e \approx (<e, 1, S>)$ are the ordinal sums with only one summand. For more details about ordinal sums see Schweizer and Sklar [14]. Note that for any given positive constant c, the additive generator f_c,

$$f_c(x) = \begin{cases} c \cdot f(x) & \text{if } x \leq e \\ f(x) & \text{otherwise} \end{cases}$$

leads to an associative compensatory operator C_c such that C_c and $C = C_1$ coincide on the squares $[0, e]^2$ and $[e, 1]^2$.

On the remainder of the unit square, small values of c make C_c close to MAX, while large values of c make C_c close to MIN.

Let **U** be a uni-norm. Then its dual **D(U)** is again a uni-norm with neutral element 1- e. If **U** is conjunctive, then **D(U)** is disjunctive, and vice versa. When neglecting the corners (0, 1) and (1, 0), the only known self-dual uni-norms are the associative compensatory operator **C**, **C** = **D(C)**, with generator f symmetric with respect to the point (0.5, 0), i. e., $f(x) + f(1 - x) = 0$ for each $x \in\,]0, 1\,[$. An example of a self-dual uni-norm is then $\mathbf{C}^{(1)}$ from Example 5.

The idempotency property is fulfilled for arbitrary $\mathbf{U}_c$ and $\mathbf{U}_d$ based on the t-norm $\mathbf{T}_M$ and the t-conorm $\mathbf{S}_M$.

Recall that the associativity of a uni-norm **U** allows an immediate extension of the binary operation **U** to an n-ary operation.

Example 5.

Let $f^{(\lambda)}(x) = \log\dfrac{\lambda x}{1 - x}$, $\lambda \in\,]0, \infty\,[$.

Then $C^{(\lambda)}(x, y) = (f^{(\lambda)})^{-1}(f^{(\lambda)}(x) + f^{(\lambda)}(y)) = \dfrac{\lambda x y}{\lambda x y + (1 - x)(1 - y)}$,

whenever $(x, y) \notin \{(0,1), (1, 0)\}$. Further,

$e^{(\lambda)} = 1/(1 + \lambda)$, $T^{(\lambda)} = T^H_{(1+\lambda)/\lambda}$, $S^{(\lambda)} = S^H_{1+\lambda}$ (Hamacher's t-norms and t-conorms), and

$$C^{(\lambda)}(x_1, \ldots, x_n) = \dfrac{\lambda^{n-1} \prod_{i=1}^{n} x_i}{\lambda^{n-1} \prod_{i=1}^{n} x_i + \prod_{i=1}^{n}(1 - x_i)}.$$

Recall that $\mathbf{C}^{(1)} = \mathbf{L}_{T_P}$ was introduced already in the previous section, i. e., that $\mathbf{C}^{(1)}$ is a convex non-linear compensatory operator, too.

Figure 3 contains the contour plots of $\mathbf{C}^{(0.1)}$, $\mathbf{C}^{(10)}$, $\mathbf{C}_1^{(1)}$, $\mathbf{C}_{0.1}^{(1)}$, $\mathbf{C}_{10}^{(1)}$.

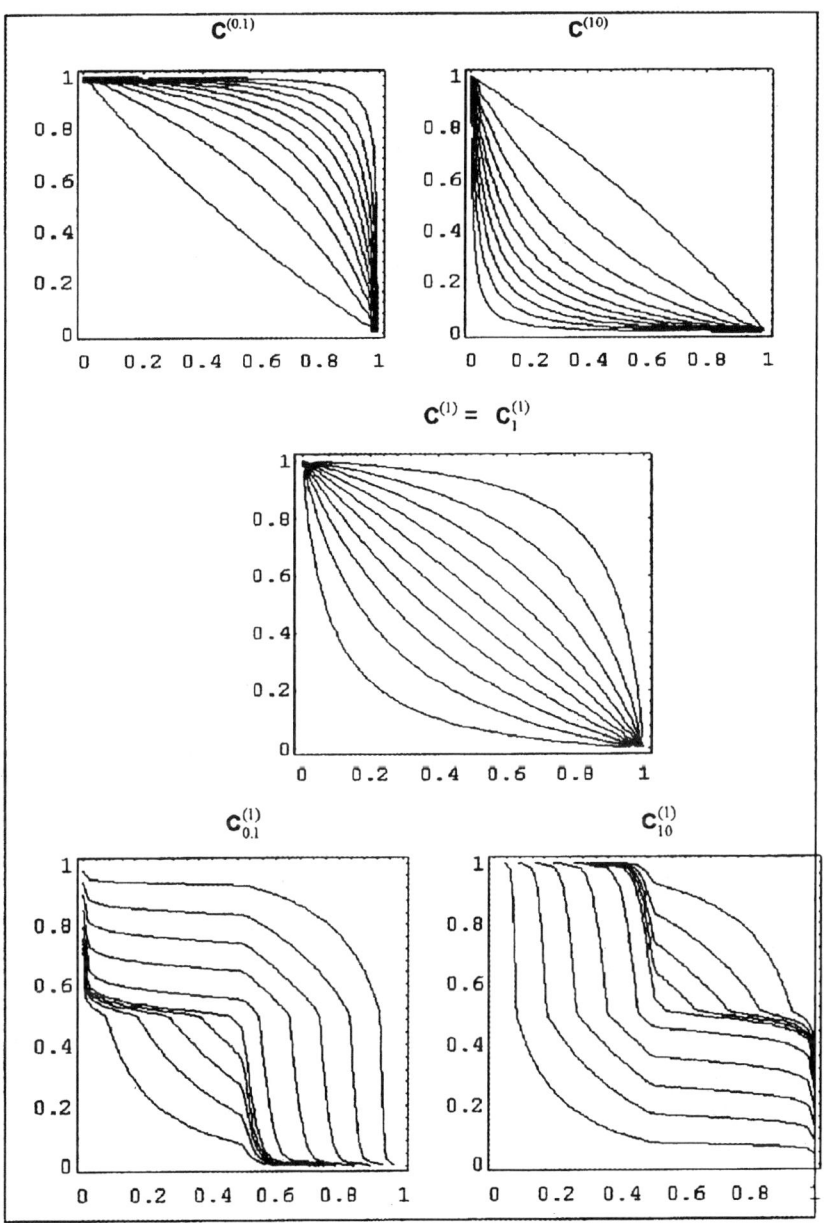

Fig. 3

Example 6.

	A	B	C	D	E
$C^{(1)}$	0.012	0.988	0.500	$2.3*10^{-8}$	1.000
$C^{(1)}_{0.1}$	0.012	0.988	0.878	0.555	1.000
$C^{(1)}_{10}$	0.012	0.988	0.122	$3.22*10^{-9}$	0.445
$C^{(0.1)}$	0.0012	0.890	0.091	$2.3*10^{-17}$	0.041
$C^{(10)}$	0.1098	0.999	0.909	0.9587	1.000
U_c	0.000	1.000	0.100	0.000	0.100
U_d	0.000	1.000	0.900	0.900	1.000

where U_c and U_d are determined by $T = T_L$, $S = S_L$ and $e = 0.5$.

7. Aggregations with additive generators

Associative compensatory operators combine strict t-norms and strict t-conorms in an ordinal sum-like construction. Similarly we can combine arbitrary t-norm and t-conorm with continuous additive generators. However, the associativity is violated up to the case of strict t-norm and t-conorm. Let g be a continuous additive generator of the t-norm **T** and let h be a continuous additive generator of the t-conorm **S**. Take an element $e \in]0,1[$ and put

$$f(x) = \begin{cases} -g\left(\dfrac{x}{e}\right) & \text{if } x \leq e \\ h\left(\dfrac{x-e}{1-e}\right) & \text{otherwise} \end{cases}.$$

If either **T** or **S** is not strict on the whole unit square $[0,1]^2$ we can define an aggregation operator $C : [0,1]^2 \to [0,1]$, $C(x, y) = f^{(-1)}(f(x) + f(y))$, where $f^{(-1)}$ is the pseudo-inverse of f, i. e.,

$$f^{(-1)}(x) = \begin{cases} 1 & \text{if } x > f(1) \\ f^{-1}(x) & \text{if } f(0) \leq x \leq f(1) \\ 0 & \text{if } x < f(0) \end{cases}.$$

For n input values $\{x_1, \ldots, x_n\}$ we define

$$C(x_1, \ldots, x_n) = f^{(-1)}\left(\sum_{i=1}^{n} f(x_i)\right).$$

C is not associative, but it is non-decreasing, commutative, continuous and e is its neutral element. **C** coincide on the square $[0, e]^2$ with the t-norm $T_e \approx$ (<0, e, **T**>) and on the square $[e,1]^2$ with the t-conorm $S_e \approx$ (<e, 1, **S**>), and on the remainder of the unit square is between MIN and MAX.

Vice versa, arbitrary continuous strictly increasing mapping $f : [0, 1] \to [-\infty, +\infty]$ with $f(e) = 0$ generates an aggregation operator **C**. Corresponding **T** and **S** are generated by the additive generator $g(x) = -f(x\,e)$ and $h(x) = f(\,e + (1 - e)\,x\,)$, respectively.

The only idempotents for a generated aggregation operator **C** are elements of the set $\{0, e, 1\}$. The dual operator **D(C)** is generated by the generator $\varphi(x) = -f(1-x)$. Similarly as in the previous section, we can define C_c with the same values as **C** on the squares $[0, e]^2$ and $[e, 1]^2$, and on the remainder of the domain close to MAX for c small and close to MIN for c large.

Example 7.

Take $T = T_L$, $S = S_L$, $e = 0.5$. Then $g(x) = 1 - x$, $h(x) = x$ and $f(x) = 2x - 1$. Further,

$$f^{(-1)}(x) = \begin{cases} 0 & \text{if } x < -1 \\ \dfrac{x+1}{2} & \text{if } -1 \le x \le 1 \\ 1 & \text{if } x > 1 \end{cases}.$$

Consequently,

$C(x, y) = /\,x + y - 0.5\,/$, where $/\,x\,/ = \max(0, \min(x, 1))$.

Further,

$$C(x_1, \ldots, x_n) = /\,0.5 + \sum_{i=1}^{n}(x_i - 0.5)\,/$$

Figure 4 contains the contour plots of **C**.

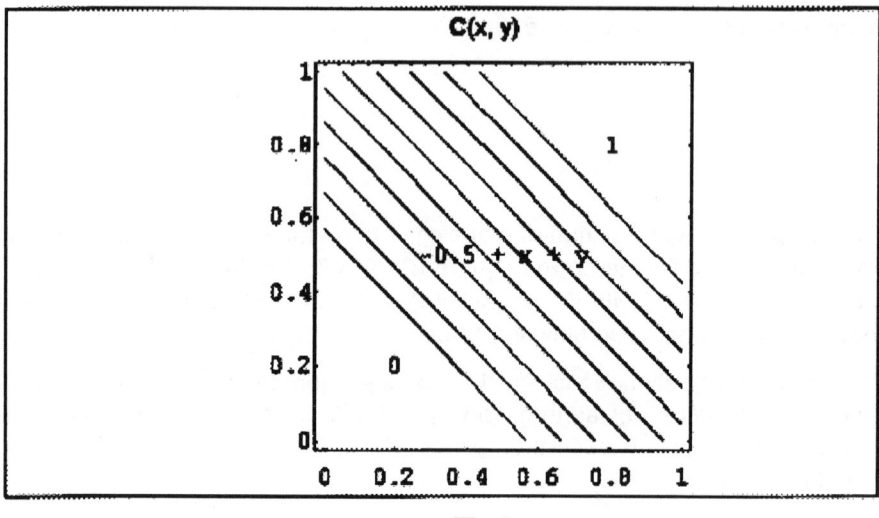

Fig. 4

Recall another interesting property of generated aggregation operators **C**, the so called pseudo-Archimedean property. In the t-norms and t-conorms case, the Archimedean property can be described as follows:

I. the only idempotents are the trivial idempotents 0 and 1;

II. for arbitrary element $x \in [0, 1]$, $\{x^{(n)}\}$ is a monotone sequence with idempotent limit, where $x^{(1)} = x$ and $x^{(n)} = T(x^{(n-1)}, x)$ for $n > 1$ (similarly for **S**).

In the case of a generated aggregation operator **C**, the only idempotents are 0, 1 and the neutral element e. For arbitrary $x \in [0,1]$, the sequence $\{x^{(n)}\}$, $x^{(1)} = x$ and $x^{(n)} = C(\underbrace{x, ..., x}_{n\text{-times}})$ is a monotone and its limit is some of idempotents. It is easy to see, that for $x < e$ the corresponding limit is 0, for $x > e$ it is 1 and for e it is e.

Further note that generated aggregation operators form a proper subclass of Yager's MICA operator with fixed identity e (so called FIMICA operators [20]).

For the bounded generator f, i. e., when starting from a nilpotent t-norm **T** and a nilpotent t-conorm **S**, the corresponding operator **C** is open in the sense of Yager and Rybalov [22], i. e., for any $A = \{x_1, ..., x_n\}$ there is $B = \{y_1, ..., y_m\}$ so that $C(A) \neq C(A, B)$. More, B can be always chosen so that $C(A, B) = t$ for arbitrary chosen value $t \in [0,1]$.

Example 8.

Let $C_{(i)}$, i =1, 2, 3, 4, be generated by $f_1(x) = x - 0.85$, $f_2(x) = x - 0.5$ (see also Example 7), $f_3(x) = x - 0.15$,

$$f_4(x) = \begin{cases} \log(3x) & \text{if } x \leq \dfrac{1}{3} \\ 3x - 1 & \text{if } x > \dfrac{1}{3} \end{cases}.$$

Then

	A	B	C	D	E
$C_{(1)}$	0.000	0.950	0.150	0.000	0.550
$C_{(2)}$	0.000	1.000	0.500	0.000	1.000
$C_{(3)}$	0.050	1.000	0.850	0.450	1.000
$C_{(4)}$	0.030	1.000	0.499	$3.6*10^{-5}$	1.000

8. Limit generated aggregation operators

Starting from the Example 5, the pointwise limits

$C^{(0)} = \lim\limits_{\lambda \to 0^+} C^{(\lambda)}$ and $C^{(\infty)} = \lim\limits_{\lambda \to \infty} C^{(\lambda)}$ are defined (up to the corners (0, 1) and (1,0)) by

$$C^{(0)}(x, y) = \begin{cases} 1 & \text{if } \max(x, y) = 1 \\ 0 & \text{otherwise} \end{cases}$$

and

$$C^{(\infty)}(x, y) = \begin{cases} 0 & \text{if } \min(x, y) = 0 \\ 1 & \text{otherwise} \end{cases}.$$

It is easy to see that for each generated aggregation operator C we have $C^{(0)} \leq C \leq C^{(\infty)}$, i. e., $C^{(0)}$ and $C^{(\infty)}$ are the boundaries for generated aggregation operators.

Let **C** be a generated aggregation operator with additive generator f and neutral element e. We have already introduced a new generated aggregation operator $\mathbf{C}_c$ generated by f_c,

$$f_c(x) = \begin{cases} c \cdot f(x) & \text{if } x \leq e \\ f(x) & \text{otherwise} \end{cases}.$$

Then $\lim_{c \to 0^+} \mathbf{C}_c = \mathbf{U}_d$ and $\lim_{c \to +\infty} \mathbf{C}_c = \mathbf{U}_c$, (up to the corners (0,1) and (1, 0)), where $\mathbf{U}_d$ and $\mathbf{U}_c$ are disjunctive and conjunctive uni-norm, respectively, defined by means of $\mathbf{T}_e \approx (<0, e, \mathbf{T}>)$, $\mathbf{S}_e \approx (<e, 1, \mathbf{S}>)$ and e, see sections 6, 7. Note that **T** is a t-norm generated by additive generator g, $g(x) = -f(e\,x)$, and **S** is a t-conorm generated by additive generator h, $h(x) = f(e + (1-e)x)$.

Let g be an additive generator of some t-norm **T**. Then also g^λ, $\lambda \in]0, \infty[$, is an additive generator of some t-norm $\mathbf{T}^{(\lambda)}$. By Dombi [3]

$$\lim_{\lambda \to +\infty} \mathbf{T}^{(\lambda)} = \mathbf{T}_M.$$

Similarly we can show that

$$\lim_{\lambda \to 0^+} \mathbf{T}^{(\lambda)} = \mathbf{T}_W.$$

An analog claim is true for the t-conorms. Now, we try to generalize the above results to the case of generated aggregation operators. Let f be an additive generator of some aggregation operator **C** with neutral element e. For $\lambda \in]0, \infty[$, put

$$f^{[\lambda]}(x) = \begin{cases} -(-f(x))^\lambda & \text{if } x \leq e \\ f^\lambda(x) & \text{otherwise} \end{cases}.$$

Then $f^{[\lambda]}$ generates an aggregation operator $\mathbf{C}^{[\lambda]}$. Put

$$\mathbf{C}^{[0]} = \lim_{\lambda \to 0^+} \mathbf{C}^{[\lambda]} \quad \text{and} \quad \mathbf{C}^{[\infty]} = \lim_{\lambda \to +\infty} \mathbf{C}^{[\lambda]}.$$

The commutativity and monotonicity of $C^{[0]}$ and $C^{[\infty]}$ is obvious as well as that e is their neutral element. However, the associativity of both limit operators is violated. For two incoming arguments x, y ∈]0, 1 [it is

$$C^{[0]}(x, y) = \begin{cases} 0 & \text{if } x < e, y < e \\ 1 & \text{if } x > e, y > e \\ y & \text{if } x = e \\ x & \text{if } y = e \\ e & \text{otherwise} \end{cases}$$

and

$$C^{[\infty]}(x, y) = \begin{cases} \min(x, y) & \text{if } f(x) + f(y) < 0 \\ e & \text{if } f(x) + f(y) = 0 \\ \max(x, y) & \text{if } f(x) + f(y) > 0 \end{cases}.$$

So, e. g., if $C = C^{(1)}$ from Example 5, i. e.,

$$C(x, y) = \frac{xy}{xy + (1 - x)(1 - y)}, \text{ we have}$$

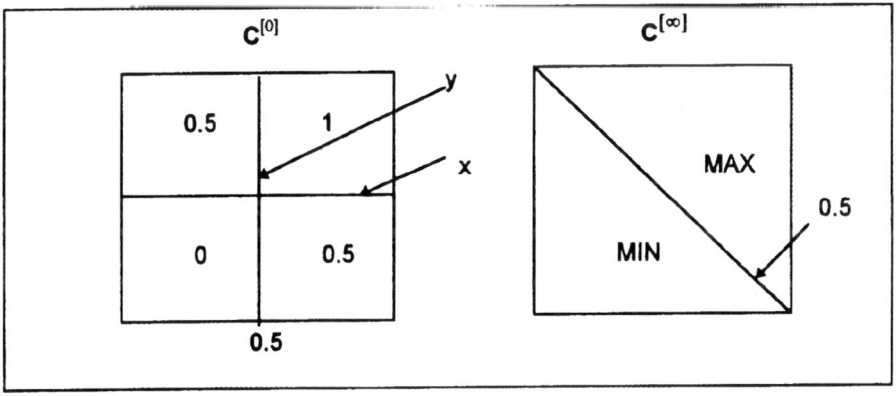

Fig. 5

Let $A = \{x_1, ..., x_n\}$, n > 2 be the set of inputs x_i ∈]0, 1 [, i = 1,...,n. Then $C^{[\infty]}$ should be obtained recursively,

$$C^{[\infty]}(A) = \begin{cases} min\ (A) & \text{if } f(max(A)) + f(min(A)) < 0 \\ max\ (A) & \text{if } f(max(A)) + f(min(A)) > 0, \\ C^{[\infty]}(A^*) & \text{if } f(max(A)) + f(min(A)) = 0 \end{cases}$$

where A* is (n - 2) - tuple arising from the n - tuple A deleting one maximal and one minimal input value from A.

Note that if card A* =1, A* = { x }, then $C^{[\infty]}$ (x) = x.

For the complementary limit aggregation operator $C^{[0]}$, denote by s the sum of sign(x_i), where sign(x_i) is 1 if x > 0, it is 0 if x = 0 and it is -1 if x < 0. Then

$$C^{[0]}(A) = \begin{cases} 0 & \text{if } s < -1 \\ e & \text{if } s = 0 \\ 1 & \text{if } s > +1 \\ \alpha & \text{if } s = -1 \\ \beta & \text{if } s = 1 \end{cases}$$

where

$$\alpha = f^{(-1)}\left(-\prod_{i=1}^{n}|f(x_i)|^{-sign(f(x_i))}\right)$$

and

$$\beta = f^{(-1)}\left(\prod_{i=1}^{n}|f(x_i)|^{sign(f(x_i))}\right).$$

Recall that 0^0 =1 by convention.

Example 9.

Starting from $C^{(1)}$ from Example 5, we get $C_1^{[\infty]}$ and $C_1^{[0]}$. Starting from C from Example 7, we get $C_2^{[\infty]}$ and $C_2^{[0]}$. Then

	A	B	C	D	E	F	G
$C_1^{[0]}$	0.000	1.000	0.500	0.000	1.000	0.027	0.631
$C_1^{[\infty]}$	0.100	0.900	0.500	0.000	1.000	0.100	0.100
$C_2^{[0]}$	0.000	1.000	0.500	0.000	1.000	0.000	0.650
$C_2^{[\infty]}$	0.100	0.900	0.500	0.000	1.000	0.100	0.100

where $F = \{0.1, 0.2, 0.7\}$ and $G = \{0.1, 0.7, 0.8\}$.

9. Conclusions

Several types of aggregation operators based on triangular norms and conorms were introduced and investigated. All of them can be applied to process the evidence under fuzziness. The majority of them violate the associativity property. However, for practical applications the lack of associativity seldom causes difficulties, up to, possibly, computing complexity. Recall, e. g., the non-associative arithmetical mean In several types of non-associative aggregation operators, the quasi-associativity is fulfilled [21], allowing to reduce the computing complexity. So, e. g., by the arithmetical mean, we have that $\bar{x} = \dfrac{\text{Sum } x_i}{\text{Card } x_i}$, where both Sum and Card are associative.

The suggested aggregation operators can be applied to aggregate pointwisely the fuzzy inputs (generalizing the fuzzy intersection and/or the fuzzy union).

Another possible way of application is the generalization of the extension principle (e. g. in the calculus of fuzzy numbers).

Finally recall that the original Menger's [10] requirements to a t-norm (commutativity, monotonicity, idempotency of 0 and 1 and $C(1, a) > 0$ for each $a > 0$) include not only nowadays t-norms and t-conorms, but also each of introduced aggregative operators.

Acknowledgement: This work has been supported by grants VEGA 95/5305/471, VEGA 1/4064/97 and GAČR 402/96/0414.

References

[1] von Altrock, C., Krause, B. and Zimmermann, H.-J.: Advanced fuzzy logic of a model car in extreme situations. Fuzzy Sets and Systems 48 (1992), 41 - 52.

[2] Dombi, J.: Basic concepts for a theory of evaluation: The aggregative operator. Europ. J. Oper. Research 10 (1982), 282 - 293.

[3] Dombi, J.: A general class of fuzzy operators, the de Morgan class of fuzzy operators and fuzziness measures induced by fuzzy operators. Fuzzy Sets and Systems 8 (1982),149 -163.

[4] Fodor, J. C., Yager, R. R. and Rybalov, A.:structure of uni-norms. Technical Report #MII-1501, Machine Intelligence Institute, Iona College, New Rochelle, NY,1994.

[5] Klement, E. P.: Construction of fuzzy σ-algebras using triangular norms. J. Math. Anal. Appl. 85 (1982), 543 - 565.

[6] Klement, E. P., Mesiar, R. and Pap, E.: Additive generators of t-norms which are not necessarily continuous. Proc. EUFIT'96, Aachen,1996, 70 - 73.

[7] Klement, E. P., Mesiar, R. and Pap, E.: On the relationship of associative compensatory operators to triangular norms and conorms. Int. Journal of Uncertainty, Fuzziness and Knowledge - Base systems, Vol. 4, No. 2 (1996), 129 - 144.

[8] Luhandjula, M. K.: Compensatory operators in fuzzy linear programming with multiple objectives. Fuzzy Sets and Systems 8 (1982), 245 - 252.

[9] Ling, C. M.: Representation of associative functions. Publ. Math. Debrecen 12 (1965),189 - 212.

[10] Menger, K.: Statistical metrics. Proc. Nat. Acad. Sci. U. S. A. 8 (1942), 535 - 537.

[11] Mesiar, R.: Compensatory operators based on triangular norms and conorms. Proc. EUFIT'95, ed. H.-J. Zimmermann, Aachen,1995,131-135.

[12] Píš, P., Mesiar, R.: Fuzzy model of inexact reasoning in medical consulting systems. IFSA-EC and EURO-MC Workshop, Warszawa 1986, 302 - 310.

[13] Schweizer, B. and Sklar, A.: Statistical metric spaces. Pacific J. Math. 10 (1960), 313 - 334.

[14] Schweizer, B. and Sklar, A.: Probabilistic metric spaces. North-Holland, Amsterdam,1983.

[15] Turksen, I. B: Interval-valued fuzzy sets and `compensatory AND'. Fuzzy Sets and Systems 51 (1992), 295 - 307.

[17] Yager, R. R.: Aggregation operators and fuzzy systems modeling. Fuzzy Sets and Systems 67 (1994),129 -146.

[18] Yager, R. R.: Aggregating operators and fuzzy systems modelling. Techn. Report #MII-1401, Machine Intelligence Institute, Iona College, New Rochelle, NY,1994.

[19] Yager, R. R. and Rybalov, A.: Uniform aggregation operators. Technical Report #MII-1501, Machine Intelligence Institute, Iona College, New Rochelle, NY,1994.

[20] Yager, R. R. and Rybalov, A.: Full reinforcement operators in aggregation techniques. Techn. Report #MII-1502, Machine Intelligence Institute, Iona College, New Rochelle, NY,1995.

[21] Yager, R. R.: On a class of weak triangular norm operators. Techn. Report #MII-1528, Machine Intelligence Institute, Iona College, New Rochelle, NY, 1996.

[22] Yager, R. R. and Rybalov, A.: A note on the incompatibility of openness and associativity. Techn. Report #MII-1529, Machine Intelligence Institute, Iona College, New Rochelle, NY,1996.

[23] Zimmermann, H.-J. and Zysno, P.: Latent connectives in human decision making. Fuzzy Sets and Systems 4 (1980), 37 - 51.

Aggregation Functions Defined by *t*-Norms and *t*-Conorms

János Fodor* and Tomasa Calvo**

* Department of Mathematics
University of Agricultural Sciences (GATE)
Páter Károly u. 1 H-2103 Gödöllő, Hungary
E-mail: jfodor@cs.elte.hu or jfodor@mszi.gau.hu

** Department of Mathematics and Computer Science
University of the Balearic Islands
Ctra. de Valldemossa, km 7.5
E-07071 Palma de Mallorca, Spain
E-mail: DMITCSO@PS.UIB.ES

Abstract

We introduce a class of aggregation functions by the help of continuous *t*-norms and *t*-conorms. The general functional forms of such aggregations is determined. Associativity and idempotency are also studied separately.

Keywords: aggregation functions, triangular norms, triangular conorms, associativity, idempotency.

1 Introduction

The problem of aggregating fuzzy sets in a meaningful way has been of central interest since the late 1970s. In most cases, the aggregation operators are defined on a pure axiomatic basis and are interpreted either as logical connectives (such as *t*-norms and *t*-conorms) or as averaging operators allowing a compensation effect (such as the arithmetic mean).

On the other hand, it can be observed by some empirical tests that the above-mentioned classes of operators differ from those ones that people use in practice (see Zimmermann and Zysno (1980)). Therefore, it is important to find operators that are, in a sense, mixtures of the previous ones, and allow some degree of compensation. A class of such operators was introduced and investigated by Mayor and Trillas (1986).

The main aim of the present paper is to extend their approach by introducing so called T-S aggregation functions, where T is a continuous *t*-norm and S is a continuous *t*-conorm. Informally speaking, a T-S aggregation function is a symmetric and nondecreasing function from the unit square to the unit interval, and its value at points $(x,0) \in [0,1]^2$ and $(x,1) \in [0,1]^2$ can be expressed by the help of its value at $(0,1) \in [0,1]^2$, and by using T and S, respectively. For different T-S pairs, this broad class includes, among others, t-norms, t-conorms, the median and the arithmetic mean.

The paper is organized as follows. In Section 2 we introduce T-S aggregation functions, give the weakest and strongest members of the family for fixed T and S, and show some examples. In Section 3 we establish a characterization theorem, which indicates the general (typical) forms of T-S aggregation functions. In Section 4 we study associative T-S aggregations, and obtain a subclass of operations that, in their structure, are similar to the median, but are more general. In Section 5 the idempotency of continuous T-S aggregation functions is investigated. Apart from the case when T is the product and S is its dual, the most general continuous idempotent T-S aggregation function is the median.

2 T-S aggregation functions

Mayor and Trillas (1986) introduced the following class of binary operations on the unit interval.

Definition 1 A function $F : [0,1]^2 \to [0,1]$ is called an *aggregation function* if the following conditions hold:

(i) $F(x,y) = F(y,x)$ for all $x,y \in [0,1]$;

(ii) $F(x,y) \leq F(u,v)$ when $x \leq u$, $y \leq v$;

(iii) $F(x,0) = F(1,0)x$ for all $x \in [0,1]$;

(iv) $F(x,1) = (1 - F(1,0))x + F(1,0)$ for all $x \in [0,1]$.

Using the notation $k := F(1,0)$, conditions (iii) and (iv) can be viewed at least two different ways. First, (iii) requires that the value $F(x,0)$ is the weighted arithmetic mean of x and 0; in the same way, $F(x,1)$ is the weighted arithmetic mean of x and 1. Second, the value kx is the value of the product t-norm at k and x, and $(1-k)x+k$ is the value of its dual t-conorm at k and x. This way of thinking makes it possible to extend the definition of Mayor and Trillas (1986) by applying any continuous t-norm and t-conorm as follows.

Definition 2 Let T be a continuous t-norm and S be a continuous t-conorm. A function $F : [0,1]^2 \to [0,1]$ is called a T-S *aggregation function* if the following conditions hold:

(i) $F(x,y) = F(y,x)$ for all $x,y \in [0,1]$;

(ii) $F(x,y) \leq F(u,v)$ when $x \leq u$, $y \leq v$;

(iii) $F(x,0) = T(F(1,0),x)$ for all $x \in [0,1]$;

(iv) $F(x,1) = S(F(1,0),x)$ for all $x \in [0,1]$.

Clearly, we obtain Definition 1 when $T(x,y) = xy$ and $S(x,y) = x+y-xy$ ($x,y \in [0,1]$).

Proposition 1 *Suppose T is a continuous t-norm, S is a continuous t-conorm and $k \in [0,1]$ is fixed. Then the weakest T-S aggregation function F_w with $F_w(0,1) = k$ is given by*

$$F_w(x,y) = \begin{cases} S(k, \min(x,y)) & \text{if } \max(x,y) = 1 \\ T(k, \max(x,y)) & \text{otherwise} \end{cases}$$

and the strongest one F_s with $F_s(0,1) = k$ is of form

$$F_s(x,y) = \begin{cases} S(k, \min(x,y)) & \text{if } \min(x,y) > 0 \\ T(k, \max(x,y)) & \text{otherwise} \end{cases}.$$

Proof. It is easy by Definition 2.
■

Before studying T-S aggregation functions in detail, consider the following examples.

Example 1 For given continuous t-norm T and continuous t-conorm S, define two functions F_1, F_2 as follows:

$$F_1(x,y) = S(T(k, S_1(x,y)), T_1(x,y)), \quad x, y \in [0,1], \tag{1}$$
$$F_2(x,y) = T(S(k, T_1(x,y)), S_1(x,y)), \quad x, y \in [0,1], \tag{2}$$

where T_1 is a t-norm, S_1 is a t-conorm and $k \in [0,1]$. It is easy to see that both F_1 and F_2 are T-S aggregation functions, with $F_1(1,0) = F_2(1,0) = k$.
If we choose $k = 0$ then $F_1 = T_1$, and in case of $k = 1$ we have $F_2 = S_1$. Thus, any t-norm and any t-conorm is a T-S aggregation function.

Example 2 Consider the particular case of (1) and (2) with $T = T_1 = \min$ and $S = S_1 = \max$. Then, after simple calculation, we obtain that

$$F_1(x,y) = F_2(x,y) = \text{med}(x,y,k),$$

that is, the well-known median operator characterized by Fung and Fu (1975).

Example 3 Let $T(x,y) = xy$, $S(x,y) = x + y - xy$, $T_1(x,y) = \max(x+y-1, 0)$, $S_1(x,y) = \min(x+y, 1)$ and $k = 1/2$ in (1) and (2). Then we obtain

$$F_1(x,y) = F_2(x,y) = \frac{x+y}{2}, \quad x, y \in [0,1].$$

By these examples, it might be clear that the class of T-S aggregation functions contains a wide variety of diverse binary operations.

3 Characterizations

In this section we show that any T-S aggregation function is of form similar to (1) and (2).

For any t-norm T and for any t-conorm S their residuals I_T and J_S, called R-implications and R-coimplications, respectively (see Fodor and Roubens (1994)), are defined as usual:

$$I_T(x,y) = \sup\{z \mid T(x,z) \leq y\}, \quad J_S(x,y) = \inf\{t \mid S(x,t) \geq y\},$$

for $x,y \in [0,1]$. Main properties that we will use in the construction are listed as follows.

Lemma 1 *For any t-norm T and any t-conorm S the following conditions are satisfied:*

(i) I_T and J_S are nonincreasing in their first place and are nondecreasing in their second place;

(ii) $I_T(1,x) = J_S(0,x) = x$ for all $x \in [0,1]$;

(iii) $x \leq y$ implies $I_T(x,y) = 1$ and $J_S(y,x) = 0$.

(iv) If, in addition, T and S are continuous then for all $x,y \in [0,1]$ we have

$$T(x, I_T(x,y)) = \min(x,y), \quad S(x, J_S(x,y)) = \max(x,y).$$

Proof. For (i)–(iii) we refer to Fodor and Roubens (1994).

To prove (iv), consider two cases. If $x \leq y$ then $I_T(x,y) = 1$, so

$$T(x, I_T(x,y)) = T(x,1) = x = \min(x,y).$$

If $x > y$ then there exists $z \in [0,1]$ such that $y = T(x,z)$. Then $z \leq I_T(x,y)$, by definition of I_T, so $y = T(x,z) \leq T(x, I_T(x,y))$. On the other hand, $T(x, I_T(x,y)) \leq y$ is equivalent to $I_T(x,y) \leq I_T(x,y)$, which is obviously true. This completes the proof for T. The case of S is similar. ∎

Now we are able to state the main theorem of this section.

Theorem 1 *Let T be a continuous t-norm and S be a continuous t-conorm. A function $F : [0,1]^2 \to [0,1]$ is a T-S aggregation function with $k = F(1,0)$ if and only if there exist two symmetric, nondecreasing functions $G, H : [0,1]^2 \to [0,1]$ with $G \geq H$, $G(x,1) = 1$, $H(x,0) = 0$ and $T(k,x) = T(k,G(x,0))$, $S(k,x) = S(k,H(x,1))$ for all $x \in [0,1]$ such that either $F = F_1$, or $F = F_2$, where*

$$F_1(x,y) = S(T(k,G(x,y)), H(x,y)), \tag{3}$$
$$F_2(x,y) = T(S(k,H(x,y)), G(x,y)) \tag{4}$$

for all $x,y \in [0,1]$.

Proof. Assume first that F is given by (3). Clearly, then F is nondecreasing and symmetric. Moreover,

$$F(1,0) = S(T(k, G(1,0)), H(1,0)) = S(T(k,1), 0) = k.$$

In addition,

$$F(x,0) = S(T(k, G(x,0)), H(x,0)) = S(T(k,x), 0) = T(k,x),$$

and

$$F(x,1) = S(T(k, G(x,1)), H(x,1)) = S(k, H(x,1)) = S(k,x).$$

These conditions imply that F is a T-S aggregation function with $k = F(1,0)$.

The proof that (4) defines a T-S aggregation function is similar.

To prove the converse statement, assume that F is a T-S aggregation function with $k = F(1,0)$. We want to find functions G and H, with the prescribed properties, such that (3) and/or (4) hold.

Denote I_T the residual implication defined by T, and J_S the residual coimplication defined by S. Then define G and H by

$$G(x,y) = I_T(k, F(x,y)), \quad H(x,y) = J_S(k, F(x,y)), \quad x, y \in [0,1].$$

By Lemma 1, thus defined G and H are symmetric and nondecreasing functions. Moreover,

$$G(x,1) = I_T(k, F(x,1)) = I_T(k, S(k,x)) = 1,$$

because $k \leq S(k,x)$. Similarly,

$$H(x,0) = J_S(k, F(x,0)) = J_S(k, T(k,x)) = 0,$$

since $k \geq T(k,x)$. We can write

$$\begin{aligned} T(k, G(x,0)) &= T(k, I_T(k, F(x,0))) \\ &= T(k, I_T(k, T(k,x))) = \min(k, T(k,x)) \\ &= T(k,x), \end{aligned}$$

since T is a continuous t-norm. Next, by continuity of S, we have

$$\begin{aligned} S(k, H(x,1)) &= S(k, J_S(k, F(x,1))) \\ &= S(k, J_S(k, S(k,x))) = \max(k, S(k,x)) \\ &= S(k,x). \end{aligned}$$

We still have to prove that $G \geq H$. This follows trivially, because if $x, y \in [0,1]$ are such that $k \leq F(x,y)$ then $G(x,y) = 1$, while $k > F(x,y)$ implies $H(x,y) = 0$.

Finally, we prove that F can be written in the form of (4). Consider two cases:

Case 1: $x, y \in [0,1]$ are such that $k \leq F(x,y)$.

Then $G(x,y) = I_T(k, F(x,y)) = 1$, so we have

$$S(T(k, G(x,y)), H(x,y)) = S(k, J_S(k, F(x,y))) = \max(k, F(x,y)) = F(x,y).$$

Case 2: $x, y \in [0,1]$ are such that $k > F(x,y)$.

Then $H(x,y) = J_S(k, F(x,y)) = 0$, so we have

$$S(T(k, G(x,y)), H(x,y)) = T(k, I_T(k, F(x,y))) = \min(k, F(x,y)) = F(x,y),$$

which completes the proof.

∎

Notice that by the construction of G and H in the previous proof, both forms of (3) and (4) give the same F. Naturally, if we start from some G and H, we cannot guarantee in general that the T-S aggregations defined by (3) and (4) are equal, except when $T = \min$ and $S = \max$.

Proposition 2 *Let $T = \min$ and $S = \max$, G and H be functions having properties required in Theorem 1, and F_1, F_2 be functions defined by (3) and (4), respectively. Then $F_1(x,y) = F_2(x,y) = \mathrm{med}(k, G(x,y), H(x,y))$ for all $x, y \in [0,1]$.*

Proof. By $G \geq H$ we have

$$\begin{aligned}
F_1(x,y) &= \max(\min(k, G(x,y)), H(x,y)) \\
&= \min(\max(k, H(x,y)), \max(G(x,y), H(x,y))) \\
&= \mathrm{med}(k, G(x,y), H(x,y)) \\
&= \min(\max(k, H(x,y)), G(x,y)) \\
&= F_2(x,y).
\end{aligned}$$

∎

For the general case, we prove the following result.

Theorem 2 *Let T be a continuous t-norm different from the minimum, and S be a continuous t-norm different from the maximum. Assume G and H are functions satisfying conditions of Theorem 1, and define F_1, F_2 by (3), (4), respectively. Then $F_1 = F_2$ if and only if there exist G and H such that $\min(1 - G(x,y), H(x,y)) = 0$ for any $x, y \in [0,1]$.*

Proof. First suppose that G and H satisfy $\min(1 - G(x,y), H(x,y)) = 0$ for any $x, y \in [0,1]$. Consider two cases.

If $x, y \in [0,1]$ are such that $G(x,y) = 1$ then $F_1(x,y) = S(k, H(x,y)) = F_2(x,y)$.

If $x, y \in [0,1]$ are such that $H(x,y) = 0$ then $F_1(x,y) = T(k, G(x,y)) = F_2(x,y)$.

Therefore, in both cases we have $F_1 = F_2$.

To prove the converse, assume that for given G, H we have $F_1 = F_2$. Then we also know that functions $G_1(x,y) = I_T(k, F_1(x,y))$ and $H_1(x,y) = J_S(k, F_1(x,y))$ are such that they define the same F_1 and F_2, and satisfy $\min(1 - G_1(x,y), H_1(x,y)) = 0$ for any $x, y \in [0,1]$. ∎

4 Associativity

In this section we investigate the associativity of T-S aggregation functions. We have the following general result.

Theorem 3 *Let F be a T-S aggregation function with $k = F(1,0)$. Then F is associative if and only if one of the following cases occurs:*

(a) $k = 0$ *and F is a t-norm;*

(b) $k = 1$ *and F is a t-conorm;*

(c) $0 < k < 1$, $T = \min$, $S = \max$ *and there exist a t-norm T_1 and a t-conorm S_1 such that*

$$F(x,y) = \begin{cases} kS_1(x/k, y/k) & \text{if } x,y \leq k \\ k + (1-k)T_1\left(\dfrac{x-k}{1-k}, \dfrac{y-k}{1-k}\right) & \text{if } x,y \geq k \\ k & \text{otherwise} \end{cases} \quad (5)$$

Proof. It is obvious that if F is a t-norm or a t-conorm then it is an associative T-S aggregation function for any t-norm T and t-conorm S.

If F has the form (5) then it is obviously associative and nondecreasing function. Moreover,

$$F(x,0) = \begin{cases} x \text{ if } x \leq k \\ k \text{ if } x > k \end{cases} = \min(x, k),$$

and similarly, $F(x,1) = \max(x, k)$. That means that F is a min-max aggregation function.

To prove the converse, suppose F is an associative T-S aggregation function for some t-norm T and t-conorm S. If $k = 0$ then F is a t-norm, while $k = 1$ implies that F is a t-conorm.

Suppose $0 < k < 1$. First we prove that $F(x,0) = \min(k, x)$. Indeed, F is associative, so we have for all $x, y, z \in [0,1]$ that

$$F(x, F(y,z)) = F(F(x,y), z). \quad (6)$$

Let $x = y = 0$ in (6). Then, by definition of a T-S aggregation function, we get

$$T(k, T(k, z)) = T(k, z) \quad \text{for all } z \in [0,1]. \quad (7)$$

If $z = 1$, this implies $T(k,k) = k$. Then, by monotonicity, we have for any $z \geq k$ that
$$k = T(k,k) \leq T(k,z) \leq T(k,1) = k,$$
and this implies that $T(k,z) = k = \min(k,z)$ for $z \geq k$.

Assume now that $z < k$. Because T is continuous, there exists $y \in [0,1]$ such that $T(k,y) = z$ From (7) then we obtain
$$T(k,z) = z = \min(k,z) \quad \text{for all } z < k.$$

In a similar way, we can prove that $S(k,z) = \max(k,z)$ for all $z \in [0,1]$. Next we prove that $F(x,k) = k$ for all $x \in [0,1]$. Indeed, we have
$$F(x,k) = F(x, F(1,0)) = F(F(x,1), 0)$$
$$= T(k, S(k,x)) = \min(k, \max(k,x))$$
$$= k.$$

Assume now that $x \leq k \leq y$. Then $F(x,y) \leq F(k,y) = k$, and $F(x,y) \geq F(x,k) = k$, which imply $F(x,y) = k$ in the present case.

Finally, define two functions T_1 and S_1 by F as follows:
$$T_1(x,y) = \frac{F(k + (1-k)x, k + (1-k)y) - k}{1-k}, \quad x,y \in [0,1],$$
$$S_1(x,y) = \frac{F(kx, ky)}{k}, \quad x,y \in [0,1].$$

Obviously, both T_1 and S_1 are associative, nondecreasing and commutative operations, since F is associative, nondecreasing and commutative. Moreover,
$$T_1(x,1) = \frac{F(k + (1-k)x, 1) - k}{1-k} = \frac{\max(k, k + (1-k)x) - k}{1-k} = x,$$
and
$$T_1(x,0) = \frac{F(k + (1-k)x, k) - k}{1-k} = 0.$$

Next,
$$S_1(x,1) = \frac{F(kx, k)}{k} = 1,$$
and
$$S_1(x,0) = \frac{F(kx, 0)}{k} = \frac{\min(k, kx)}{k} = x.$$

These properties imply that T_1 is a t-norm and S_1 is a t-conorm. Moreover, we obviously have (5). ■

Corollary 1 *The most general associative and idempotent T-S aggregation is the median.*

5 Idempotency

In this section we study the idempotency of T-S aggregation functions defined from a t-norm and a t-conorm. That is, we consider the following two main classes:

$$F_1(x,y) = S(T(k, S_1(x,y)), T_1(x,y)), \quad x,y \in [0,1], \tag{8}$$
$$F_2(x,y) = T(S(k, T_1(x,y)), S_1(x,y)), \quad x,y \in [0,1], \tag{9}$$

where T_1 is a t-norm and S_1 is a t-conorm, and determine conditions so that either F_1 or F_2 is idempotent: $F_1(x,x) = x$ or $F_2(x,x) = x$ for all $x \in [0,1]$.

We distinguish the usual cases as follows, starting with the extremes 'min' and 'max'.

Theorem 4 *Suppose $T = \min$ and $S = \max$. Then a min-max aggregation function F with $k = F(1,0)$ is idempotent if and only if $F(x,y) = \mathrm{med}(x,y,k)$, $x,y \in [0,1]$.*

Proof. As we have seen in Proposition 2, in the present case $F_1 = F_2 = F$ with $F(x,y) = \mathrm{med}(k, S_1(x,y), T_1(x,y))$ for $x,y \in [0,1]$.

If there exists $x \in [0,1]$ such that $T_1(x,x) < k < S_1(x,x)$ then we have $k = F(x,x) = x$, a contradiction. Thus, for all $x \in [0,1]$ we have either $T_1(x,x) \leq S_1(x,x) \leq k$, or $k \leq T_1(x,x) \leq S_1(x,x)$.

In the first case it follows that $x \leq k$, and $S_1(x,x) = x$, while in the second case we have $x \geq k$ and $T_1(x,x) = x$. These together imply that $S_1(x,y) = \max(x,y)$ for $x,y \leq k$, and $T_1(x,y) = \min(x,y)$ for $x,y \geq k$.

These properties imply that $F(x,k) = k$ for any $x \in [0,1]$, whence we obtain that $F(x,y) = \mathrm{med}(x,y,k)$. ■

Now we turn to the study the nilpotent case.

Theorem 5 *Let $T(x,y) = \max(x+y-1, 0)$ and $S(x,y) = \min(x+y, 1)$, $x,y \in [0,1]$, in (8) and (9). Then F_1 and F_2 are continuous and idempotent if and only if either $F_1 = F_2 = \min$ or $F_1 = F_2 = \max$.*

Proof. F_1 and F_2 are idempotent if and only if for all $x \in [0,1]$ we have

$$F_1(x,x) = \min(\max(k + S_1(x,x) - 1, 0) + T_1(x,x), 1) = x,$$
$$F_2(x,x) = \max(\min(k + T_1(x,x), 1) + S_1(x,x) - 1, 0) = x.$$

These equalities are certainly hold for $x = 0$ and for $x = 1$. Thus, assume that $x \in (0,1)$.

If $k = 0$ then the above equations are of the form for $x \in (0,1)$:

$$T_1(x,x) = x,$$
$$T_1(x,x) + S_1(x,x) - 1 = x.$$

The first equation implies that $T_1 = \min$, and this together with the second equation imply that $S_1 = S_s$, the strongest t-conorm. Therefore, we have that $F_1 = F_2 = \min$.

If $k = 1$ then we have for all $x \in (0,1)$ that
$$T_1(x,x) + S_1(x,x) = x,$$
$$S_1(x,x) = x.$$

Then it follows that $S_1 = \max$ and $T_1 = T_w$, the weakest t-norm. Hence, we have $F_1 = F_2 = \max$.

Suppose $0 < k < 1$. If there is an $x \in [0,1]$ such that $S_1(x,x) < 1-k$ then we have
$$x = T_1(x,x) = \max(0, k + T_1(x,x) + S_1(x,x) - 1),$$
which is possible in the present case if and only if $T_1(x,x) = 0 = x$. Similarly, $T_1(x,x) > 1 - k$ implies $x = 1$.

Therefore, we have the inequality $T_1(x,x) \leq 1-k \leq S_1(x,x)$ for all $x \in (0,1)$, whence we have
$$T_1(x,y) \leq 1 - k \leq S_1(x,y) \quad \text{for all } x, y \in (0,1).$$

This inequality implies that the partial mapping $T_1(x,.)$ of T_1 cannot be continuous at the point 1 when $x > 1 - k$. Similarly, $S_1(x,.)$ cannot be continuous at 0 when $x < 1 - k$, whence discontinuity of F_1 and F_2 at the corresponding points follow.

Thus, we have shown that $0 < k < 1$ implies that F_1 and F_2 are not continuous, which completes the proof. ∎

Although the only continuous idempotent W-W' aggregation functions are 'min' and 'max', we show some other interesting non-continuous classes in the following examples when $0 < k < 1$.

Example 4 Define T_1 and S_1 as follows:
$$T_1(x,y) = \begin{cases} \min(x,y) & \text{if } \max(x,y) = 1 \\ \min\{\max(x-k,0), \max(y-k,0)\} & \text{otherwise} \end{cases},$$

and
$$S_1(x,y) = \begin{cases} \max(x,y) & \text{if } \min(x,y) = 0 \\ \max\{\min(1+x-k,1), \min(1+y-k,1)\} & \text{otherwise} \end{cases}.$$

Then we have $T_1(x,y) \leq 1 - k \leq S_1(x,y)$ for $x,y \in (0,1)$, and
$$F_1(x,y) = F_2(x,y) = \begin{cases} \text{med}(x,y,k) & \text{if } x,y \in (0,1) \\ \max(k+x-1,0) & \text{if } y = 0 \\ \min(k+x,1) & \text{if } y = 1 \end{cases},$$

which is a non-continuous version of the median. ∎

Example 5 Now we define T_1 and S_1 by the following formulas:

$$T_1(x,y) = \begin{cases} \min(x,y) & \text{if } \max(x,y) = 1 \text{ or } x,y \leq 1-k \\ 1-k & \text{otherwise} \end{cases},$$

and

$$S_1(x,y) = \begin{cases} \max(x,y) & \text{if } \min(x,y) = 0 \text{ or } x,y \geq 1-k \\ 1-k & \text{otherwise} \end{cases}.$$

Then we have $T_1(x,y) \leq 1-k \leq S_1(x,y)$ for $x,y \in (0,1)$, and

$$F_1(x,y) = F_2(x,y) = \begin{cases} \min(x,y) & \text{if } 0 < x,y \leq 1-k \\ \max(x,y) & \text{if } 1 > x,y \geq 1-k \\ \min(k+x,1) & \text{if } y = 1 \\ \max(x+k-1,0) & \text{if } y = 0 \\ 1-k & \text{otherwise} \end{cases}.$$

Although this operation is not associative, it is linked with the so-called *uninorms* (for details, see Yager and Rybalov (1996), Fodor et al. (1996)). ■

Example 6 Let $k = 1/2$ and T_1 and S_1 be defined as follows:

$$T_1(x,y) = \begin{cases} \min(x,y) & \text{if } \max(x,y) = 1 \\ \dfrac{xy}{2} & \text{otherwise} \end{cases},$$

and

$$S_1(x,y) = \begin{cases} \max(x,y) & \text{if } \min(x,y) = 0 \\ 1 - \dfrac{(1-x)(1-y)}{2} & \text{otherwise} \end{cases}.$$

Then again we have $T_1(x,y) \leq 1/2 \leq S_1(x,y)$ for $x,y \in (0,1)$, and

$$F_1(x,y) = F_2(x,y) = \begin{cases} \dfrac{x+y}{2} & \text{if } x,y \in (0,1) \\ \min(x+1/2,1) & \text{if } y = 1 \\ \max(x-1/2,0) & \text{if } y = 0 \end{cases}.$$

■

Finally we consider the case when $T(x,y) = xy$ and $S(x,y) = x+y-xy$ for any $x,y \in [0,1]$. Then the equations to be solved are given as follows:

$$kS_1(x,x) + T_1(x,x) - kS_1(x,x)T_1(x,x) = x, \quad \text{for all } x \in [0,1],$$
$$kS_1(x,x) + T_1(x,x)S_1(x,x) - kS_1(x,x)T_1(x,x) = x, \quad \text{for all } x \in [0,1].$$

If $k = 0$ then it follows that $T_1 = \min$ and $S_1 = S_s$ introduced in the previous proof. Hence, in this case $F_1 = F_2 = \min$.

Similarly, if $k = 1$ then $S_1 = \max$ and $T_1 = T_w$, whence $F_1 = F_2 = \max$.

Obviously, if $0 < k < 1$ then the above equations imply that $T_1(x,x)(1 - S_1(x,x)) = 0$ for all $x \in [0,1]$, whence we have only one equation to be solved, and it can be written in the following form:

$$kS_1(x,x) + (1-k)T_1(x,x) = x, \quad \text{for all } x \in [0,1],$$

where now T_1 is a nilpotent t-norm and S_1 is a nilpotent t-conorm and $0 < k < 1$. If, in addition, we assume that both T_1 and S_1 are Archimedean and S_1 is the N-dual of T_1 with some strong negation N, then the last equation was completely solved by Mayor (1994). We cite his result now.

Theorem 6 *For any given k with $0 < k < 1$, there exists a unique nilpotent Archimedean t-norm T_k such that its corresponding aggregation operator F_k is idempotent. In particular, F_k is a quasi-arithmetic mean of the form*

$$F_k(x,y) = \varphi_k^{-1}\left(\frac{\varphi_k(x) + \varphi_k(y)}{2}\right),$$

where φ_k is a particular automorphism of the unit interval.

Note that the automorphism φ_k is the inverse of the solution of the so-called De Rham system of functional equations for any fixed $0 < k < 1$. For more details on that system we refer to De Rham (1956).

6 Acknowledgment

The paper has been written when J. Fodor was a visiting professor at the University of the Balearic Islands, Palma de Mallorca. His research has also been supported in part by OTKA (National Scientific Research Fund, Hungary) I/6–14144.

7 References

1. G. De Rham, Sur quelques courbes définies par des équations fonctionelles, *Rend. Sem. Mat. Univ. Politec. Torino* **16** (1956) 101–113.
2. J. Fodor and M. Roubens, *Fuzzy Preference Modelling and Multicriteria Decision Support* (Kluwer, Dordrecht, 1994).
3. J.C. Fodor, R.R. Yager and A. Rybalov, Structure of uni-norms, *Inf. Sci.* (to appear).
4. L.W. Fung and K.S. Fu, "An axiomatic approach to rational decision making in a fuzzy environment", in: *Fuzzy Sets and Their Applications to Cognitive and Decision Processes*, eds. L.A. Zadeh et al. (Academic Press, New York, 1975) pp. 227–256.
5. C.H. Ling, Representation of associative functions, *Publ. Math. Debrecen* **12** (1965) 182–212.

6. G. Mayor, On a family of quasi-arithmetic means, *Aeq. Math.* **48** (1994) 137–142.
7. G. Mayor and E. Trillas, On the representation of some aggregation functions, *Proc. ISMVL* (1986) 110–114.
8. R.R. Yager and A. Rybalov, Uninorm aggregation operators, *Fuzzy Sets and Systems* **80** (1996) 111–120.
9. H.-J. Zimmermann and P. Zysno, Latent connectives in human decision making, *Fuzzy Sets and Systems* **4** (1980) 37–51.

2

AGGREGATION IN DECISION MAKING AND CONTROL

Fuzzy Integral as a Flexible and Interpretable Tool of Aggregation

Michel GRABISCH

Thomson-CSF, Central Research Laboratory

Domaine de Corbeville, 91404 Orsay Cedex, France

Abstract

The fuzzy integral with respect to a fuzzy measure has been used in many applications of multicriteria evaluation. We present here its properties for aggregation and its situation among common aggregation operators. The concept of Shapley value and interaction index, which are well rooted in a theory of representation of fuzzy measures, can afford a semantical analysis of the aggregation operation, and facilitate the use of fuzzy integral in practical problems.

1 Introduction: a Historical Perspective

Since its origin (Sugeno, 1974 [35]), it has been felt that fuzzy integral with respect to a fuzzy measure could be favourably used in problems of evaluation involving several criteria. The argument which was put ahead at this time was that fuzzy measure is more flexible than probability, which is stuck into its additivity property: the importance of two criteria in the probability framework can be nothing else than the sum of the individual importances, while with fuzzy measures, it can be greater or lower, allowing the modelling of interaction phenomena between criteria. Many applications in Japan were conducted along this line, during the eighties [19, 31, 41, 18, 43] (see also a summary in [14], and a short description in [11]), showing that the method was effectively successfull and could bring new insights into the problem of aggregation.

Despite its richness and appeal, the method appeared to be difficult to use and the interpretation of results not as intuitive as expected. This was due to essentially two facts:

- a problem involving n criteria requires the determination of 2^n coefficients (the fuzzy measure). The difficulty lies more in the proper identification of these coefficients than in the computational complexity.

- although it was felt that fuzzy measure can model interaction among criteria, the exact way to proceed was not clear. In fact, in the above cited works, there was no explicit reference to known concepts of multicriteria decision making, or multiattribute utility theory, which could have been the basis for a consistent view of the interaction phenomenon.

In the nineties, more fundamental research has been done in order to cope with the problem of complexity and interpretation of fuzzy measures, to situate them into the contexts of multicriteria decision making, multiattribute utility theory and game theory, and many new ideas have been put forward. Some of the most significant advances are listed below:

- referring to cooperative game theory, where importance of coalitions of players are modelled by fuzzy measures (which are called there *characteristic form of a game*), Murofushi proposed the use of the Shapley value [34] in order to model the importance of criteria [25].

- borrowing concepts from multiattribute utility theory [20], Murofushi and Soneda defined the interaction index between two criteria [26]. This definition replaced the rather *ad hoc* definitions of Ishii and Sugeno [19].

- Murofushi and Sugeno gave a clear interpretation of the meaning of independance of criteria, relating to the concept of mutual preferential independence used in multiattribute utility theory [20]. They showed that if the utility function is a Choquet integral, then there is equivalence between additivity of the fuzzy measure and mutual preferential independence [29].

- the author has tried to imbed fuzzy integral in a true context of multicriteria decision making, showing its properties for aggregation, some characterizations results, and its relation to common aggregation operators [8, 10].

- a powerful (although suboptimal) and quick algorithm for identification of fuzzy measures has been proposed by the author [9]. It allows the identification using few learning data.

- in order to cope with the exponential complexity of fuzzy measures, the author has introduced the concept of k-order additive fuzzy measure, which can range freely between the additive case (simple but too poor) and the general case (rich but too complicated) [12, 13]. Also, he generalizes the Shapley value and the interaction index of Murofushi and Soneda to an interaction index among any subset of criteria.

- several many other ideas, which will not be discussed here, including the hierarchical decomposition of Choquet integral of Fujimoto *et al.* [7, 37, 36], and some new generalisation of the fuzzy integral by Imaoka [17].

In order to solve the problem of exponential complexity of fuzzy measures, Fujimoto has looked for conditions for a Choquet integral to be decomposed into several Choquet integrals with respect to fuzzy measures defined on subsets of criteria. His results has been applied to practical cases by Sugeno and Kwon [38, 39].

Imaoka has found a common representation of Sugeno and Choquet integrals, based on cumulative distribution functions. His work gives new directions for a general definition of the fuzzy integral, a subject which has been often researched in the past by e.g. Kruse [22], Weber [44], Ichihashi et al. [16], Murofushi and Sugeno [27], and Mesiar [23, 24].

This paper will develop what is thought to be fundamental in a practical application of fuzzy integral to multicriteria decision problem. We will concentrate mainly on interpretation and identification issues, introducing the concept of k-order additive measure and general interaction index.

As this is explained elsewhere in this book, we will not present at length the framework for multicriteria decision making, and we will restrict ourself to a quick presentation of fundamental properties for aggregation, which are necessary for the paper to be self-contained. For the sake of simplicity, we will consider that we aggregate scores with respect to criteria into a single score representing the global value of the alternative to be evaluated, avoiding the presentation of fuzzy preference relations which constitute a better framework for decision (see Fodor and Roubens [6]). More specifically, we will denote by $X = \{1, \ldots, n\}$ the set of criteria. An alternative a will be represented by a n-dimensional vector of scores in $[0, 1]$, denoted $[a_1 \cdots a_n]^T$, where a_i is the score of a with respect to the criterion i. Fuzzy measures and integrals will be then defined on a finite set, thus avoiding measure-theoretic intricacies. The reader is referred to the monographs of Denneberg [3], Grabisch et al. [14], Sugeno and Murofushi [40], and Wang and Klir [42] for a more complete treatment.

Throughout the paper, *min* and *max* will be denoted by $\wedge$ and $\vee$ respectively.

2 Basic Definitions of Fuzzy Integrals

Let us denote by $X = \{x_1, \ldots, x_n\}$ the set of criteria, and by $\mathcal{P}(X)$ the power set of X, i.e. the set of all subsets of X.

Definition 1 *A fuzzy measure on X is a set function $\mu : \mathcal{P}(X) \longrightarrow [0, 1]$, satisfying the following axioms.*

(i) $\mu(\emptyset) = 0, \mu(X) = 1$.

(ii) $A \subset B$ implies $\mu(A) \leq \mu(B)$, for $A, B \in \mathcal{P}(X)$.

$\mu(A)$ can be viewed as the weight of importance of the set of criteria A. Thus, in addition to the weights on criteria taken separately which are often used in multicriteria decision making methods, weights on any combination of criteria are also defined.

A fuzzy measure is said to be *additive* if $\mu(A \cup B) = \mu(A) + \mu(B)$ whenever $A \cap B = \emptyset$, *superadditive* (resp. *subadditive*) if $\mu(A \cup B) \geq \mu(A) + \mu(B)$ (resp. $\mu(A \cup B) \leq \mu(A) + \mu(B)$) whenever $A \cap B = \emptyset$. If a fuzzy measure is additive, then it suffices to define the n coefficients (weights) $\mu(\{x_1\}), \ldots, \mu(\{x_n\})$ to define entirely the measure. In general, one needs to define the $2^n - 2$ coefficients corresponding to the 2^n subsets of X, except $\emptyset$ and X itself.

Another particular case of fuzzy measure is the *0-1 fuzzy measure*, whose values are either 0 or 1. An example of 0-1 fuzzy measure is common in game theory: a fuzzy measure is said to be a *unanimity game* u_A for subset $A \subset X$ when $u_A(B) = 1$ if and only if $B \supset A$, and is zero otherwise. For any fuzzy measure μ, the dual fuzzy measure μ^* is defined by $\mu^*(A) = 1 - \mu(A^c)$ for any $A \subset X$.

We introduce now the concept of discrete fuzzy integrals, viewed as aggregation operators. For this reason, we will adopt a connective-like notation instead of the usual integral form, and the integrand will be a set of n values $a_1, \ldots, a_n$ in $[0, 1]$.

Definition 2 *Let μ be a fuzzy measure on X. The* discrete Sugeno integral *of $a_1, \ldots, a_n$ with respect to μ is defined by :*

$$\mathcal{S}_\mu(a_1, \ldots, a_n) := \bigvee_{i=1}^{n} (a_{(i)} \wedge \mu(A_{(i)})),$$

where $a_{(i)}, i = 1, \ldots, n$ indicates a permutation of $a_i, i = 1, \ldots, n$ such that $a_{(1)} \leq a_{(2)} \leq \cdots \leq a_{(n)}$, and $A_{(i)} := \{x_{(i)}, \ldots, x_{(n)}\}$.

Another definition was proposed later by Murofushi and Sugeno [28], using a concept introduced by Choquet in capacity theory [2].

Definition 3 *Let μ be a fuzzy measure on X. The* discrete Choquet integral *of $a_1, \ldots, a_n$ with respect to μ is defined by*

$$\mathcal{C}_\mu(a_1, \ldots, a_n) := \sum_{i=1}^{n} (a_{(i)} - a_{(i-1)}) \mu(A_{(i)}),$$

with the same notations as above, and $a_{(0)} = 0$.

An equivalent expression of (3) is

$$\mathcal{C}_\mu(a_1, \ldots, a_n) = \sum_{i=1}^{n} a_{(i)} [\mu(A_{(i)}) - \mu(A_{(i+1)})], \tag{1}$$

with $A_{(n+1)} = \emptyset$.

Sugeno and Choquet integrals are essentially different in nature, since the former is based on non linear operators (*min* and *max*), and the latter on usual linear operators. Both compute a kind of distorted average of $a_1, \ldots, a_n$. More general definitions exist but will not be considered here (see the introduction above, and also [8, 14]).

3 The Fuzzy Integral as an Aggregation Operator

3.1 Properties for Aggregation

We begin this section by giving a short list of properties that should be verified by a "good" aggregation operator. These properties are divided into *mathematical* properties and *behavioral* properties. Usually, attention is given only to the mathematical ones, but in multicriteria decision problems, one should be careful also on the so-called behavioral properties. A thorough survey of properties related to aggregation can be found in [6, 5].

In this paragraph, $\mathcal{H}$ denotes an aggregation operator. When necessary, the number of arguments appears in superscript, e.g. $\mathcal{H}^{[3]}(a_1, a_2, a_3)$.

mathematical properties

- worst and best alternatives
$$\mathcal{H}(0, \ldots, 0) = 0, \quad \mathcal{H}(1, \ldots, 1) = 1.$$
A stronger requirement is idempotence (or unanimity, agreement):
$$\mathcal{H}(a, a, \ldots, a) = a, \forall a.$$

- continuity: $\mathcal{H}$ is a continuous function of $a_1, \ldots, a_n$.

- monotonicity (non decreasingness) with respect to each argument:
$$a'_i > a_i \Rightarrow \mathcal{H}(a_1, \ldots, a'_i, \ldots, a_n) \geq \mathcal{H}(a_1, \ldots, a_i, \ldots, a_n).$$
This means that increasing a partial score cannot decrease the result.

- neutrality (or commutativity, anonymity)
$$\mathcal{H}(a_1, \ldots, a_n) = \mathcal{H}(a_{\sigma(1)}, \ldots, a_{\sigma(n)}),$$
for all permutation σ in $\{1, \ldots, n\}$. This is required when combining criteria of equal importance or anonymous expert's opinions.

- compensativeness
$$\bigwedge_{i=1}^{n} a_i \leq \mathcal{H}(a_1, \ldots, a_n) \leq \bigvee_{i=1}^{n} a_i.$$

- associativity. Suppose an operator $\mathcal{H}$ is defined for two arguments. Associativity permits to extend it for more arguments unambiguously.
$$\mathcal{H}^{[3]}(a_1, a_2, a_3) = \mathcal{H}^{[2]}(\mathcal{H}^{[2]}(a_1, a_2), a_3) = \mathcal{H}^{[2]}(a_1, \mathcal{H}^{[2]}(a_2, a_3)).$$

- stability for the same ordinal transformation. Suppose ϕ is a continuous, strictly increasing mapping from **R** to **R**. Then
$$\mathcal{H}(\phi(a_1), \ldots, \phi(a_n)) = \phi(\mathcal{H}(a_1, \ldots, a_n)).$$
A particular case which is useful in practice is the case of positive linear transformation:
$$\mathcal{H}(ra_1 + t, \ldots, ra_n + t) = r\mathcal{H}(a_1, \ldots, a_n) + t,$$
with $r > 0$, $t \in \mathbf{R}$. These properties are closely related to the choice of a measurement scale.

- decomposability (Kolmogoroff [21])

$$\mathcal{H}^{[n]}(a_1, a_k, a_{k+1}, \ldots, a_n) = \mathcal{H}^{[n]}(\underbrace{a, \ldots, a}_{k \text{ times}}, a_{k+1}, \ldots, a_n),$$

where $a = \mathcal{H}^{[k]}(a_1, \ldots, a_k)$.

behavioral properties

- possibility of expressing weights of importance on criteria if this is necessary.

- possibility of expressing the behavior of the decision maker. This refers to any typical behaviors which can be observed on decision makers. Most common in multicriteria decision making are disjuntive and conjunctive behaviors. A disjunctive behavior is a tolerant behavior, where the satisfaction of one of the criteria, or few criteria, entails the global satisfaction. On the opposite, a conjunctive behavior is an intolerant behavior, where the satisfaction of all the criteria are necessary to obtain the global satisfaction. Other examples are veto attitudes on some criteria, which mean that the non satisfaction of these criteria entails the global non satisfaction.

- possibility of expressing various effects of interaction between criteria. For example, possible interactions between criteria are *redundancy* (two criteria are redundant if they express more or less the same thing) and *support* or *reinforcement* (two criteria with little importance when taken separately, become very important when considered jointly).

- possibility of an easy semantical interpretation, i.e. being able to relate the values of parameters defining $\mathcal{H}$ to the behavior implied by $\mathcal{H}$ in a decision making point of view. This point is crucial and forbids the use of a "black box" methodology, e.g. neural nets.

3.2 Common Aggregation Operators

We give a short list of common aggregation operators, belonging to the class of averaging operators (i.e. being compensative, monotone and idempotent), since they are the most suitable for multicriteria decision problems.

quasi-arithmetic means. the easiest way to aggregate is the simple arithmetic mean $1/n \sum_i a_i$. Many other means exist, such as geometric, harmonic means, etc. In fact all these common means belong to the family of quasi-arithmetic means, defined as follows.

$$M_f(a_1, \ldots, a_n) = f^{-1} \left[\frac{1}{n} \sum_{i=1}^{n} f(a_i) \right] \tag{2}$$

where f is any continuous strictly monotonic function. This family has been characterized by Kolmogoroff [21], as being the class of all decomposable commutative averaging operators. It is easy to extend the definition by introducing weights of importance $w_1, \ldots, w_n$ on criteria, with the constraint $\sum_{i=1}^{n} w_i = 1$.

$$M^f_{w_1,\ldots,w_n}(a_1,\ldots,a_n) = f^{-1}\left[\sum_{i=1}^{n} w_i f(a_i)\right] \quad (3)$$

median.

$$\text{med}(a_1,\ldots,a_n) = \begin{cases} a_{(\frac{n+1}{2})}, & \text{if } n \text{ is odd} \\ \frac{1}{2}(a_{(\frac{n}{2})} + a_{(\frac{n}{2}+1)}), & \text{if } n \text{ is even.} \end{cases} \quad (4)$$

where parentheses () around the index show that elements have been arranged in increasing order, i.e. $a_{(1)} \leq a_{(2)} \leq \cdots \leq a_{(n)}$. Weights of importance on criteria can be modelled by duplicating the corresponding elements in the list. *Order statistics*, which give as output the kth value of the ordered list, are a generalisation of the median.

weighted minimum and maximum. They have been introduced in [4] in the framework of possibility theory, and are a generalisation of *min* and *max*. They are defined as follows.

$$\text{wmin}_{w_1,\ldots,w_n}(a_1,\ldots,a_n) = \bigwedge_{i=1}^{n} [(1 - w_i) \vee a_i] \quad (5)$$

$$\text{wmax}_{w_1,\ldots,w_n}(a_1,\ldots,a_n) = \bigvee_{i=1}^{n} [w_i \wedge a_i] \quad (6)$$

where weights are normalized so that $\bigvee_{i=1}^{n} w_i = 1$.

ordered weighted averaging operators (OWA). They have been introduced by Yager [45, 46].

$$\text{OWA}_{w_1,\ldots,w_n}(a_1,\ldots,a_n) = \sum_{i=1}^{n} w_i a_{(i)} \quad (7)$$

where $\sum_{i=1}^{n} w_i = 1$, and $a_{(1)} \leq a_{(2)} \leq \cdots \leq a_{(n)}$ as above. Taking all weights equal to $1/n$ leads to the arithmetic mean, while the *min* operator (resp. *max*, the median, order statistics) can be recovered by taking all weights equal to 0 except $w_1 = 1$ (resp. $w_n = 1$, $w_{\frac{n+1}{2}} = 1$, $w_k = 1$).

3.3 Properties of Fuzzy Integral for Aggregation

The following properties of fuzzy integrals show their adequacy to aggregation in multicriteria decision problems.

- the Sugeno and Choquet integral are idempotent, continuous, monotonic, and compensative operators.

- the Choquet integral is stable under positive linear transformations. The Sugeno integral does not share this property, but satisfies a similar property with *min* and *max* replacing product and sum. In this sense, it can be said that the Choquet integal is suitable for cardinal aggregation (where numbers have a real meaning), while the Sugeno integral seems to be more suitable for ordinal aggregation (where only order makes sense).

- if μ is a 0-1 fuzzy measure, then the Choquet and the Sugeno integral take the following form [30]:

$$\mathcal{C}_\mu(a_1,\ldots,a_n) = \mathcal{S}_\mu(a_1,\ldots,a_n) = \bigvee_{A:\mu(A)=1} \left(\bigwedge_{x_i \in A} a_i \right).$$

- commutative fuzzy integrals are obtained only when the fuzzy measure satisfies $\mu(A) = \mu(B)$ if $|A| = |B|$, where $|A|$ denotes the cardinal of A.

- Sugeno integrals with respect to a fuzzy measure satisfying $\mu(A) = \alpha, \forall A \neq \emptyset, X$ are associative.

These are only mathematical properties. The whole section 4 will be devoted entirely to the behavioral properties of the Choquet integral, since this is a much more complex subject.

3.4 Relation of Fuzzy Integral with Common Operators

The main results are summarized below (see Grabisch *et al.* [8, 14] and Murofushi and Sugeno [30]).

- the Choquet integral with respect to an additive measure μ coincides with a weighted arithmetic mean, whose weights w_i are $\mu(\{x_i\})$.

- any OWA operator with weights $w_1, \ldots, w_n$ is a Choquet integral, whose fuzzy measure μ is defined by

$$\mu(A) = \sum_{j=0}^{i-1} w_{n-j}, \quad \forall A \text{ such that } |A| = i,$$

where $|A|$ denotes the cardinal of A. Reciprocally, any commutative Choquet integral is such that $\mu(A)$ depends only on $|A|$, and coincides with an OWA operator, whose weights are $w_i = \mu(A_{n-i+1}) - \mu(A_{n-i}), i = 2, \ldots, n$, and $w_1 = 1 - \sum_{i=2}^{n} w_i$. A_i denotes any subset such that $|A_i| = i$.

- the Sugeno and the Choquet integral contains all order statistics, thus in particular, min, max and the median. The fuzzy measure μ corresponding to the k order statistic is defined by $\mu(A) = 0$ if $|A| \leq n - k$, $\mu(A) = 1$ otherwise.

- weighted minimum and weighted maximum are particular cases of Sugeno integral. The fuzzy measure corresponding to $\text{wmax}_{w_1,\ldots,w_n}$ is defined by

$$\mu(A) = \bigvee_{x_i \in A} \mu(\{x_i\}),$$

while for $\text{wmin}_{w_1,\ldots,w_n}$

$$\mu(A) = 1 - \bigvee_{x_i \notin A} \mu(\{x_i\}).$$

In the above, $\mu(\{x_i\}) = w_i$. It can be noticed that μ is a possibility measure in the case of wmax, and a necessity measure for wmin.

- any arbitrary combination of minimum and maximum can be represented by a Choquet or a Sugeno integral with respect to a suitable 0-1 fuzzy measure. For example, $(a_1 \wedge a_3) \vee (a_2 \wedge a_4)$ is represented by a fuzzy measure μ with $\mu(\{x_1, x_3\}) = \mu(\{x_2, x_4\}) = 1$, the other coefficients being 0 (this is obvious from the property of 0-1 fuzzy measures given in section 3.3).

Figure 1 gives a summary of all set relations between various aggregation operators and fuzzy integrals.

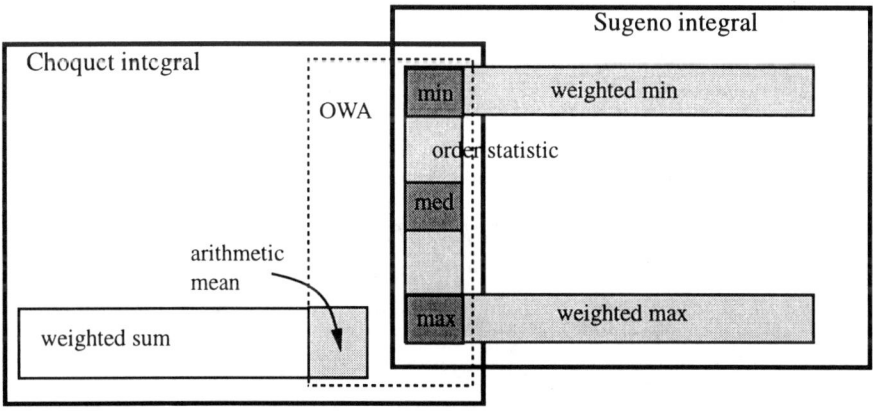

Figure 1: Set relations between various aggregation operators and fuzzy integrals

4 Behavioral Analysis of Fuzzy Measure and Choquet Integral

This section is devoted to a deep analysis of the Choquet integral and fuzzy measures, in terms of behavior in multicriteria decision making. Recent advances on this subject have permitted to make a significant progress in the understanding of the mechanism of aggregation of the Choquet integral.

In the sequel, the set of criteria will be denoted as $X = \{1, \ldots, n\}$ to avoid heavy notations. Also $\mu(\{i\}), \mu(\{i,j\}), \mu(A)$ and $\mu(A \cup B)$ will be often denoted by μ_i, μ_{ij}, μ_A and μ_{AB} respectively, and similarly for any set function.

4.1 The Representation of Importance and Interaction

In the context of cooperative game theory, Shapley has introduced the following index, representing the importance of a player in the game, based on axiomatic considerations.

Definition 4 *(Shapley [34]) Let μ be a fuzzy measure on X. The Shapley index for every $i \in X$ is defined by*

$$v_i := \sum_{k=0}^{n-1} \gamma_k \sum_{K \subset X \setminus i, |K|=k} (\mu_{iK} - \mu_K), \tag{8}$$

with $\gamma_k := \frac{(n-k-1)!k!}{n!}$. *The* Shapley value *of μ is the vector $v(\mu) := [v_1 \cdots v_n]^t$.*

The Shapley index v_i can be interpreted as a kind of weighted average value of the contribution of element i alone in all coalitions. It satisfies $\sum_{i=1}^{n} v_i = 1$, so that the Shapley value is a kind a sharing out of the total value of the game μ to the individual players. Murofushi proposed in [25] to use the Shapley value in the context of multicriteria decision making.

Murofushi and Soneda, relying on concepts of multiattribute utility theory, have defined an interaction index between two criteria.

Definition 5 *(Murofushi and Soneda [26]) Let μ be a fuzzy measure on X. The* interaction index *of elements i, j is defined by*

$$\mathcal{I}_{ij} := \sum_{k=0}^{n-2} \xi_k \sum_{K \subset X \setminus \{i,j\}, |K|=k} (\mu_{ijK} - \mu_{iK} - \mu_{jK} + \mu_K), \tag{9}$$

with $\xi_k := \frac{(n-k-2)!k!}{(n-1)!}$.

The interaction index $\mathcal{I}_{ij}$ can be interpreted as a kind of average value of the *added value* given by putting i and j together, all coalitions being considered. The following cases can happen:

- $\mathcal{I}_{ij} > 0$: i and j are said to be *complementary*, or to have a *positive synergy*, since put together they become very important.

- $\mathcal{I}_{ij} < 0$: i and j are said to be *substitutive*, or to have a *negative synergy* since adding one or the either or both to a coalition is not very significant.

- $\mathcal{I}_{ij} = 0$: i and j have no interaction, they are said to be *independent*.

We will come back to this interpretation later.

Relying on the analogy between (8) and (9), let us define interaction indexes for more than two elements.

Definition 6 *(Grabisch [12]) Let μ be a fuzzy measure on X. The* interaction index *of elements among $A \subset X$, $A \neq \emptyset$, is defined by:*

$$\mathcal{I}(A) := \sum_{k=0}^{n-|A|} \Xi_k^{|A|} \sum_{K \subset X \setminus A, |K|=k} \sum_{L \subset A} (-1)^{|A|-|L|} \mu_{LK}, \tag{10}$$

with $\Xi_k^p := \frac{(n-k-p)!k!}{(n-p+1)!}$.

Clearly, $v_i = \mathcal{I}(\{i\})$, $i = 1, \ldots, n$, and we recover (9) substituting A by $\{i,j\}$ in (10).

Since (10) is defined for every $A \subset X$, $A \neq \emptyset$, putting $\mathcal{I}_\emptyset = 0$, we have defined a set function on X, constructed from μ. In fact, as this will be explained in the next section, $\mathcal{I}$ is another representation of a fuzzy measure, very useful for the decision maker. This is made clear in the next section.

4.2 Three Representations of a Fuzzy Measure

Including the usual measure form of a fuzzy measure, there are three possible representations of a fuzzy measure :

the (usual) measure representation: the set function $\mu : \mathcal{P}(X) \longrightarrow [0,1]$, determined by $\mu(A)$, for all $A \subset X$. These are 2^n coefficients in $[0,1]$, which can be put into a lattice isomorphic to the lattice of subsets ordered by inclusion.

the Möbius representation: the Möbius inversion formula is known in the evidence theory of Shafer [33], to compute the basic probability assignment from a belief function. This formula is in fact known in combinatorics since a long time (see e.g. Rota [32]), and can be applied to any set function. Let us denote m the set function from $\mathcal{P}(X)$ to **R** which is the transformation of μ by the Möbius transform. It is defined by:

$$m(A) := \sum_{B \subset A} (-1)^{|A \setminus B|} \mu(B), \tag{11}$$

for any $A \subset X$. This is again a set of 2^n coefficients, but in **R** and without any lattice structure.

the interaction representation: it is given by the formula (10), for any $\emptyset \neq A \subset X$, and $\mathcal{I}(\emptyset) := 0$. This is also a set function, whose values range in **R**, and without any lattice structure.

Of course, these are not the only possible alternative representations of a fuzzy measure, but these ones present a particular interest in a decision making framework.

We give below relations between these different representations.

- passage from m to μ (this is known as the zeta transform):

$$\mu(A) = \sum_{B \subset A} m(B), \forall A \subset X. \tag{12}$$

- passage from m to $\mathcal{I}$:

$$\mathcal{I}_A = \sum_{B \subset X \setminus A, |B| \leq n - |A|} \frac{1}{|B| + 1} m_{AB}, \quad \forall A \subset X, A \neq \emptyset. \tag{13}$$

- passage from $\mathcal{I}$ to m:

$$m_A = \sum_{j=1}^{n-|A|+1} \alpha_j \sum_{B \subset X \setminus A, |B| = j-1} \mathcal{I}_{AB}, \quad \forall A \subset X, A \neq \emptyset, \tag{14}$$

where the coefficients α_j are obtained in a recursive way by the formula

$$\alpha_j := -\sum_{l=1}^{j-1} \frac{\alpha_l}{j - l + 1} C_{j-1}^{j-l},$$

and $\alpha_1 := 1$. First values are $\alpha_2 = -1/2, \alpha_3 = 1/6, \alpha_4 = 0, \alpha_5 = -1/30$, etc.

The set of equations (10), (11), (12), (13) and (14) allows to pass freely from one representation to another.

4.3 k-order Additive Fuzzy Measures

In the introduction, it was remarked that the complexity of fuzzy measures was exponential, thus limiting their application. Many people have tried to overcome this difficulty by using decomposable measures, which require only n coefficients to be defined: these are for examples probability measures (additive measures), λ-measures introduced by Sugeno, etc. As a matter of fact, additive measures, as well as decomposable measures, are far too simple when considered as a modelling of strength of coalitions. In [12], the author has proposed the concept of *k-order additive fuzzy measures*, which can range freely from purely additive measures to general fuzzy measures. In fact, this concept is related to the Möbius transform.

Let us consider another way of introducing the Möbius representation. In the field of complexity analysis, pseudo-Boolean functions are often used. A *pseudo-Boolean function* is simply a real valued function $f : \{0,1\}^n \longrightarrow \mathbf{R}$ (see e.g. [15]). It is easy to see that a fuzzy measure is a particular case of pseudo-Boolean function: simply remark that for any $A \subset X$, A is equivalent to a point $(x_1,\ldots,x_n)$ in $\{0,1\}^n$ such that $x_i = 1$ iff $i \in A$. It can be shown that any pseudo-Boolean function can be put under a multilinear polynomial in n variables:

$$f(x) = \sum_{T \subset X}\left[a_T \prod_{i \in T} x_i \right], \qquad (15)$$

with $a_T \in \mathbf{R}$ and $x = (x_1,\ldots,x_n) \in \{0,1\}^n$. Clearly, $a_T = m(T)$ for every $T \subset X$ (see formula (12)). For this reason, m will be called either the *Möbius representation* or the *multilinear representation* of μ.

Looking at (15), one can notice that additive measures have a linear representation $f(x) = \sum_{i=1}^{n} a_i x_i$. By extension, we call *k-order additive fuzzy measures* or simply *k-additive measures* any fuzzy measure of which the multilinear expression is of degree k, i.e. $m(A) = 0$, for all A such that $|A| > k$. For example, a 2-additive measure is expressed by:

$$f(x) = a_1 x_1 + a_2 x_2 + \cdots + a_n x_n + a_{12} x_1 x_2 + \cdots + a_{(n-1)n} x_{n-1} x_n.$$

Turning this into the usual measure representation, one can verify easily that only the following coefficients $\mu_1,\ldots,\mu_n, \mu_{12},\ldots,\mu_{(n-1)n}$ are necessary to define the whole 2-additive measure. Indeed:

$$\mu(K) = \sum_{\{i,j\} \subset K} \mu_{ij} - (|K|-2) \sum_{i \in K} \mu_i$$

Thus, k-additive measures can be represented by a limited set of coefficients $\mu(A)$ or $m(A)$, $|A| \leq k$, i.e. at most $\sum_{i=1}^{k} C_n^i$. Furthermore, looking at the interaction representation (10), we see that this representation uses the same number of coefficients as the multilinear representation. Remark also that for any k-additive measure, $m(A) = \mathcal{I}(A)$ for all A such that $|A| = k$.

4.4 Main Properties

We give now some useful properties, focusing on the interaction representation.

Property 1 *(Chateauneuf and Jaffray [1]) A set of 2^n coefficients m_T, $T \subset X$ corresponds to the Möbius representation of a fuzzy measure if and only if*
 (i) $m_\emptyset = 0$, $\sum_{A \subset X} m_A = 1$,
 (ii) $\sum_{i \in B \subset A} m_B \geq 0$, *for all* $A \subset X$, *for all* $i \in A$.

A set of necessary and sufficient conditions on the $\mathcal{I}_A$ coefficients to ensure monotonicity is much more difficult to obtain. Partial results are given in [12].

Property 2 *Let μ be a fuzzy measure. Then the interaction index $\mathcal{I}_{ij}$ ranges in $[-1, +1]$. $\mathcal{I}_{ij} = 1$ if and only if $\mu = u_{\{i,j\}}$, the unanimity game for the pair i,j. Similarly, $\mathcal{I}_{ij} = -1$ if and only if $\mu = u^*_{\{i,j\}}$, the dual measure of the unanimity game for the pair i,j.*

Let us denote by $\mu^{i \wedge j}$ and $\mu^{i \vee j}$ the measures $u_{\{i,j\}}$ and $u^*_{\{i,j\}}$ respectively.

Property 3 *$\mu^{i \wedge j}$ and $\mu^{i \vee j}$ are 2-additive measures, whose Möbius representation is given by*
$$m_{ij}^{i \wedge j} = 1, \quad m_A^{i \wedge j} = 0 \text{ for } A \neq \{i,j\},$$
$$m_i^{i \vee j} = m_j^{i \vee j} = 1, \quad m_{ij}^{i \vee j} = -1 \text{ and } 0 \text{ for other subsets.}$$
Moreover, for both measures, $\mathcal{I}_{kl} = 0$ whenever $\{k,l\} \neq \{i,j\}$, and $v(i) = v(j) = 1/2$, $v(k) = 0$ for all $k \neq i, j$. The Choquet integrals with respect to $\mu^{i \wedge j}$ and $\mu^{i \vee j}$ are given by
$$\mathcal{C}_{\mu^{i \wedge j}}(t_1, \ldots, t_n) = t_i \wedge t_j, \quad \mathcal{C}_{\mu^{i \vee j}}(t_1, \ldots, t_n) = t_i \vee t_j.$$

Property 4 *Let μ, μ^* be a pair of dual measures, and denote by $v(i)$, $\mathcal{I}_{ij}$, $v(i)^*$, $\mathcal{I}^*_{ij}$, $\mathcal{I}_A$, $\mathcal{I}^*_A$ their respective Shapley and interaction indexes for two elements and more. Then*
$$v(i)^* = v(i), \quad \mathcal{I}^*_{ij} = -\mathcal{I}_{ij}, \quad \mathcal{I}^*_A = (-1)^{|A|+1} \mathcal{I}_A.$$

Property 5 *Any convex combination of fuzzy measures is a fuzzy measure, and the three representations are invariant under convex combinations. More specifically, let $\mu^1, \ldots, \mu^p$ be a family of fuzzy measures, with $m^i, \mathcal{I}^i, i = 1, \ldots, p$ their representations, and $\alpha_1, \ldots, \alpha_p \in [0,1]$ so that $\sum_{i=1}^p \alpha_i = 1$. Then $\mu = \sum_{i=1}^p \alpha_i \mu^i$ is a fuzzy measure whose Möbius and interaction representations are given by*
$$m_A = \sum_{i=1}^p \alpha_i m_A^i, \quad \mathcal{I}_A = \sum_{i=1}^p \alpha_i \mathcal{I}_A^i, \quad \forall A \subset X.$$

Property 6 *(Chateauneuf and Jaffray [1]) Let μ be a fuzzy measure, with m its Möbius representation, and $t_1, \ldots, t_n$ a set of positive real numbers. Then the expression of the Choquet integral in terms of m is given by:*
$$\mathcal{C}_\mu(t_1, \ldots, t_n) = \sum_{A \subset X} m_A \inf_{i \in A} t_i.$$

Using (14), we can obtain the corresponding expression for the interaction representation. A more interesting expression will be given in the next paragraph.

4.5 Application to the Behavioral Analysis of Fuzzy Measures

4.5.1 Importance and Interaction of Criteria

The interaction representation gives a clear understanding of the behavior of fuzzy measures in multicriteria decision problems. We have already stressed that the Shapley index v_i of i expresses the global importance of i, all coalitions being considered. On the other hand, the interaction index $\mathcal{I}_{ij}$ of a pair i, j expresses the amount of positive or negative synergy between i and j.

We can have a deeper insight into the meaning of v_i and $\mathcal{I}_{ij}$ if we use a Choquet integral as aggregation operator, in the case of 2-additive measures, which represents a nice compromise between richness of modelling and simplicity. Recall that in this case we can represent a fuzzy measure only by its Shapley value and by a set of interaction indexes $\mathcal{I}_{ij}$, for any pair (i, j). It is possible to express the Choquet integral with respect to a 2-additive measure using only the Shapley value and the $\mathcal{I}_{ij}$ indices. The following relation can be obtained, using properties 3 and 5:

$$\begin{aligned} \mathcal{C}_\mu(t_1,\ldots,t_n) &= \sum_{\mathcal{I}_{ij}>0}(t_i \wedge t_j)\mathcal{I}_{ij} + \sum_{\mathcal{I}_{ij}<0}(t_i \vee t_j)|\mathcal{I}_{ij}| \\ &+ \sum_{i=1}^n t_i(v(i) - \frac{1}{2}\sum_{j\neq i}|\mathcal{I}_{ij}|), \end{aligned} \tag{16}$$

Moreover, we have $v(i) - \frac{1}{2}\sum_{j\neq i}|\mathcal{I}_{ij}| \geq 0$, $i = 1,\ldots,n$. Looking at this equation, we see that the Choquet integral has in fact three components: a conjunctive component for all pairs (i, j) such that $\mathcal{I}_{ij} > 0$, a disjunctive component for all pairs (i, j) such that $\mathcal{I}_{ij} < 0$, and a linear component. The following become clear:

- a positive $\mathcal{I}_{ij}$ implies a conjunctive behavior between i and j. This means that the simultaneous satisfaction of criteria i and j is significant for the global score, but a unilateral satisfaction has no effect.

- a negative $\mathcal{I}_{ij}$ implies a disjunctive behavior, which means that the satisfaction of either i or j is sufficient to have a significant effect on the global score.

- the Shapley value acts as a weight vector in a weighted arithmetic mean. This represents the linear part of Choquet integral.

4.5.2 Veto and Favor Criteria

Another important topic in multicriteria decision making is the concept of *veto*. Suppose $\mathcal{H}$ is an aggregation operator. A criterion i is a veto for $\mathcal{H}$ if for any n-uple $(t_1,\ldots,t_n)$ of scores,

$$\mathcal{H}(t_1,\ldots,t_n) = t_i \wedge \mathcal{G}(t_1,\ldots,t_n).$$

This means that if the score on criterion i is high, it has no effect on the evaluation, but if it becomes low, the global score will become low too, whatever the values of the

other scores are. Similarly, criterion i is said to be a *favor* if for any n-uple $(t_1, \ldots, t_n)$ of scores,
$$\mathcal{H}(t_1, \ldots, t_n) = t_i \vee \mathcal{G}(t_1, \ldots, t_n).$$
In this case, low values of t_i have no effect on the global score, but a high value of t_i entails a high global score.

Let us take for $\mathcal{H}$ the Choquet integral and show that these veto and favor effects can be represented by a suitable fuzzy measure. We will show the following property:

Property 7 *For the Choquet integral, a veto effect on i is obtained by any fuzzy measure $\mu^{i\wedge}$ satisfying $\mu^{i\wedge}(A) = 0$ whenever $i \notin A$. Similarly, a favor effect on i is obtained by any fuzzy measure $\mu^{i\vee}$ satisfying $\mu^{i\vee}(A) = 1$ whenever $i \in A$. Moreover, $(\mu^{i\wedge})^*$ has a favor effect for i and $(\mu^{i\vee})^*$ a veto effect for i.*

Let us show that $\mu^{i\wedge}$ has a veto effect for the Choquet integral. Using (1) and the definition of $\mu^{i\wedge}$, we have, supposing without loss of generality that $t_i = t_{(k)}$,

$$\begin{aligned}
\mathcal{C}_{\mu^{i\wedge}}(t_1, \ldots, t_n) &= \sum_{j=1}^{k-1} t_{(j)}(\mu_{A_{(j)}} - \mu_{A_{(j+1)}}) + t_i \mu_{A_{(k)}} \\
&\leq \sum_{j=1}^{k-1} t_i(\mu_{A_{(j)}} - \mu_{A_{(j+1)}}) + t_i \mu_{A_{(k)}} \\
&= t_i.
\end{aligned}$$

Thus we have shown that $\mathcal{C}_{\mu^{i\wedge}}(t_1, \ldots, t_n) \leq t_i, \forall t_1, \ldots, t_n$, which is equivalent to say that i is a veto for the Choquet integral. Results on favor effect can be shown similarly. Remark also that interpreting fuzzy measures as strength of coalitions, the definitions of $\mu^{i\wedge}$ and $\mu^{i\vee}$ seem to be natural.

Lastly, it is interesting to look at the interaction representation of $\mu^{i\wedge}$ and $\mu^{i\vee}$. Denoting $\mathcal{I}_{jk}^{i\wedge}$ and $\mathcal{I}_{jk}^{i\vee}$ their respective interaction indices, it is easy to show that (use (9))

$$\mathcal{I}_{ik}^{i\wedge} \geq 0, \quad \mathcal{I}_{ik}^{i\vee} \leq 0, \quad \forall j, k \neq i.$$

No particular result can be obtained for other coefficients $\mathcal{I}_A$. This property shows that a veto or a favor effect can be detected on any μ simply by inspection of the $\mathcal{I}_{ij}$ indices.

It is possible to generalize the concept of veto to several criteria. For example, a veto for criteria 1 and 2, which means $\mathcal{H}(t_1, \ldots, t_n) \leq t_1 \wedge t_2$, is obtained by any fuzzy measure $\mu^{1\wedge 2\wedge}$ such that $\mu^{1\wedge 2\wedge}(A) = 0$ whenever $\{1, 2\} \not\subset A$. Then $\mathcal{I}_{12i} \geq 0, \forall i$ (similarly for favour effect and more than two criteria).

5 Identification of Fuzzy Measures

The parameters involved in a fuzzy integral are the 2^n coefficients of the fuzzy measure. The problem of identification is thus far more complex than for OWA and ordinary weighted operators, due to the exponential growing of the number of parameters. Of course, the use of k-additive measures can considerably reduce the complexity of the problem. We can distinguish essentially three approaches (see [14] for more details).

5.1 Identification Based on the Semantics

It consists of guessing the coefficients of μ, on the basis of semantical considerations. The interaction representation is clearly the natural framework for doing this. This could be for example:

- *importance of criteria*. This can be properly done by use of the Shapley value.
- *interaction between criteria*. The interaction index (def. 5) is suitable for this.
- *symmetric criteria*. Two criteria x_i, x_j are symmetric if they can be exchanged without changing the aggregation mode. Then, $\mu(A \cup \{i\}) = \mu(A \cup \{j\})$, $\forall A \subset X - \{i, j\}$. This reduces the number of coefficients.
- *veto and favor effect*. They can be expressed by a suitable sign of $\mathcal{I}_{ij}$.

This approach is practicable only for low values of n, and above all, if one has at his disposal an expert or decision maker who is able to tell the relative importance of criteria, and the kind of interaction between them, if any. This could be the case, in application of design of new products, where the marketing "defines" what should be the ideal product, in terms of aggregation of criteria.

5.2 Identification Based on Learning Data

Considering the fuzzy integral model as a system, one can identify its parameters by minimizing an error criterion, provided learning data are available. Suppose that $(z_k, y_k), k = 1, \ldots, l$ are learning data where $z_k = [z_{k,1} \cdots z_{k,n}]^t$ is a n dimensional input vector, containing the partial scores of object k with respect to criteria 1 to n, and y_k is the global score of object k. Then, one can try to identify the best fuzzy measure μ so that the squared error criterion is minimized.

$$E^2 = \sum_{k=1}^{l} (C_\mu(z_{k,1}, \ldots, z_{k,n}) - y_k)^2. \tag{17}$$

It can be shown [14] that (17) can be put under a quadratic program form, that is

minimize $\frac{1}{2}\mathbf{u}^t \mathbf{D}\mathbf{u} + \mathbf{c}^t\mathbf{u}$
under the constraint $\mathbf{A}\mathbf{u} + \mathbf{b} \geq 0$

where $\mathbf{u}$ is a $(2^n - 2)$ dimensional vector containing all the coefficients of the fuzzy measure μ (except $\mu(\emptyset)$ and $\mu(X)$ which are fixed), $\mathbf{D}$ is a $(2^n - 2)$ dimensional square matrix, $\mathbf{c}$ a $(2^n - 2)$ dimensional vector, $\mathbf{A}$ a $n(2^{n-1} - 1) \times (2^n - 2)$ matrix, and $\mathbf{b}$ a $n(2^{n-1} - 1)$ dimensional vector.

If there are too few learning data (there must be at least $n!/[(n/2)!]^2$ data: see [14] for details), matrices may be ill-conditioned. Moreover, the constraint matrix $\mathbf{A}$ is a sparse matrix, and become sparser as n grows, causing bad behavior of the algorithm. For all these reasons, including memory problems, time of convergence, the solution given by a quadratic program is not always reliable in practical situations.

Some authors have proposed alternatives to quadratic programming under the form of "heuristic" algorithms taking advantage of the peculiar structure of fuzzy measures, but often suboptimal. One of the most powerful in terms of performance, time and memory has been proposed by Grabisch [9]. The basic idea is that, in the absence of any information, the most reasonable way of aggregation is the arithmetic mean (provided the problem is cardinal), i.e. a Choquet integral with respect to an additive equidistributed fuzzy measure. Any input of information tends to move away the fuzzy measure from this equilibrium point. This means that, in case of few data, coefficients of the fuzzy measure which are not concerned with the data are kept as near as possible to the equilibrium point. This solves the problem of having too few data.

5.3 Combining Semantics and Learning Data

Obviously, the combination of semantical considerations, which are able to reduce the complexity and provide guidelines, with learning data should lead to more efficient algorithms. An attempt in this direction is given by Yoneda *et al.* [47]. The basic ideas are the following.

- objective: minimize the distance to the additive equidistributed fuzzy measure,

- constraints: the usual constraints implied by the monotonicity of fuzzy measures, and constraints coming from semantical considerations, which can be translated into constraints on coefficients of the interaction representation. These could be about relative importance of criteria, redundancy, support between criteria, veto, etc.

The constraints being linear, this leads to a quadratic program as before.

6 Conclusion

This paper has attempted to show the richness of fuzzy measures and integrals as a new mean of aggregation. Emphasis has been put on the interpretation of fuzzy measures in terms of importance and interaction among criteria, and of veto and favor, which are understandable to a decision maker.

As it can be noticed, the concept of k-additive measure, which is a generalization of the concept of additivity, is well suited to the Choquet integral, since the Choquet integral becomes additive for additive measures. Now, a similar development has yet to be done for the Sugeno integral and possibility measure. This opens the road to k-order decomposable fuzzy measures.

References

[1] A. Chateauneuf and J.Y. Jaffray. Some characterizations of lower probabilities and other monotone capacities through the use of Möbius inversion. *Mathematical Social Sciences*, 17:263–283, 1989.

[2] G. Choquet. Theory of capacities. *Annales de l'Institut Fourier*, 5:131–295, 1953.

[3] D. Denneberg. *Non-Additive Measure and Integral*. Kluwer Academic, 1994.

[4] D. Dubois and H. Prade. Weighted minimum and maximum operations in fuzzy set theory. *Information Sciences*, 39:205–210, 1986.

[5] J. Fodor and M. Roubens. On meaningfulness of means. *Journal of Computational and Applied Mathematics*.

[6] J.C. Fodor and M. Roubens. *Fuzzy Preference Modeling and Multi-Criteria Decision Aid*. Kluwer Academic Publisher, 1994.

[7] K. Fujimoto. *On hierarchical decomposition of Choquet integral model*. PhD thesis, Tokyo Institute of Technology, 1995.

[8] M. Grabisch. Fuzzy integral in multicriteria decision making. *Fuzzy Sets & Systems*, 69:279–298, 1995.

[9] M. Grabisch. A new algorithm for identifying fuzzy measures and its application to pattern recognition. In *Int. Joint Conf. of the 4th IEEE Int. Conf. on Fuzzy Systems and the 2nd Int. Fuzzy Engineering Symposium*, pages 145–150, Yokohama, Japan, march 1995.

[10] M. Grabisch. The application of fuzzy integrals in multicriteria decision making. *European J. of Operational Research*, 89:445–456, 1996.

[11] M. Grabisch. Fuzzy measures and integrals: a survey of applications and recent issues. In D. Dubois, H. Prade, and R. Yager, editors, *Fuzzy Sets Methods in Information Engineering: A Guided Tour of Applications*. J. Wiley & Sons, 1996.

[12] M. Grabisch. k-order additive fuzzy measures. In *6th Int. Conf. on Information Processing and Management of Uncertainty in Knowledge-Based Systems (IPMU)*, Granada, Spain, july 1996.

[13] M. Grabisch. k-order additive fuzzy measures and their application to multicriteria analysis. In *2nd World Automation, Int. Symp. on Soft Computing for Industry*, Montpellier, France, may 1996.

[14] M. Grabisch, H.T. Nguyen, and E.A. Walker. *Fundamentals of Uncertainty Calculi, with Applications to Fuzzy Inference*. Kluwer Academic, 1995.

[15] P.L. Hammer and R. Holzman. On approximations of pseudo-Boolean functions. *ZOR - Methods and Models of Operations Research*, 36:3–21, 1992.

[16] H. Ichihashi, H. Tanaka, and K. Asai. Fuzzy integrals based on pseudo-addition and multiplication. *J. Math. Anal. Appl.*, 130:354–364, 1988.

[17] H. Imaoka. A proposal of opposite-Sugeno integral and a uniform expression of fuzzy integrals. In *Int. Joint Conf. 4th IEEE Int. Conf. on Fuzzy Systems and 2nd Int. Fuzzy Engineering Symp.*, pages 583–590, Yokohama, march 1995.

[18] K. Inoue and T. Anzai. A study on the industrial design evaluation based upon non-additive measures. In *7th Fuzzy System Symp.*, pages 521–524, Nagoya, Japan, June 1991. in japanese.

[19] K. Ishii and M. Sugeno. A model of human evaluation process using fuzzy measure,. *Int. J. Man-Machine Studies*, 22:19–38, 1985.

[20] R.L. Keeney and H. Raiffa. *Decision with Multiple Objectives*. Wiley, New York, 1976.

[21] A. Kolmogoroff. Sur la notion de moyenne. *Atti delle Reale Accademia Nazionale dei Lincei Mem. Cl. Sci. Fis. Mat. Natur. Sez.*, 12:323–343, 1930.

[22] R. Kruse. Fuzzy integrals and conditional fuzzy measures. *Fuzzy Sets & Systems*, 10:309–313, 1983.

[23] R. Mesiar. Choquet-like integrals. *J. of Mathematical Analysis and Application*, 194:477–488, 1995.

[24] R. Mesiar and J. Šipoš. A theory of fuzzy measures: integration and its additivity. *Int. J. General Systems*, 23:49–57, 1994.

[25] T. Murofushi. A technique for reading fuzzy measures (i): the Shapley value with respect to a fuzzy measure. In *2nd Fuzzy Workshop*, pages 39–48, Nagaoka, Japan, october 1992. in japanese.

[26] T. Murofushi and S. Soneda. Techniques for reading fuzzy measures (iii): interaction index. In *9th Fuzzy System Symposium*, pages 693–696, Sapporo, Japan, may 1993. In japanese.

[27] T. Murofushi and M. Sugeno. Fuzzy t-conorm integrals with respect to fuzzy measures : generalization of Sugeno integral and Choquet integral. *Fuzzy Sets & Systems*, 42:57–71, 1991.

[28] T. Murofushi and M. Sugeno. A theory of fuzzy measures. Representation, the Choquet integral and null sets. *J. Math. Anal. Appl.*, 159(2):532–549, 1991.

[29] T. Murofushi and M. Sugeno. Non-additivity of fuzzy measures representing preferential dependence. In *2nd Int. Conf. on Fuzzy Systems and Neural Networks*, pages 617–620, Iizuka, Japan, july 1992.

[30] T. Murofushi and M. Sugeno. Some quantities represented by the Choquet integral. *Fuzzy Sets & Systems*, 56:229–235, 1993.

[31] T. Onisawa, M. Sugeno, Y. Nishiwaki, H. Kawai, and Y. Harima. Fuzzy measure analysis of public attitude towards the use of nuclear energy. *Fuzzy Sets & Systems*, 20:259–289, 1986.

[32] G.C. Rota. On the foundations of combinatorial theory i. theory of Möbius functions. *Zeitschrift für Wahrscheinlichkeitstheorie und Verwandte Gebiete*, 2:340–368, 1964.

[33] G. Shafer. *A Mathematical Theory of Evidence*. Princeton Univ. Press, 1976.

[34] L.S. Shapley. A value for n-person games. In H.W. Kuhn and A.W. Tucker, editors, *Contributions to the Theory of Games, Vol. II*, number 28 in Annals of Mathematics Studies, pages 307–317. Princeton University Press, 1953.

[35] M. Sugeno. *Theory of fuzzy integrals and its applications*. PhD thesis, Tokyo Institute of Technology, 1974.

[36] M. Sugeno, K. Fujimoto, and T. Murofushi. A hierarchical decomposition of Choquet integral model. *Int. J. of Uncertainty, Fuzziness and Knowledge-Based Systems*.

[37] M. Sugeno, K. Fujimoto, and T. Murofushi. Hierarchical decomposition theorems for Choquet integral models. In *Int. Joint Conf. 4th IEEE Int. Conf. on Fuzzy Systems and 2nd Int. Fuzzy Engineering Symp.*, pages 2245–2252, Yokohama, Japan, march 1995.

[38] M. Sugeno and S.H. Kwon. A clusterwise regression-type model for subjective evaluation. *J. of Japan Society for Fuzzy Theory and Systems*, 7(2):291–310, 1995.

[39] M. Sugeno and S.H. Kwon. A new approach to time series modeling with fuzzy measures and the Choquet integral. In *Int. Joint Conf. of the 4th IEEE Int. Conf. on Fuzzy Systems and the 2nd Int. Fuzzy Engineering Symp.*, pages 799–804, Yokohama, Japan, march 1995.

[40] M. Sugeno and T. Murofushi. *Fuzzy measure theory*, volume 3 of *Course on fuzzy theory*. Nikkan Kōgyō, 1993. in japanese.

[41] K. Tanaka and M. Sugeno. A study on subjective evaluations of color printing images. In *4th Fuzzy System Symposium*, pages 229–234, Tokyo, Japan, may 1988. in japanese.

[42] Z. Wang and G.J. Klir. *Fuzzy measure theory*. Plenum, 1992.

[43] T. Washio, H. Takahashi, and M. Kitamura. A method for supporting decision making on plant operation based on human reliability analysis by fuzzy integral. In *2nd Int. Conf. on Fuzzy Logic and Neural Networks*, pages 841–845, Iizuka, Japan, July 1992.

[44] S. Weber. $\perp$-decomposable measures and integrals for archimedean t-conorms $\perp$. *J. Math. Anal. Appl.*, 101:114–138, 1984.

[45] R.R. Yager. On ordered weighted averaging aggregation operators in multicriteria decision making. *IEEE Trans. Systems, Man & Cybern.*, 18:183–190, 1988.

[46] R.R. Yager. Connectives and quantifiers in fuzzy sets. *Fuzzy Sets & Systems*, 40:39–75, 1991.

[47] M. Yoneda, S. Fukami, and M. Grabisch. Interactive determination of a utility function represented by a fuzzy integral. *Information Sciences*, 71:43–64, 1993.

Using Priorities in Aggregation Connectives

Antoine Kelman
LAFORIA, University of Paris-VI
4 Place Jussieu
75230 Paris Cedex 05, France
kelman@laforia.ibp.fr

Ronald R. Yager
Machine Intelligence Institute
Iona College
New Rochelle, New York 10801, USA
ryager@iona.edu

Abstract : We present new multicriteria aggregation methods based on OWA operators and on the management of priorities between different criteria. These methods are parameterized in order to be used in various different application contexts. For each method, we present a learning process and a neural representation.

1. Introduction

Many decision problems require the consideration of multiple criteria. The search for a solution in these situations inevitably leads to a multicriteria aggregation problem, where using the local satisfactions to each criterion we must to compute a global satisfaction function.

Fuzzy set theory from its early inception has provided a fertile medium for addressing these types of problems. We recall that if X is a set of alternatives in a decision problem we can represent each criteria as fuzzy subset over X. In these fuzzy subsets the membership grades correspond to the satisfaction of the alternative to the criteria the fuzzy subset is representing. Once having represented the criteria as fuzzy subsets over the space of alternatives the problem of combining the multiple criteria to form the global satisfaction function becomes an issue of aggregating fuzzy subsets. Considerable research has been done on the aggregation of fuzzy subsets [1]. These aggregation methods span both a logical inspired ones such as the union and intersection as well as numerically inspired ones such as the mean.

An important benefit of using the fuzzy set framework in the multi-criteria aggregation problem is the ability to be able to provide a semantical meaning for the various operations used in the aggregation process. This makes it easier to model a decision makers desired imperative for combining multiple criteria in the fuzzy set framework than in other frameworks.

In combining multiple criteria the two most extreme behaviors are the conjunctive one (where we want to satisfy all the criteria at the same time) and the disjunctive one (where only at least one criterion has to be satisfied). In fuzzy logic, such operators are available with the use of the intersection and union of the criteria fuzzy sets.

In this work we consider situations in which there exists a priority with respect to our desire to satisfy the individual criteria. In this environment we develop a new aggregation method, first for the case of two criteria then in the more general case of M criteria. For each method that we present, we show a neural representation and possible learning process for the associated parameters based on observations.

2. Review of Some Fuzzy Set Aggregation Methods

In this section we provide a brief review of some established fuzzy set aggregation methods which play a role in the application fuzzy subsets to multi-criteria decision making.

2.1. Triangular t-norms and t-conorms

These conjunctive (t-norms) and disjunctive (t-conorms) operators play an important role in the fuzzy sets theory. Here, we only recall their definition and a few of the useful properties that we shall need later on, more details can be found in [2-3].

A *triangular norm (t-norm)* is a mapping $T : [0,1] \times [0,1] \rightarrow [0,1]$ such that $\forall\ x, y, z, t \in [0,1]$:

i) $T(x, y) = T(y, x)$ (commutativity)
ii) $T(x, T(y, z)) = T(T(x, y), z)$ (associativity)
iii)   If $x \leq z$ and $y \leq t$, $T(x, y) \leq T(z, t)$ (monotonicity)
iv) $T(x, 1) = x$ (one is neutral element)

t-norms are used to implement fuzzy subset intersection operators, the actual choice of t-norm is context dependent. The t-norm that was originally used by Zadeh's was the Min and is notable because of the property that if T is any t-norm, $\forall\ x, y \in [0,1]$, $T(x, y) \leq \min(x, y)$. Another useful property related to these operators is that if T is any t–norm, $\forall\ x \in [0,1]$, $T(x, 0) = 0$.

We can also recall a few t-norms that are frequently used :

$T(x, y) = \min(x, y)$: Zadeh
$T(x, y) = x\,y$: probabilistic
$T(x, y) = \max(0, x + y - 1)$: Lukasiewicz

In a similar manner we define t-conorms. A *triangular conorm (t-conorm)* is also a mapping $\perp: [0,1] \times [0,1] \rightarrow [0,1]$ such that $\forall\ x, y, z, t \in [0,1]$:

i) $\perp(x, y) = \perp(y, x)$ (commutativity)
ii) $\perp(x, \perp(y, z)) = \perp(\perp(x, y), z)$ (associativity)
iii)   If $x \leq z$ and $y \leq t$, $\perp(x, y) \leq \perp(z, t)$ (monotonicity)
iv) $\perp(x, 0) = x$ (zero is the neutral element)

The t-conorms are logical disjunction operators used to implement fuzzy subset unions. The t-conorm that was originally used by Zadeh was Max and is notable because of the property that if $\perp$ is any t-conorm, $\forall\ x, y \in [0,1]$, $\perp(x, y) \geq \max(x, y)$. Another useful property related to these operators is that if $\perp$ is a t-conorm, $\forall\ x \in [0,1]$, $\perp(x, 1) = 1$.

We can recall the following often used t-conorms:

$\perp(x, y) = \max(x, y)$: Zadeh
$\perp(x, y) = x + y - xy$: probabilistic sum
$\perp(x, y) = \min(x + y, 1)$: Lukasiewicz

The intersection and union, triangular norms and conorms, where the first aggregation operators used in fuzzy logic. However, in order to get a softer behavior and more closely model the aggregation of multiple criteria we must consider more complex methods.

2.2. The OWA Operators

Here we are interested in aggregation operators that lie between the extremes of conjunctive and disjunctive aggregation described above [5-9]. We are particularly interested in the OWA operators which provide a mean for implementing linguistic quantifier guided aggregations. We recall linguistic quantifiers correspond to terms such as *most, some, about half*. They generalize the concept of existential and universal quantifiers. The linguistic quantifiers were first introduced by Zadeh [4] where he represents them as fuzzy subsets. Several types of linguistic quantifiers can be defined. Here we shall only consider the non-decreasing ones. Using these quantifiers we can express aggregation imperatives that lie between the conjunctive case, requiring all criteria be satisfied, and the disjunctive case, requiring any criteria be satisfied. Thus using linguistic quantifiers we can consider situations where we desire *most* of the criteria be satisfied. In order to formally implement these quantifier guided aggregations we must use the OWA operators introduced by Yager [7].

Definition : An mapping F from $[0,1]^n \to [0,1]$ is an OWA operator of dimension n if there exists a weighting vector $W = \begin{bmatrix} w_1 \\ w_2 \\ \vdots \\ w_n \end{bmatrix}$ associated with F for which :

- $w_i \in [0, 1]$
- $\sum_{i=1}^{n} w_i = 1$,

and such that $F(a_1, a_2, \cdots, a_n) = \sum_{i=1}^{n} w_i b_i$ where b_i is the i-th largest element of the $\{a_1, a_2, ..., a_n\}$.

We can show that OWA operators are idempotent and monotonic. The OWA operators are effectively mean-type aggregation operators [1] as we can show that:

$$T(a_1, a_2, ..., a_n) \leq F(a_1, a_2, ..., a_n) \leq \perp(a_1, a_2, ..., a_n)$$

where T and $\perp$ are respectively any t-norm and t-conorm.

As we indicated before, the OWA operators can be used in aggregation of fuzzy subsets guided by linguistic quantifiers. Let us consider a linguistic quantifier Q such as *most*. As suggested by Zadeh this quantifier can be expressed as a fuzzy subset on the unit interval where the membership grade $Q(r)$ indicates the degree to which the proportion r satisfies the concept *most*. Assume we have a collection of n criteria, represented as fuzzy subsets, A_i. The process for obtaining the aggregation of these criteria under the imperative that *most* criteria should be satisfied by an optimal alternative is as follows. Find the weights of the associated OWA operator as:

$$w_i = Q(i/n) - Q((i-1)/n), \text{ for } i = 1 \text{ to } n.$$

Then for each alternative x, the overall decision function is $D(x) = F(a_1, a_2, \ldots, a_n)$ where F is an OWA aggregation using the above weights and $a_i = A_i(x)$.

Hence we can see that OWA operators and linguistic quantifiers are closely and simply related.

2.3. Priority and Second Order Criteria

The concept of second order criteria was introduced in fuzzy set theory by Yager [10] in the framework of considering prioritized aggregation of criteria. Here, we explain that concept in a two-criteria case. Let A and B be two criteria which are defined by their membership functions on the universe of discourse X. Our objective here is to model the knowledge that in aggregating these two criteria A has a priority over B. By the idea of priority in the aggregation process we mean that the optimal solution has to satisfy the criterion A and if possible B (if that doesn't interfere with the satisfaction to A). Let us denote as D the global criterion. We have then that

$$D = A \text{ and } (B, \text{ if } B \text{ is compatible with } A)$$

The second part of the conjunction can be expressed as a rule (R):

(R) : if B is compatible with A then B *is a relevant criterion*

The compatibility is measured by the possibility of having A and B simultaneously, that is:

$$Poss(A,B) = \max_{x \in X} \left(A(x) \wedge B(x) \right)$$

If we represent (R) in a logical way we have:

$$(R) = \neg(Poss(A, B)) \vee B = (1 - Poss(A, B)) \vee B$$

So finally the expression of the global criterion is

$D(x) = A(x) \wedge ((1 - Poss(A, B)) \vee B(x))$, which can also be expressed as:
$D(x) = (A(x) \wedge (1 - Poss(A, B))) \vee (A(x) \wedge B(x))$.

We can also mention a few of the major properties of this operator.

Property : If $Poss(A, B) = 0$ then $D(x) = A(x)$

This means that if the second order criterion B is completely incompatible with the higher priority criterion, A, then that second order criterion is not important in the choice of the optimal solution.

Property : if $Poss(A, B) = 1$ then $D(x) = A(x) \wedge B(x)$

This implies that if the two criteria are fully compatible we shall try to find a solution that satisfy them simultaneously.

It is possible to introduce the idea of a degree of priority of A over B. For example as suggested in [10] we can consider

$$D(x) = A(x) \wedge (B(x) \vee ((1 - Poss(A, B)) \wedge \alpha))$$

where α denotes a degree of priority of A over B. We note if $\alpha = 1$ we get

$$D(x) = A(x) \wedge (B(x) \vee ((1 - Poss(A, B)))),$$

which was our original formulation where we have the situation in which A has complete priority over B. On the contrary, if $\alpha = 0$, $D(x) = A(x) \wedge B(x)$: A and B have the same importance.

We now present our aggregation methods, beginning with the two-criteria case. The presented operators can be used in multi-criteria aggregation but also in other fields as for example information fusion, when the information comes from multiple sources. It is in the spirit of a greater adaptability that the third method will be presented (cf § 5).

3. Aggregation of Two Criteria

Let us assume that the set of alternatives in which we shall search for the optimal solution is the set $X = \{x_1, x_2, ..., x_n\}$. The two criteria that are to be aggregated will be denoted as A and B as in § 2.3. If one of the criteria is more important, in the sense of priority, we could use the method that we recalled in § 2.3. But here we shall consider a more general formulation where the priority relationship can be variable. In this spirit we shall consider the following formulation of our overall objective function

A has a priority over B with a degree α
or
B has a priority over A with a degree β

Using these considerations we obtain the following aggregation operator

$D_1(x) = \quad [A(x) \wedge (B(x) \vee ((1 - Poss(A, B)) \wedge \alpha))]$
$\vee [B(x) \wedge (A(x) \vee ((1 - Poss(B, A)) \wedge \beta))]$

The operators $\wedge$ and $\vee$ can be chosen from among the t-norms and t-conorms operators. Some of the properties of D_1 will change with that choice.

3.1. Properties

In order to study the behavior of our new operator, we shall first look at a few properties that are associated with extreme values of the parameters.

Property :   If $\alpha = 1$ and $\beta = 0$, A has a priority over B.

                 If $\beta = 1$ and $\alpha = 0$, B has a priority over A.

<u>Proof</u> :

If $\alpha = 1$ and $\beta = 0$, $D_1(x) = \left[A(x) \wedge \left(B(x) \vee \left(1 - Poss(A, B)\right)\right)\right] \vee \left[B(x) \wedge A(x)\right]$

as $x \wedge 1 = x$ and $x \wedge 0 = 0$ for any t-norm $\wedge$ (cf § 2.1).

Hence, $D_1(x) = [A(x) \wedge B(x)] \vee [A(x) \wedge (1 - Poss(A, B))] \vee [B(x) \wedge A(x)]$
hence, $D_1(x) = [A(x) \wedge B(x)] \vee [A(x) \wedge (1 - Poss(A, B))]$, what is the expression of the fact that A has a priority over B (cf § 2.3)

Here, we see that the expression of a complete priority is possible. However, some more interesting results can be obtained when the parameters are equal.

Property :   If $\alpha = \beta = 0$, $D_1(x) = A(x) \wedge B(x)$

$\alpha = \beta = 0$ means that neither A nor B are considered to have a priority. This case is a pure conjunction : the optimal solution has to satisfy both criteria.

Property : If $\alpha = \beta = 1$, $D_1(x) = [A(x) \wedge B(x)] \vee [A(x) \wedge (1 - Poss(A, B))]$
$\vee [B(x) \wedge (1 - Poss(A, B))]$

Here, A and B are both very important in the choice of a final solution. We can notice that we obtain an aggregation method that is very close to one suggested by Dubois and Prade [11] :

$$D_1(x) = \left[\frac{A(x) \wedge B(x)}{Poss(A, B)}\right] \vee \left[A(x) \wedge \left(1 - Poss(A, B)\right)\right] \vee \left[B(x) \wedge \left(1 - Poss(A, B)\right)\right]$$

Our aggregation method gives results lying between the conjunction and the disjunction. Its behavior changes with the value of the compatibility between the two criteria :

• If $Poss(A, B) = 0$, the two criteria are not compatible and we have :

$$D_1(x) = \left[A(x) \wedge B(x)\right] \vee A(x) \vee B(x)$$

If we also have $\wedge = \min$ and $\vee = \max$, $D_1(x) = A(x) \vee B(x)$

The method is then a pure disjunction : as the two criteria cannot be satisfied at the same time, it is sufficient to satisfy one of them.
- If $Poss(A, B) = 1$, the two criteria are fully compatible and we have :

$$D_1(x) = [A(x) \wedge B(x)]$$

The method is then a pure conjunction : we look for a solution that satisfies both criteria at the same time.

We just looked at the behaviors of our methods for extreme values of the parameters : the bigger a parameter is, the more important is the criterion to which the parameter is associated. If $\alpha = 0$, for example, A has no priority over B which means that A is at best as important as B : satisfying A is not our first priority. Each criterion is compared to the other via its associated parameter. The bigger α is, the more A is supposed to be discriminating about the choice of the final solution.

Let us show a few algebraic properties :

Theorem : If $\wedge = \min$ and $\vee = \max$, D_1 is *idempotent*, which means that
if $A(x) = B(x) = y$ then $D_1(x) = y$

Proof :

$$D_1(x) = \left[y \wedge \left(y \vee \left((1 - Poss(A,B)) \wedge \alpha \right) \right) \right] \vee \left[y \wedge \left(y \vee \left((1 - Poss(A,B)) \wedge \beta \right) \right) \right]$$

now, $y \vee \left((1 - Poss(A,B)) \wedge \alpha \right) \geq y$ as $\vee = \max$

therefore $y \wedge \left(y \vee \left((1 - Poss(A,B)) \wedge \alpha \right) \right) = y$ as $\wedge = \min$

and identically, $y \wedge \left(y \vee \left((1 - Poss(A,B)) \wedge \beta \right) \right) = y$

hence, $D_1(x) = y \vee y = y$ as $\vee = \max$.

Theorem : Limit conditions :   if $A(x) = B(x) = 0$ then $D_1(x) = 0$
if $A(x) = B(x) = 1$ then $D_1(x) = 1$

Proof : • if $A(x) = B(x) = 0$, $\forall y$, $A(x) \wedge y = 0 \wedge y = 0$, due to the properties of t-norms (cf § 2.1). Identically, $B(x) \wedge y = 0 \wedge y = 0$ hence $D_1(x) = 0 \vee 0 = 0$.
- if $A(x) = B(x) = 1$, $\forall y$, $B(x) \vee y = 1 \vee y = 1$, due to the properties of t-conorms. Hence, $D_1(x) = (1 \wedge 1) \vee (1 \wedge 1) = 1$.

Theorem : Monotonicity : if $A(x) \geq A(y)$ and $B(x) \geq B(y)$ then $D_1(x) \geq D_1(y)$

Proof : t-norms and t-conorms are monotonic (cf § 2.1) hence we have :

$$\underbrace{B(x) \vee \left(\left(1 - Poss(A,B)\right) \wedge \alpha\right)}_{C_\alpha} \geq B(y) \vee C_\alpha$$

hence, $A(x) \wedge \left(B(x) \vee C_\alpha\right) \geq A(y) \wedge \left(B(y) \vee C_\alpha\right)$ as $\wedge$ is monotonic,

Identically, $B(x) \wedge \left(A(x) \vee C_\beta\right) \geq B(y) \wedge \left(A(y) \vee C_\beta\right)$,

hence, $D_1(x) \geq D_1(y)$ as $\vee$ is monotonic.

Finally we can show that the presented method is actually a hybrid aggregation mean-type method as we have the following theorem:

Theorem: $\quad \forall\, x \in X, A(x) \wedge B(x) \leq D_1(x) \leq A(x) \vee B(x)$

<u>Proof</u>: • $B(x) \vee C_\alpha \geq \max\left(B(x), C_\alpha\right) \geq B(x)$ as $\vee$ is a t-conorm (cf § 2.1)

hence, $A(x) \wedge \left(B(x) \vee C_\alpha\right) \geq A(x) \wedge B(x)$ as $\wedge$ is monotonic.

Now, $D_1(x) \geq \max\left(A(x) \wedge \left(B(x) \vee C_\alpha\right),\ B(x) \wedge \left(A(x) \vee C_\beta\right)\right)$

therefore, $D_1(x) \geq A(x) \wedge B(x)$.

• $A(x) \wedge \left(B(x) \vee C_\alpha\right) \leq \min\left(A(x) \wedge \left(B(x) \vee C_\alpha\right)\right) \leq A(x)$ as $\wedge$ is a t-norm (cf § 2.1)

Identically, $B(x) \wedge \left(A(x) \vee C_\beta\right) \leq \min\left(B(x) \wedge \left(A(x) \vee C_\beta\right)\right) \leq B(x)$

hence, $D_1(x) \leq A(x) \vee B(x)$ as $\vee$ is monotonic.   .

Hereby, we see that the suggested method has all the main properties of an aggregation one [6]. The tuning parameters can be semantically understood, which enables us to use human expert knowledge.

3.2. Neural Representation

In order to be able to use techniques and hardware related to artificial neural networks, we present a neural representation of the latter method.

According to the results in [12-13], we find that the following fuzzy operators (t-norms and t-conorms) are easily implemented in a neural framework:

- $\wedge$ = product $\Leftrightarrow a \wedge b = a \times b$
- $\vee$ = Lukasiewicz's t-conorm $\Leftrightarrow a \vee b = \min(1, a + b)$

Using these operators, we get:

$$Poss(A,B) = \bigvee_{x \in X} (A(x) \wedge B(x)) = min\left(1, \sum_{i=1}^{N} A(x_i) \cdot B(x_i)\right)$$

We shall denote as f the function :
$$f : [0, \infty[\to [0,1]$$
$$x \mapsto min(1, x)$$

Hence we have : $Poss(A,B) = f\left(\sum_{i=1}^{N} A(x_i) \cdot B(x_i)\right)$

Noting Id the identity function, we get the following network :

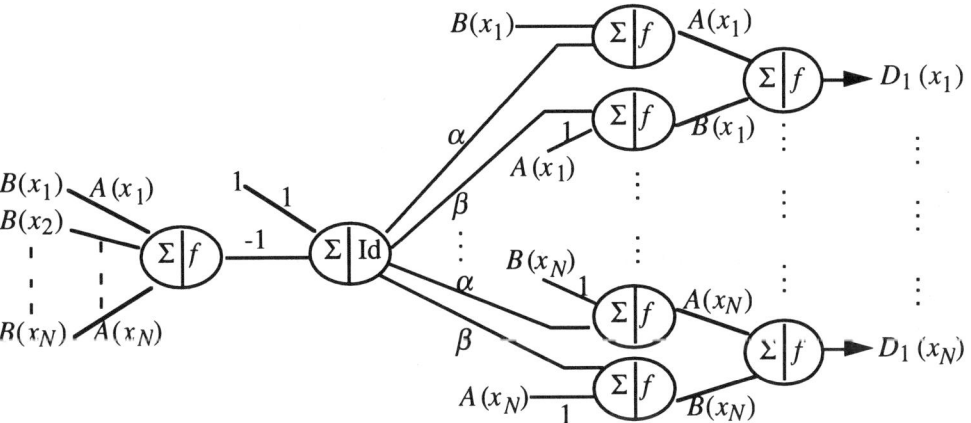

Figure 3.2.1 : Neural representation of D_1

We notice that the parameters of the method are the weights of the network. These parameters are to be learned and it here appears that α and β can easily be depending on x. Therefore we can generalize the method, and we obtain :

$$D_2(x_i) = \left[A(x_i) \wedge \left(B(x_i) \vee ((1 - Poss(A,B)) \wedge \alpha_i)\right)\right]$$
$$\vee \left[B(x_i) \wedge \left(A(x_i) \vee ((1 - Poss(A,B)) \wedge \beta_i)\right)\right]$$

That way, the parameters can be tuned more precisely : the way the criterion A has a priority over B doesn't have to be homogeneous on the whole universe X. We could change the degree of priority with the values of x. That is why it is more efficient if the parameter is not a constant.

This new method has the same properties that the first one except for the monotonicity that doesn't hold here. The neural representation is as easy as before :

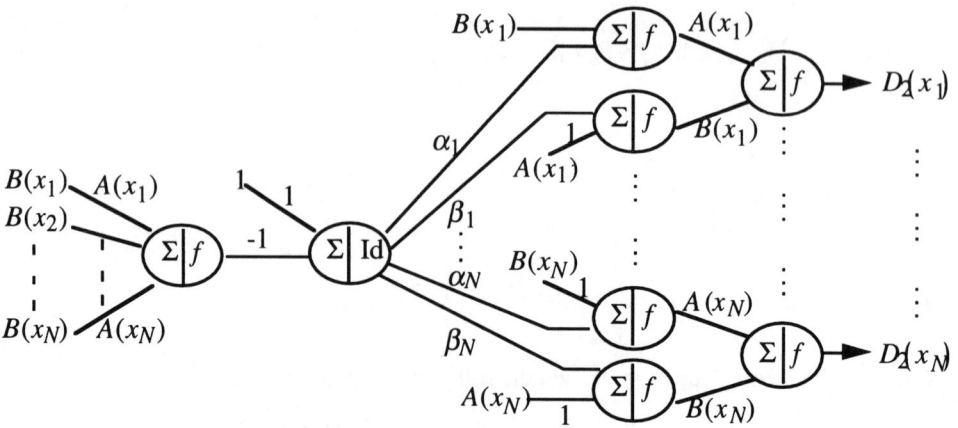

Figure 3.2.2 : Neural representation of D_2

Now we can step to the learning process of the parameters α_i and β_i based on the latter representation.

3.3. Learning Process for the Parameters

We shall use a backpropagation algorithm which is actually a gradient descent method. Therefore we have to use differentiable firing functions for each of our neurons. We shall use an approximation of the function $f: x \to \min(1, x)$ by assuming that :

$$f(x) = \frac{1}{1 + e^{-7(x-0.5)}} \quad [12].$$

The derivative of f is then defined for any point. It will be denoted as f', as usual.

We try to minimize the error : $E = \frac{1}{2} \sum_{i=1}^{N} (D_2(x_i) - T_i)^2$, where T_i is the expected value for the output $D_2(x_i)$. We note $A(x_i) = A_i$, $B(x_i) = B_i$ and $1 - Poss(A,B) = \overline{P_{AB}}$.

At each step of the learning process we have :

$$\Delta \alpha_i(t+1) = \alpha_i(t+1) - \alpha_i(t) = -\eta \cdot \frac{\partial E}{\partial \alpha_i} \quad \text{where } \eta \text{ is the descent step.}$$

$$\frac{\partial E}{\partial \alpha_i} = (D_2(x_i) - T_i) \cdot \frac{\partial D_2(x_i)}{\partial \alpha_i} \quad \text{as } D_2(x_i) \text{ only depends on } \alpha_i.$$

As $D_2(x_i) = f\left(A_i \cdot f\left(B_i + \alpha_i \overline{P_{AB}}\right) + B_i \cdot f\left(A_i + \beta_i \overline{P_{AB}}\right)\right)$, we have :

$$\frac{\partial D_2(x_i)}{\partial \alpha_i} = A_i \overline{P_{AB}} \cdot f'\left(B_i + \alpha_i \overline{P_{AB}}\right) \cdot f'\left(A_i f\left(B_i + \alpha_i \overline{P_{AB}}\right) + B_i f\left(A_i + \beta_i \overline{P_{AB}}\right)\right)$$

Identically, $\Delta \beta_i(t+1) = \beta_i(t+1) - \beta_i(t) = -\eta \cdot \dfrac{\partial E}{\partial \beta_i}$

and $\dfrac{\partial E}{\partial \beta_i} = (D_2(x_i) - T_i) \cdot \dfrac{\partial D_2(x_i)}{\partial \beta_i}$

and $\dfrac{\partial D_2(x_i)}{\partial \beta_i} = B_i \overline{P_{AB}} \cdot f'\left(A_i + \beta_i \overline{P_{AB}}\right) \cdot f'\left(A_i f\left(B_i + \alpha_i \overline{P_{AB}}\right) + B_i f\left(A_i + \beta_i \overline{P_{AB}}\right)\right)$.

Furthermore, since the α_i and the β_i are subject to constraints: $\alpha_i \in [0,1]$ and $\beta_i \in [0,1]$, we shall bound the learning process in the following way:

$$\begin{cases} \text{If } \alpha_i(t) + \Delta \alpha_i(t+1) \geq 1 \text{ then } \alpha_i(t+1) = 1 \\ \text{If } \alpha_i(t) + \Delta \alpha_i(t+1) \leq 0 \text{ then } \alpha_i(t+1) = 0 \end{cases} \text{ and identically for } \beta_i$$

We have a learning process for the tuning parameters of the D_2 method. The learning can be done with an examples base. For these examples the results are of course to be known. They can be guessed by a human expert if he is competent enough in the field of the application.

4. M Criteria Aggregation

Based on the latter paragraph, it seems to be interesting to extend the applicability of the method to cases where we could have several criteria. According to the spirit of the method, we should consider every alternative of priorities between the M criteria. Assuming that each criterion stands at a single level of priority, we would have $(M!)$ ways of ordering the criteria and also $(M!)$ parameters (or priority degrees) for the method. As this seems way too much (especially when the number of criteria is big), we shall use another approach: in the two criteria case, we were comparing each criterion to the other one, which actually represented every other criterion. Hence, here, we can decide to compare each criterion with the $(M-1)$ remaining criteria, or more exactly with a part of these $(M-1)$ criteria.

The universe of discourse will still be represented by $X = \{x_1, x_2, \cdots, x_N\}$. The global criterion will be denoted as C (defined by the $C(x)$). The method is the following:

Let us consider M criteria (or information sources) denoted $S_j(x), j = 1 \cdots M$. The global criterion is defined by :

$$C(x) = \bigvee_{j=1}^{M} C_j(x) \text{ with } C_j(x) = S_j(x) \wedge \left[Q_j(x) \vee \left(\left(1 - Poss(S_j, Q_j)\right) \wedge \alpha_j \right) \right],$$

where Q_j is a linguistic quantifier or an OWA operator applied to the criteria $(S_k)_{k \neq j}$. We can see C_j as a partial criterion related to S_j. C_j allows us to compare S_j with some of the remaining criteria, from the priority point of view. The quantifier Q_j will be considered as an OWA operator as we saw that the OWA can simply model linguistic quantifiers.

Generalizing the method D_2 (cf § 3.2), it seems natural to consider the priority degrees as functions of x. That way, we get the new global criterion that will be denoted as $D(x)$:

$$D(x) = \bigvee_{j=1}^{M} D_j(x) \text{ with } D_j(x_i) = S_j(x_i) \wedge \left[Q_j(x_i) \vee \left(\left(1 - Poss(S_j, Q_j)\right) \wedge \alpha_{ji} \right) \right]$$

where Q_j is an OWA operator on the $(S_k(x_i))_{k \neq j}$.

4.1. Properties

As in § 3, we first present numerical properties in order to understand the behavior of the method. Theoretic properties will then be presented.

Let us consider $D_j(x_i) = S_j(x_i) \wedge \left[Q_j(x_i) \vee \left(\left(1 - Poss(S_j, Q_j)\right) \wedge \alpha_{ji} \right) \right]$. This formula, that was initially introduced by referring to the priorities (cf § 2.1), can also be interpreted in terms of weights : if $\alpha_{ji} = 1$, $D_j(x_i)$ is the weighted minimum (if $\wedge$ = min and $\vee$ = max) of $S_j(x_i)$ and $Q_j(x_i)$ whose respective weights would be 1 and $Poss(S_j, Q_j)$ [14]. No matter how we interpret $D_j(x_i)$, we see that $Q_j(x_i)$ only interfere if it has a big value : when $\vee$ = max, $\left(\left(1 - Poss(S_j, Q_j)\right) \wedge \alpha_{ji} \right)$ is the threshold from which the information given by $Q_j(x_i)$ are taken as relevant. The smaller α_{ji} is, the more $Q_j(x_i)$ is taken into account in the calculus.

The fact that α_{ji} depends on x_i allows to privilege such or such source of information depending on the object we consider. That can be interesting when one of the sources is very discriminating for one object (or for a subset of objects) but not on the whole set X.

As in § 3, we find some limit properties :

Property : If $\forall i,j$, $\alpha_{ji} = 0$ and $\forall j$, $Q_j = \min$ then, if $\wedge = \min$, $D(x_i) = \bigwedge_j S_j(x_i)$

Proof : here we want to express a conjunction : no source has any priority.
If $\forall i,j$, $D_j(x_i) = S_j(x_i) \wedge Q_j(x_i) = S_j(x_i) \wedge \bigwedge_{k \neq j} S_k(x_i) = \bigwedge_l S_l(x_i)$

hence $D(x_i) = \bigvee_j \left(\bigwedge_l S_l(x_i) \right) = \bigwedge_l S_l(x_i)$: conjunction.

Property : If the sources or criteria are totally incompatible and that $\forall i,j$, $\alpha_{ji} = 1$ and $\forall j$, $Q_j = \max$ then, if $\wedge = \min$ and $\vee = \max$, $D(x_i) = \bigvee_j S_j(x_i)$.

Proof : we are here in the case of a disjunction. As the criteria are not compatible, $Poss(S_i, S_j) = 0$, $\forall i \neq j$, which means that $\forall i \neq j$ and $\forall x$, $S_i(x) \wedge S_j(x) = 0$.
Hence, as $\forall j$ and $\forall x$, $\exists k \neq j$ / $Q_j(x) = S_k(x)$, we have $Poss(S_j, Q_j) = Poss(S_j, S_k) = 0$,
hence $D_j(x) = S_j(x) \vee (Q_j(x) \vee 1) = S_j(x)$, hence $D(x) = \bigvee_j S_j(x)$.

We see that, once again, the method can be fully conjunctive or fully disjunctive under appropriate conditions.

Now, let us look at the algebraic properties :

Theorem : If $\wedge = \min$ and $\vee = \max$, D is *idempotent*, which means that :
if $\forall j$, $S_j(x_i) = a$ then $D(x_i) = a$.

Proof : if $\forall j$, $S_j(x_i) = a$ then $D_j(x_i) = a$ as the OWA are idempotent.
Now, $a \vee \left((1-Poss(S_j, Q_j)) \wedge \alpha_{ji} \right) \geq a$ as $\vee = \max$, therefore $D_j(x_i) = a$,
hence, $D(x_i) = \bigvee_j a = a$.

Theorem : limit conditions : If $\forall j$, $S_j(x_i) = 0$ then $D(x_i) = 0$
If $\forall j$, $S_j(x_i) = 1$ then $D(x_i) = 1$.

Proof : $\wedge$ and $\vee$ are any t-norms and t-conorms.

- If $\forall j$, $S_j(x_i) = 0$, $D_j(x_i) = 0$ as $0 \wedge y = 0$ hence $D(x_i) = 0$.
- If $\forall j$, $S_j(x_i) = 1$, $Q_j(x_i) = 1$ as $0 \wedge y = 0$ hence $D_j(x_i) = 1 \wedge 1 = 1$ hence $D(x_i) = 1$.

Theorem : Monotonicity : If $\forall j$, $\alpha_{ji} = \alpha_j$ (not depending on x_i) then D is monotonic, which means that : if $\forall j$, $S_j(x) \geq S_j(y)$ then $D(x) \geq D(y)$.

Proof : The OWA are monotonic hence, $\forall j$, $Q_j(x) \geq Q_j(y)$,

hence, $Q_j(x) \vee \left(\left(1\text{-}Poss(S_j, Q_j)\right) \wedge \alpha_j\right) \geq Q_j(y) \vee \left(\left(1\text{-}Poss(S_j, Q_j)\right) \wedge \alpha_j\right)$,

and, as $S_j(x) \geq S_j(y)$, we have : $D_j(x) \geq D_j(y)$ hence $\bigvee_j D_j(x) \geq \bigvee_j D_j(y) \Leftrightarrow D(x) \geq D(y)$.

D is not generally monotonic (if α_{ji} is not a constant).

Theorem : $\forall x \in X$, $\bigwedge_j S_j(x) \leq D(x) \leq \bigvee_j S_j(x)$

Proof : $\forall i, j$, $D_j(x_i) \leq S_j(x_i)$ as $\wedge \leq \min$, hence $D(x_i) \leq \bigvee_j S_j(x_i)$.

More, $Q_j(x_i) \geq \bigwedge_{k \neq j} S_k(x_i)$ according to the properties of the OWA, hence, as $Q_j(x_i) \vee c \geq Q_j(x_i)$, we have

$D_j(x_i) \geq S_j(x_i) \wedge Q_j(x_i) \geq S_j(x_i) \wedge \bigwedge_{k \neq j} S_k(x_i) \geq \bigwedge_k S_k(x_i)$, hence $D(x_i) \geq \bigwedge_k S_k(x_i)$.

Here again, the presented method is an aggregation one in the usual way. Conjunction and disjunction are limit cases. The parameters allow us to move between these two limit cases. As the parameters are more numerous than before, a learning process is very useful.

4.2. Learning Process for the Parameters

The tuning parameters are the α_{ji} and the Q_j, that are assumed to be OWA operators.

As in § 3.2, we choose the following operators : $\wedge = $ product $\Leftrightarrow a \wedge b = a \times b$ and $\vee = $ Lukasiewicz's t-conorm $\Leftrightarrow a \vee b = \min(1, a+b)$. the function $f(x) = \dfrac{1}{1+e^{-7(x-0.5)}}$ will still be considered as a derivable approximation of $f : x \to \min(1, x)$.

The OWA operators are defined by their weighting vectors. These weights are constrained : they are non-negative and their sum is 1. In order to integrate these constraints to our learning process, we shall take the following representation [9] :

The weight n° k of Q is : $\dfrac{e^{\lambda_{jk}}}{\sum_{k=1}^{M-1} e^{\lambda_{jk}}} = \omega_{jk}$. That way, the ω_{jk} will be learned by learning the λ_{jk}. Finally we have :

$$D(x_i) = \bigvee_{j=1}^{M} D_j(x_i) = f\left(\sum_{j=1}^{M} D_j(x_i)\right)$$

and $D_j(x_i) = S_j(x_i) \cdot f\left(\dfrac{\sum_{k=1}^{M-1} b_{jik} \cdot e^{\lambda_{jk}}}{\sum_{k=1}^{M-1} e^{\lambda_{jk}}} + \alpha_{ji}\left(1 - f\left(\sum_{l=1}^{N} S_l(x_i) Q_l(x_i)\right)\right)\right)$,

where the $\{b_{jik}\}$ are the $\{S_l(x_i) / l \neq j\}$ arranged in ascending order : b_{jik} is the k-th biggest element of $\{S_l(x_i) / l \neq j\}$.

The error that we want to minimize is : $\left\{E = \dfrac{1}{2} \sum_{i=1}^{N} (D(x_i) - T_i)^2\right\}$ where T_i is the expected output for $D(x_i)$.

We have : $\Delta\alpha_{ji}(t+1) = \alpha_{ji}(t+1) - \alpha_{ji}(t) = -\eta \cdot \dfrac{\partial E}{\partial \alpha_{ji}}$ where η is the descent step.

Now, $\dfrac{\partial E}{\partial \alpha_{ji}} = (D(x_i) - T_i) \cdot \dfrac{\partial D(x_i)}{\partial \alpha_{ji}}$ for if $l \neq i$, $\dfrac{\partial D(x_l)}{\partial \alpha_{ji}} = 0$,

therefore $\dfrac{\partial D(x_i)}{\partial \alpha_{ji}} = \dfrac{\partial}{\partial \alpha_{ji}}\left(f\left(\sum_{j=1}^{M} D_j(x_i)\right)\right) = \left(\sum_{j=1}^{M} \dfrac{\partial D_j(x_i)}{\partial \alpha_{ji}}\right) \cdot f'\left(\sum_{j=1}^{M} D_j(x_i)\right)$

and $\dfrac{\partial D_j(x_i)}{\partial \alpha_{ji}} = S_j(x_i) \cdot \left(1 - Poss(S_j, Q_j)\right) f'\left(Q_j(x_i) + \alpha_{ji}\left(1 - Poss(S_j, Q_j)\right)\right)$

as $Poss(S_j, Q_j) = f\left(\sum_{l=1}^{N} S_j(x_l) Q_j(x_l)\right)$.

We can compute $\Delta\alpha_{ji}(t+1)$ by "backpropagation". As in § 3.3, we saturate the learning process in order to respect the constraints on α_{ji}:

$$\begin{cases} \text{If } \alpha_{ji}(t) + \Delta\alpha_{ji}(t+1) \geq 1 \text{ then } \alpha_i(t+1) = 1 \\ \text{If } \alpha_{ji}(t) + \Delta\alpha_{ji}(t+1) \leq 0 \text{ then } \alpha_i(t+1) = 0 \end{cases}$$

Identically, $\Delta\lambda_{ji}(t+1) = \lambda_{ji}(t+1) - \lambda_{ji}(t) = -\eta \cdot \dfrac{\partial E}{\partial \lambda_{ji}}$.

As, $\dfrac{\partial E}{\partial \lambda_{ji}} = \sum_{l=1}^{N} \left(D(x_l) - T_l\right) \cdot \dfrac{\partial D(x_l)}{\partial \lambda_{ji}}$

we have $\dfrac{\partial D(x_l)}{\partial \lambda_{ji}} = \dfrac{\partial D_j(x_l)}{\partial \lambda_{ji}} \cdot f'\left(\sum_{k=1}^{M} D_k(x_l)\right)$

$\dfrac{\partial D_j(x_l)}{\partial \lambda_{ji}} = S_j(x_l) \times \left[\dfrac{\partial Q_j(x_l)}{\partial \lambda_{ji}} - \alpha_{jl} \cdot \left(\sum_{k=1}^{N} S_j(x_k) \dfrac{\partial Q_j(x_k)}{\partial \lambda_{ji}}\right) \cdot f'\left(\sum_{k=1}^{N} S_j(x_k) Q_j(x_k)\right)\right]$

$\times f'\left(Q_j(x_l) + \alpha_{jl} \cdot \left(1 - Poss(S_j, Q_j)\right)\right)$

and

with $\dfrac{\partial Q_j(x_k)}{\partial \lambda_{ji}} = \dfrac{b_{jki} \cdot e^{\lambda_{ji}} \cdot \sum_{h=1}^{M-1} e^{\lambda_{jh}} - e^{\lambda_{ji}} \cdot \sum_{l=1}^{M-1} b_{jkl} \cdot e^{\lambda_{jl}}}{\left(\sum_{k=1}^{M-1} e^{\lambda_{jk}}\right)^2}$.

We see that we can also compute $\Delta\lambda_{ji}$ by backpropagation and that the learning of the weights of the OWA operators can be done that way. Although the method is quite complex, the tuning of the parameters can be done almost "automatically" with an examples base.

A neural representation is still possible though a little more complex than before.

4.3. Neural Representation

Before presenting the actual neural model, we recall the existence of a class of artificial neurons presented by [15] : the OWA-neurons.

These neurons were designed to represent easily OWA operators (cf § 2.2). If we denote as Q an OWA operator whose weighting vector is $\begin{bmatrix} v_1 \\ v_2 \\ \vdots \\ v_n \end{bmatrix}$, an OWA-neuron represented by :

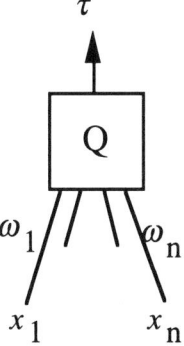

Figure 4.3.1 : OWA-neuron.

gives an output : $\tau = \sum_{j=1}^{n} b_j v_j$, where b_j is the j-th biggest of the $(\omega_i x_i)$.

These neurons allow the learning of the ω_i by backpropagation, identically that for a usual network. However, for our method, we only wish to represent a simple OWA operator, which means we want to have $\tau = \sum_{j=1}^{n} b_j v_j$, but with b_j the j-th biggest of the (x_i). Hence, we shall choose $\omega_i = 1, \forall i$.

We now present a mixed representation of $D(x_i)$, for any i. As a matter of fact, we only present fully the output of D_1. As the representation is mixed, the usual neurons will be represented by circles and the OWA-neurons by squares.

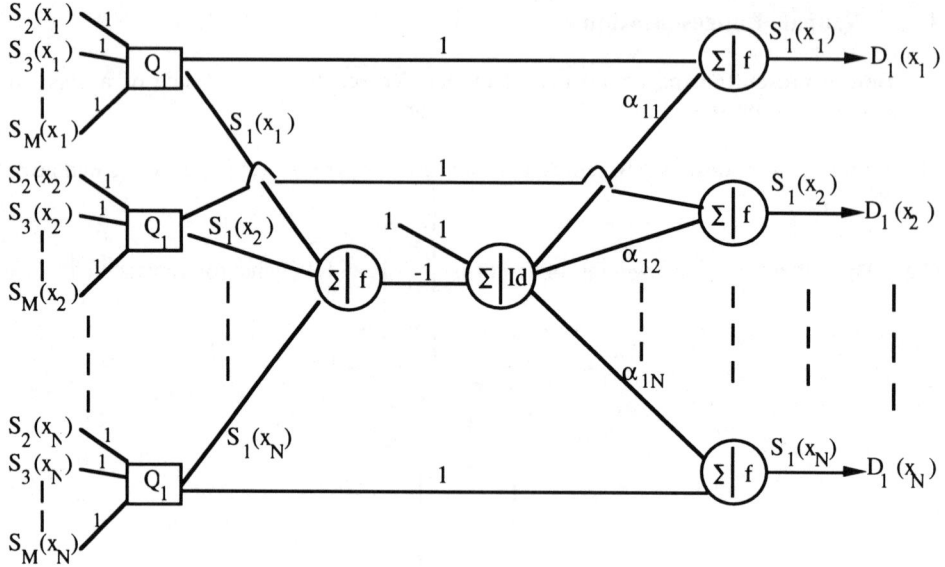

Figure 4.3.2 : Neural representation of D_1.

$D(x_i)$ can be represented in a neural framework as we have :

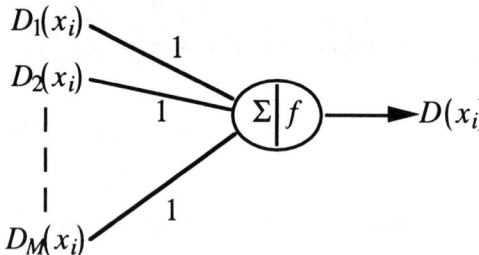

Figure 4.3.3 : Neural representation of D.

5. Toward a More Adaptive Method

5.1. Motivations and Method Presentation

When we look at the method we presented in § 4 and particularly its neural representation, it seems that it will be more intuitive to allow the parameters to depend on the objects (the x_i). First, the use of the α_{ji} can be replaced by a function we shall denote as ω [16]. In the case of two sources A and B, we shall get :

$$D_A(x) = A(x) \wedge \left[B(x) \vee \left(1 - \omega_A(Poss(A,B))\right) \right],$$

and identically $\quad D_B(x) = B(x) \wedge \left[A(x) \vee \left(1 - \omega_B(Poss(A,B))\right) \right],$

with $\quad D(x) = D_A(x) \vee D_B(x).$

Yager [16] suggests that the function ω should satisfy a few constraints. First, we shall only assume that it is a on-decreasing function, which means that the more compatible A and B are, the more the optimal solution will try to satisfy them simultaneously.

With that constraint, the function ω is more general than a single parameter α. As a matter of fact, if we want to have $(1-P) \wedge \alpha = 1 - \omega(P)$, we can choose $\omega(P) = 1 - (1-P) \wedge \alpha$. As $\wedge$ is a t-norm, it is monotonic. Hence,

if $P' \geq P$, $1 - P' \leq 1 - P$
$\Rightarrow (1-P') \wedge \alpha \leq (1-P) \wedge \alpha$
$\Rightarrow 1 - (1-P') \wedge \alpha \geq 1 - (1-P) \wedge \alpha$
$\Rightarrow \omega(P') \geq \omega(P).$

The function ω is effectively non-decreasing. Furthermore, since α depends on x, it seems natural that ω should depend on x too.

We proceed identically for the OWA operator used in § 4 : each source is compared to to a combination of the others. There is no reason why this combination should be the same for every objects : concerning object x_1, the source S_1 could be considered as having a priority on all the other sources (because, for example, S_1 is very reliable on x_1). But for the other objects $(x_i)_{i \neq 1}$, S_1 could be only averagely good and willing to be compared to some of the other sources only. We could want to express the fact that, for example, concerning x_2, S_1 is "at the same level of priority" that 20 % of the other sources, which means that the results we can get with S_1 don't hold anymore if S_1 is conflicting with more that 20 % of the remaining sources. That way, the OWA operator will depend on the considered object. We shall denote as Q_{ji} the OWA operator associated with the source S_j and the object x_i.

Finally, the $Poss(S_j, Q_{ji})$ term can also be discussed : in the case of two sources A and B the OWA operator is always equal to the identity. Hence, we have :

$$D_A(x_i) = A(x_i) \wedge \left[B(x_i) \vee \left(1 - \omega_{Ai}(Poss(A,B)) \right) \right]$$: The selection of an optimal solution takes into account the compatibility between the two sources. But once again, that compatibility could be important only on some objects. We could imagine that the sources could be very unreliable on some objects (a subset of X). On these objects, the compatibility is not very important. On the other hand, the compatibility is important when we deal with some objects on which the sources are very reliable.

All this to say that the use of $Poss(A,B)$ could be too inflexible in some cases. Hence, we propose a softer compatibility. Finally, we get :

$$D_j(x_i) = S_j(x_i) \wedge \left[Q_{ji}(x_i) \vee \left(1 - \bigvee_{l=1}^{N} \omega_{jil} \left(S_j(x_l) \wedge Q_{ji}(x_l) \right) \right) \right].$$

ω_{jil} is a non-decreasing function depending on S_j, on the object x_i and on the influence of another object x_l on the final score given to x_i :

$$\omega_{jil} = \omega(S_j, x_i, x_l).$$

We still have $D(x_i) = \bigvee_{j=1}^{M} D_j(x_i)$.

5.2. Learning Process for the Parameters

In all the methods we proposed, a learning of the tuning parameters was possible. Here, this learning is still possible and is all the more important as the method is now very complex, in order to be able be adapted to more applications. As before, we take $\wedge = $ product $\Leftrightarrow a \wedge b = a \times b$ and $a \vee b \approx f(a+b)$ with $f(x) = \dfrac{1}{1+e^{-7(x-0.5)}}$. The weights of the OWA operators are still obtained through the λ_{jik} :

$$Q_{ji}(x_l) = \frac{\sum_{k=1}^{M-1} b_{jlk} \cdot e^{\lambda_{jik}}}{\sum_{k=1}^{M-1} e^{\lambda_{jik}}} \quad,$$

where b_{jlk} is the k-th biggest element of the $\{S_h(x_l) / h \neq j\}$.

We put another constraint on the functions ω_{jil}. These functions were non-decreasing and we shall assume that they also have the shape shown on figure 5.2.1.

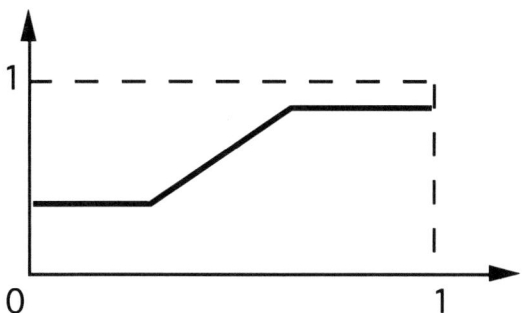

Figure 5.2.1 : shape of the functions ω_{jil}.

As we still use a gradient descent algorithm, the ω_{jil} will be approximated by the derivable functions, that are depending on 4 parameters :

$$\omega_{jil}(x) = a_{jil} + \frac{d_{jil} \cdot (1 - a_{jil})}{1 + e^{-b_{jil}(x - c_{jil})}} \quad.$$

The constraints in these parameters are :

$$a_{jil} \in [0,1], b_{jil} \geq 0, c_{jil} \in [0,1], d_{jil} \in [0,1] \quad.$$

The error we want to minimize is : $E = \frac{1}{2} \sum_{i=1}^{N} (D(x_i) - T_i)^2$, where T_i is the expected output for $D(x_i)$.

We begin by learning the parameters of the ω_{jil}.

We have : $\Delta a_{jil}(t+1) = a_{jil}(t+1) - a_{jil}(t) = -\eta \cdot \frac{\partial E}{\partial a_{jil}}$ where η is the descent step.

Now, $\dfrac{\partial E}{\partial a_{jil}} = (D(x_i) - T_i) \cdot \dfrac{\partial D(x_i)}{\partial a_{jil}}$, and $\dfrac{\partial D(x_i)}{\partial a_{jil}} = \dfrac{\partial D_j(x_i)}{\partial a_{jil}} \times f'\left(\sum\limits_{h=1}^{M} D_h(x_i)\right)$,

therefore $\dfrac{\partial D_j(x_i)}{\partial a_{jil}} = \dfrac{\partial}{\partial a_{jil}}\left(S_j(x_i) \cdot f\left[Q_{ji}(x_i) + 1 - f\left\{\sum\limits_{h=1}^{N} a_{jih} + \dfrac{d_{jih} \cdot (1 - a_{jih})}{1 + e^{-b_{jih}(S_j(x_h) Q_{ji}(x_h) - c_{jih})}}\right\}\right]\right)$

hence,

$\dfrac{\partial D_j(x_i)}{\partial a_{jil}} = -S_j(x_i) \times \left(1 - \dfrac{d_{jil}}{1 + e^{-b_{jil}(S_j(x_i) Q_{ji}(x_i) - c_{jil})}}\right) \times f'\left(\sum\limits_{h=1}^{N} \omega_{jih}(S_j(x_h) Q_{ji}(x_h))\right)$

$\times f'\left[Q_{ji}(x_i) + 1 - f\left(\sum\limits_{h=1}^{N} \omega_{jih}(S_j(x_h) Q_{ji}(x_h))\right)\right]$

We satisfy the constraints on a_{jil} by "saturating" the learning process :

$\begin{cases} \text{If } a_{jil}(t) + \Delta a_{jil}(t+1) \geq 1 \text{ then } a_{jil}(t+1) = 1 \\ \text{If } a_{jil}(t) + \Delta a_{jil}(t+1) \leq 0 \text{ then } a_{jil}(t+1) = 0 \end{cases}$

Identically, $\Delta b_{jil}(t+1) = b_{jil}(t+1) - b_{jil}(t) = -\eta \cdot \dfrac{\partial E}{\partial b_{jil}}$

and, $\dfrac{\partial E}{\partial b_{jil}} = (D(x_i) - T_i) \cdot \dfrac{\partial D_j(x_i)}{\partial b_{jil}} \times f'\left(\sum\limits_{h=1}^{M} D_h(x_i)\right)$

and,

$\dfrac{\partial D_j(x_i)}{\partial b_{jil}} = -S_j(x_i) \times \dfrac{d_{jil}(1 - a_{jil})(S_j(x_i) Q_{ji}(x_i) - c_{jil}) e^{-b_{jil}(S_j(x_i) Q_{ji}(x_i) - c_{jil})}}{\left(1 + e^{-b_{jil}(S_j(x_i) Q_{ji}(x_i) - c_{jil})}\right)^2}$

$\times f'\left(\sum\limits_{h=1}^{N} \omega_{jih}(S_j(x_h) Q_{ji}(x_h))\right) \times f'\left[Q_{ji}(x_i) + 1 - f\left(\sum\limits_{h=1}^{N} \omega_{jih}(S_j(x_h) Q_{ji}(x_h))\right)\right]$

here, the saturation is : If $b_{jil}(t) + \Delta b_{jil}(t+1) \leq 0$ then $b_{jil}(t+1) = 0$.

Identically, $\Delta c_{jil}(t+1) = c_{jil}(t+1) - c_{jil}(t) = -\eta \cdot \dfrac{\partial E}{\partial c_{jil}}$

and, $\dfrac{\partial E}{\partial c_{jil}} = (D(x_i) - T_i) \cdot \dfrac{\partial D_j(x_i)}{\partial c_{jil}} \times f'\left(\sum_{h=1}^{M} D_h(x_i)\right)$

and,

$$\dfrac{\partial D_j(x_i)}{\partial c_{jil}} = S_l(x_i) \times \dfrac{d_{jil}(1 - a_{jil}) b_{jil} e^{-b_{jil}(S_l(x_i) Q_{ji}(x_i) - c_{jil})}}{\left(1 + e^{-b_{jil}(S_l(x_i) Q_{ji}(x_i) - c_{jil})}\right)^2}$$

$$\times f'\left(\sum_{h=1}^{N} \omega_{jih}(S_l(x_h) Q_{ji}(x_h))\right) \times f'\left[Q_{ji}(x_i) + 1 - f\left(\sum_{h=1}^{N} \omega_{jih}(S_l(x_h) Q_{ji}(x_h))\right)\right]$$

and, $\begin{cases} \text{If } c_{jil}(t) + \Delta c_{jil}(t+1) \geq 1 \text{ then } c_{jil}(t+1) = 1 \\ \text{If } c_{jil}(t) + \Delta c_{jil}(t+1) \leq 0 \text{ then } c_{jil}(t+1) = 0 \end{cases}$

Finally, $\Delta d_{jil}(t+1) = d_{jil}(t+1) - d_{jil}(t) = -\eta \cdot \dfrac{\partial E}{\partial d_{jil}}$

and, $\dfrac{\partial E}{\partial d_{jil}} = (D(x_i) - T_i) \cdot \dfrac{\partial D_j(x_i)}{\partial d_{jil}} \times f'\left(\sum_{h=1}^{M} D_h(x_i)\right)$, and

$$\dfrac{\partial D_j(x_i)}{\partial d_{jil}} = - S_l(x_i) \times \dfrac{1 - a_{jil}}{1 + e^{-b_{jil}(S_l(x_i) Q_{ji}(x_i) - c_{jil})}} \times f'\left(\sum_{h=1}^{N} \omega_{jih}(S_l(x_h) Q_{ji}(x_h))\right)$$

$$\times f'\left[Q_{ji}(x_i) + 1 - f\left(\sum_{h=1}^{N} \omega_{jih}(S_l(x_h) Q_{ji}(x_h))\right)\right]$$

and, $\begin{cases} \text{If } d_{jil}(t) + \Delta d_{jil}(t+1) \geq 1 \text{ then } d_{jil}(t+1) = 1 \\ \text{If } d_{jil}(t) + \Delta d_{jil}(t+1) \leq 0 \text{ then } d_{jil}(t+1) = 0 \end{cases}$

We now learn the λ_{jik}, in order to get the OWA weights:

$\Delta \lambda_{jik}(t+1) = \lambda_{jik}(t+1) - \lambda_{jik}(t) = -\eta \cdot \dfrac{\partial E}{\partial \lambda_{jik}}$

and, $\dfrac{\partial E}{\partial \lambda_{jik}} = (D(x_i) - T_i) \cdot \dfrac{\partial D_j(x_i)}{\partial \lambda_{jik}} \times f'\left(\sum_{h=1}^{M} D_h(x_i)\right)$, and we have :

$$\frac{\partial D_j(x_i)}{\partial \lambda_{jik}} = \left[\begin{array}{c} \frac{\partial Q_{ji}(x_i)}{\partial \lambda_{jik}} \\ -f'\left(\sum_{h=1}^{N} \omega_{jih}\left(S_j(x_h)Q_{ji}(x_h)\right)\right) \cdot \sum_{h=1}^{N} \frac{d_{jih}\left(1 - a_{jih}\right) b_{jih} \frac{\partial Q_{ji}(x_h)}{\partial \lambda_{jik}} e^{-b_{jih}\left(S_j(x_h)Q_{ji}(x_h) - c_{jih}\right)}}{\left(1 + e^{-b_{jih}\left(S_j(x_h)Q_{ji}(x_h) - c_{jih}\right)}\right)^2} \\ \times S_j(x_i) \times f'\left[Q_{ji}(x_i) + 1 - f\left(\sum_{h=1}^{N} \omega_{jih}\left(S_j(x_h)Q_{ji}(x_h)\right)\right)\right] \end{array} \right]$$

with $\dfrac{\partial Q_{ji}(x_h)}{\partial \lambda_{jik}} = \dfrac{\left(b_{jhk} \cdot e^{\lambda_{jik}} \cdot \sum\limits_{l=1}^{M-1} e^{\lambda_{jil}}\right) - \left(e^{\lambda_{jik}} \cdot \sum\limits_{l=1}^{M-1} b_{jhl} \cdot e^{\lambda_{jil}}\right)}{\left(\sum\limits_{l=1}^{M-1} e^{\lambda_{jil}}\right)^2}$.

6. Conclusion

We presented a few aggregation methods based on already known methods (the OWA) and on the priority management. For each method we suggested a learning process for the tuning parameters from a examples base.

However, the learning process of the parameters is generally just a minimization of the error between the adopted model and the reality. It doesn't actually change the model. That's why we presented in § 5 a method in which the model itself is parameterized, in order to be more adaptive. A semantic interpretation of the parameters is still possible, allowing the opportunity to use human expert knowledge.

The use of a tool for managing the priorities seem to be essential to represent human reasoning where hierarchy is often met. The relative complexity of the proposed operators is necessary if we want to use a fuzzy model in fields where purely conjunctive or disjunctive aggregations would be too rigid.

7. References

[1]. Dubois, D. and Prade, H., "A review of fuzzy sets aggregation connectives", Information Sciences 36, 85 - 121, 1985.

[2]. Dubois, D. and Prade, H., Fuzzy Sets and Systems: Theory and Applications, Academic Press: New York, 1980.

[3]. Dubois, D., "Triangular norms for fuzzy sets", in Proceedings of the 2nd International Symposium of Fuzzy Sets, Linz, 39 - 68, 1981.

[4]. Zadeh, L. A., "A computational approach to fuzzy quantifiers in natural languages", Computing and Mathematics with Applications 9, 149-184, 1983.

[5]. Sugeno, M., "Theory of fuzzy integrals and its application", Doctoral Thesis, Tokyo Institute of Technology, 1974.

[6]. Grabisch, M., "Application des mesures floues et intégrales floues", Proceedings of the "Applications des ensembles flous", Nimes, France, 139-147, 1992.

[7]. Yager, R. R., "On ordered weighted averaging aggregation operators in multi-criteria decision making", IEEE Transactions on Systems, Man and Cybernetics 18, 183-190, 1988.

[8]. Grabisch, M. and Sugeno, M., "Multi-Attribute Classification using Fuzzy Integral", Proceedings of the 1st FUZZ'IEEE Congress, San Diego, USA, 47-54, 1992.

[9]. Filev, D. P. and Yager, R. R., "On the issue of OWA operator weights", Technical Report# MII-1319, Machine Intelligence Institute, Iona College, 1993.

[10]. Yager, R. R., "Nonmonotonic set theoretic operations", Fuzzy Sets and Systems 42, 173-190, 1991.

[11]. Dubois, D. and Prade, H., "Combination of information in the framework of possibility theory", in Data Fusion in Robotics and Machine Intelligence, edited by Al Abidi, M., Academic Press: New York, 1991.

[12]. Glorennec, P.-Y., "Logique Neuro-Floue", Proceedings of the "Applications des ensembles flous", Nîmes, France, 219-230, 1993.

[13]. Keller, J. M., Yager, R. R. and Tahani, H., "Neural network implementation of fuzzy logic", Fuzzy Sets and Systems 45, 1-12, 1992.

[14]. Dubois, D. and Prade, H., "Weighted minimum and maximum operations in fuzzy sets theory", Information Sciences 39, 205 - 210, 1986.

[15]. Yager, R. R., "OWA Neurons: A new class of fuzzy neurons", Proceedings Int. Joint Conference on Neural Networks, Vol. I, Baltimore, I: 226-231, 1992.

[16]. Yager, R. R., "Higher structures in multi-criteria decision making", International Journal of Man-Machine Studies 36, 553-570, 1992.

Aggregation Operators for Fuzzy Rationality Measures

Vincenzo CUTELLO[1] and Javier MONTERO[2]

[1] Department of Mathematics, University of Catania, Catania, Italy
[2] Faculty of Mathematics, Complutense University, Madrid, Spain

Abstract. Fuzzy rationality measures represent a particular class of aggregation operators. Following the axiomatic approach developed in [1,3,4,5] *rationality* of fuzzy preferences may be seen as a fuzzy property of fuzzy preferences. Moreover, several rationality measures can be aggregated into a global rationality measure. We will see when and how this can be done. We will also comment upon the feasibility of their use in real life applications. Indeed, some of the rationality measures proposed, though intuitively (and axiomatically) sound, appear to be quite complex from a computational point of view.

Keywords: aggregation rules, fuzzy preferences, decision making.

1 Introduction

Suppose we are given a finite set of alternatives X. We can consider all complete fuzzy binary relations [21] defined on X, that is, mappings

$$\mu : X \times X \to [0,1]$$

such that each value $\mu(x,y)$ represents the intensity value to which alternative y is not worse than alternative x (see also [11,16]). The completeness assumption guarantees that

$$\mu(x,y) + \mu(y,x) \geq 1$$

for all $x,y \in X$. Without such an assumption we would have to deal with possible incomparability problems.

An axiomatic characterization for the notion of *rationality* measure has been given in [4], where *consistency* was viewed as a fuzzy property of fuzzy binary preference relations. Moreover, in [5] it was studied the problem of logical combinations of such rationality measures along with some definitions of functional equivalence. Since each fuzzy preference relation can be measured according to different rationality criteria, we should be willing to check if and when we can aggregate different rationality criteria into one.

On the other hand, fuzzy rationality measures can be seen as a special class of aggregation operators. Therefore, we should be willing to investigate on the feasibility of their use in real life applications.

2 Fuzzy rationality measures

Max-min transitivity is one of the most common rationality measures in the context of fuzzy binary preference relations. However, the property of being max-min transitive is crisp, that is to say each relation either is or is not max-min transitive, despite the fact that we intuitively realize that some fuzzy preferences are much closer to max-min transitivity than others. We will see that there exist other measures which allow a fuzzy classification of fuzzy preferences. These measures verify a set of axioms which have been formally introduced in [4]. In details, rationality measures are characterized as mappings

$$\rho : \mathcal{P} \to [0,1]$$

where

$$\mathcal{P} = \bigcup_{X \text{ finite}} \mathcal{P}(X)$$

represents the family of all complete fuzzy binary relations. The following conditions were assumed in order for such a mapping to define a *rationality measure*.

(R1) $\rho(\mu) = 1$ for any μ defining a crisp strict chain on X.
(R2) Given $\mu \in \mathcal{P}$ and a permutation $\pi : X \to X$ then

$$\rho(\mu^\pi) = \rho(\mu)$$

where $\mu^\pi(x,y) = \mu(\pi(x), \pi(y))$ for all $x, y \subset X$.
(R3) For all $\mu \in \mathcal{P}$, $\rho(\mu^{-1}) = \rho(\mu)$, where $\mu^{-1}(x,y) = \mu(y,x)$ for all $x, y \in X$.
(R4) Let Y be a non-empty finite set of alternatives and let x be an extra alternative not belonging to Y. Let us consider a fuzzy preference $\mu : Y \times Y \to [0,1]$ such that $\mu(y,z) = 1, \mu(z,y) = 0, \forall y \in Y_1, \forall z \in Y_2$ for some Y_1, Y_2 partition of Y, and an extension μ' such that

$$\mu'(y,z) = \mu(y,z), \forall y, z \in Y$$
$$\mu'(y,x) = 1, \mu'(x,y) = 0, \forall y \in Y_1$$
$$\mu'(x,z) = 1, \mu'(z,x) = 0, \forall z \in Y_2$$
$$\mu'(x,x) = 1$$

Then it must be

$$\rho(\mu') \geq \rho(\mu).$$

(R5) Let $\mu \in \mathcal{P}(X)$ be fixed. Given an arbitrary ordered pair of alternatives $(\bar{x}, \bar{y})$, an arbitrary point $(\bar{a}, \bar{b}) \in [0,1] \times [0,1]$, and real numbers γ and λ such that $0 \leq \bar{a} + \lambda \cos\gamma \leq 1$, $0 \leq \bar{b} + \lambda \sin\gamma \leq 1$ and $\bar{a} + \bar{b} + \lambda(\sin\gamma + \cos\gamma) \geq 1$, we denote by $\Gamma_\mu((\bar{x},\bar{y}),(\bar{a},\bar{b}),\gamma,\lambda)$ the fuzzy preference relation defined as

$$\Gamma_\mu((\bar{x},\bar{y}),(\bar{a},\bar{b}),\gamma,\lambda)(x,y) = \begin{cases} \bar{a} + \lambda \cos\gamma & \text{if } (x,y) = (\bar{x},\bar{y}) \\ \bar{b} + \lambda \sin\gamma & \text{if } (x,y) = (\bar{y},\bar{x}) \\ \mu(x,y) & \text{otherwise} \end{cases}.$$

Let $(\bar{x},\bar{y})$, $(\bar{a},\bar{b})$, γ be fixed and let us consider the fuzzy preference relation $\Gamma^*(\lambda)$ defined as

$$\Gamma^*(\lambda)(x,y) = \Gamma_\mu((\bar{x},\bar{y}),(\bar{a},\bar{b}),\gamma,\lambda)(x,y)$$

Then, one of the following two properties must be verified by ρ.
(R5.1) there is no value λ such that

$$\rho(\Gamma^*(\lambda_1)) > \rho(\Gamma^*(\lambda))$$
$$\rho(\Gamma^*(\lambda)) < \rho(\Gamma^*(\lambda_2))$$

for some λ_1, λ_2 such that $\lambda_1 < \lambda < \lambda_2$.
(R5.2) there is no value λ such that

$$\rho(\Gamma^*(\lambda_1)) < \rho(\Gamma^*(\lambda))$$
$$\rho(\Gamma^*(\lambda)) > \rho(\Gamma^*(\lambda_2))$$

for some λ_1, λ_2 such that $\lambda_1 < \lambda < \lambda_2$.

We then have the following definition.

Definition 1. Any mapping $\rho : \mathcal{P} \to [0,1]$ is a

1. pessimistic fuzzy rationality measure if it verifies conditions (R1)-(R5.1);
2. optimistic fuzzy rationality measure if it verifies conditions (R1)-(R5.2).

A fuzzy rationality measure being both pessimistic and optimistic will be said *normal*. □

For example, max-min transitivity rationality condition is a *pessimistic* fuzzy rationality.

(Ex1)

$$\rho_{maxmin}(\mu) = \begin{cases} 1 & \text{if } \mu(x,z) \geq \min(\mu(x,y), \mu(y,z)), \forall x,y,z \in X. \\ 0 & \text{otherwise} \end{cases}$$

Some more examples of fuzzy rationality measures can be found in [4].

3 Equivalent rationality measures

As pointed out in [3], two fuzzy rationality measures can be considered equivalent if they always agree when comparing individual rationalities. In this case, it can be supposed that the discrepance between both rationality measures is just due to different underlying scales.

Definition 2. Given two fuzzy rationality measures ρ_1 and ρ_2 we will say that ρ_1 and ρ_2 are equivalent if and only if for any pair of individuals μ_1 and μ_2, $\rho_1(\mu_1) \geq \rho_1(\mu_2)$ if and only if $\rho_2(\mu_1) \geq \rho_2(\mu_2)$. □

In this particular context, it was proven that any class of equivalent fuzzy rationality measures is closed with respect to the most important logical compositions.

Theorem 3. *Let us consider* $\rho_1, \rho_2, \cdots, \rho_k$, k *equivalent fuzzy rationality measures. Let*
$$H : [0,1]^k \to [0,1]$$
be a strictly nondecreasing mapping, in such a way that
$$H(a_1, \cdots, a_k) \geq H(b_1, \cdots, b_k) \text{ if } a_j \geq b_j \text{ for all } j = 1, 2, \cdots, k$$
and
$$H(a_1, \cdots, a_k) > H(b_1, \cdots, b_k) \text{ whenever } a_j > b_j \text{ for all } j = 1, 2, \cdots, k.$$
Let also assume
$$H(1, 1, \cdots, 1) = 1.$$
Then the mapping
$$H(\rho_1, \rho_2, \cdots, \rho_k) : \mathcal{P} \to [0,1]$$
defined as
$$H(\rho_1, \rho_2, \cdots, \rho_k)(\mu) = H(\rho_1(\mu), \rho_2(\mu), \cdots, \rho_k(\mu)), \forall \mu$$
is a fuzzy rationality measure equivalent to $\rho_1, \rho_2, \cdots, \rho_k$.

Since OWA operators [18,19,20] do verify the conditions of the above theorem, they can be used to obtain new measures from finite collections of equivalent fuzzy rationality measures. T-norms and T-conorms can be used whenever they verify the above strict monotonicity condition.

4 General results on min and max operators

Max and *Min* operators have been justified in the fuzzy literature in several ways (see, e.g., [17,19]). The following results show the good performance of these two operators when restricted to pessimistic and optimistics rationality measures.

Theorem 4. *Let* ρ_1 *and* ρ_2 *be two pessimistic [resp. optimistic] fuzzy rationality measures. Then the mapping* $\min(\rho_1, \rho_2)$ *[resp.* $\max(\rho_1, \rho_2)$*] defines a pessimistic [resp. optimistic] fuzzy rationality measure.*

Proof: We will just prove the min case, since the other case is perfectly similar. Properties (R1)-(R4) are immediate. In order to check regularity, let us assume that
$$\min(\rho_1, \rho_2)(\Gamma^*(\lambda_1)) > \min(\rho_1, \rho_2)(\Gamma^*(\lambda)) < \min(\rho_1, \rho_2)(\Gamma^*(\lambda_2))$$

for some $\lambda, \lambda_1, \lambda_2$ such that $\lambda_1 < \lambda < \lambda_2$. Let us assume also that

$$\min(\rho_1, \rho_2)(\Gamma^*(\lambda)) = \rho_1(\Gamma^*(\lambda)).$$

Since ρ_1 is pessimistic but

$$\min(\rho_1, \rho_2)(\Gamma^*(\lambda_1)) \leq \rho_1(\Gamma^*(\lambda_1))$$

and

$$\min(\rho_1, \rho_2)(\Gamma^*(\lambda_2)) \leq \rho_1(\Gamma^*(\lambda_2))$$

always hold, then we have reached to a contradiction. □

Analogous closure result is obtained for OWA operators if restricted to more particular families of pessimistic or optimitic rationality measures (see [5]).

Theorem 5. *Let $\rho_1, \rho_2, \cdots, \rho_k$ be k fuzzy rationality measures, all of them pessimistic (optimistic), in such a way that -according to notation given in (R5)- every mapping $\rho_i(\Gamma^*(\lambda))$ is concave (convex) in λ. Then $H(\rho_1, \rho_2, \cdots, \rho_k)$ is also a pessimistic (optimistic) fuzzy rationality measure, if H is an OWA operator with increasing (decreasing) associated weights.*

5 The computational problem

Let us consider the fuzzy rationality measure ρ_{maxmin}. Though semantically such a rationality measure is quite unsatisfactory, since it requires a crisp property, from a computational point of view is instead a good example. Indeed, as it can easily be seen (see, e.g., [9]), for any μ the value of $\rho_{maxmin}(\mu)$ can be computed in time $\mathcal{O}(|X|^3)$. Therefore, we have a polynomial time computable fuzzy rationality measure which makes it easy to use in real life applications. This is not always the case.

The example that follows is based upon Orlovsky's choice set of unfuzzy nondominated alternatives (see, e.g., [16]). Given a fuzzy preference

$$\mu : X \times X \to [0, 1]$$

for any arbitrary non-empty subset Y of alternatives we define

$$Y^\mu_{UND} = \{x \in Y | \mu(x, y) > \mu(y, x), \forall y \in Y\}.$$

Then the following map is an optimistic fuzzy rationality measure (see also [4]):

$$\rho_N(\mu) = \begin{cases} 0 \text{ if } \exists Y^\mu_{UND} = \emptyset, \text{ for some } Y \neq \emptyset \\ \min\{\frac{1}{|Y^\mu_{UND}|} / Y^\mu_{UND} \neq \emptyset\} \text{ otherwise} \end{cases}$$

The problem of computing such a rationality measures can be rewritten in a graph-theoretical manner (see, e.g., [10]. Given μ and X we can build a graph G_μ whose vertices are the elements of X and such that there exists an edge from x to y if and only if $\mu(x, y) > \mu(y, x)$. We then have

Proposition 6. $\rho_N(\mu) = 0$ if and only if there exists a cycle in G_μ. □

As a consequence, it is relatively easy to check whether μ is absolutely irrational (acording to this particular rationality measure). However, once we know that there are no cycles in G_μ, computing the value of $\rho_N(\mu)$ becomes a computationally hard problem. Indeed, the following holds.

Proposition 7. Given a positive integer K, $\rho_N(\mu) \leq \frac{1}{K}$ if and only if G_μ has an independent set of size K or more. □

We recall that an independent set is a subset of vertices such that no two vertices in it are connected by an edge. Moreover, it is well known that the problem of deciding whether a graph has an independent set of size K or more belongs to the class of $\mathcal{NP}$-complete problems.

If we introduce the following definition

Definition 8. A fuzzy rationality measure is computationally hard if given any $0 \leq r \leq 1$ and fuzzy preference relation μ the problem of deciding whether $\rho(\mu) \geq r$ is NP-hard. □

then we have that ρ_N is computationally hard.

5.1 Polynomial Constructability

We end this section by remarking a property of the fuzzy rationality measure ρ_N which will be used later on. Given any positive integer K it is possible in polynomial time to produce a fuzzy preference relation μ such that $\rho_N(\mu) = \frac{1}{K}$. By using the graph G_μ, we build a graph on X with a maximum independent set of size K as follows:

- choose randomly a set Y of K vertices;
- add an edge from each vertex of Y to each vertex not in Y;
- if $K \geq \frac{|X|}{2}$ then stop, else recursively, build a graph on $X \setminus Y$ with an independent set of size K.

Definition 9. A fuzzy rationality measure ρ is [polynomially] constructible if there exists a [polynomial] algorithm which given any number $0 \leq r \leq 1$ decides whether there exists μ such that $\rho(\mu) = r$ and it outputs an example. □

Obviously, ρ_N is polynomially constructible.

In an attempt to bypass the hardness problem such as the one above seen, one may think to find a rationality measure, equivalent to a given computationally hard one but that instead can be computed easily. In fact, this approach is in general doomed to failure. Indeed the following holds.

Theorem 10. Let ρ be a computationally hard fuzzy rationality measure. Suppose that ρ is polynomially constructible. Then, if ρ' is equivalent to ρ, is computationally hard as well.

6 Final remarks

Consistency of fuzzy preferences use to be defined according to the main application the decision maker is thinking on. When our objective is just descriptive, we should realize that depending on the potential application our idea of consistency may be changing. In this case we should keep the information of an adequate number of rationality measures allowing a right description for all those potential purposes. Then, we may be interested in some global index, which will be some amalgamated value of those partial indices.

Computational aspects should be always addressed in the aggregation procedure of fuzzy rationality measures, but also to each rationality measure itself. Computational difficulties should be taken into account, for example, when considering Montero's rationality measure [2,12,13,14], where completeness of fuzzy preferences was also assumed. We must point out the importance of such a completeness assumption. Rationality under incomparability as modeled in [6,7,8] (see also [15]) appears as a need in future research.

Acknowledgements: This research has been partially supported by Dirección General de Investigación Científica y Técnica (Spain).

References

1. V. Cutello and J. Montero. An axiomatic approach to fuzzy rationality. In: K.C. Min, Ed., *IFSA '93* (Korea Fuzzy Mathematics and Systems Society, Seoul,1993), 634–636.
2. V. Cutello and J. Montero. A characterization of rational amalgamation operations. *International Journal of Approximate Reasoning* 8:325–344 (1993).
3. V. Cutello and J. Montero. Equivalence of Fuzzy Rationality Measures. In: H.J. Zimmermann, Ed., *EUFIT'93* (Elite Foundation, Aachen, 1993), vol. 1, 344–350.
4. V. Cutello and J. Montero. Fuzzy rationality measures. *Fuzzy sets and Systems* 62:39–54 (1994).
5. V. Cutello and J. Montero. Equivalence and Composition of Fuzzy rationality measures. *Fuzzy sets and Systems*, 1995. To Appear.
6. J.C. Fodor and M. Roubens. Preference modelling and aggregation procedures with valued binary relations. In: R. Lowen and M. Roubens, Eds., *Fuzzy Logic* (Kluwer Academic Press, Amsterdam, 1993), 29–38.
7. J.C. Fodor and M. Roubens. Valued preference structures. *European Journal of Operational Research* 79:277–286 (1994).
8. J.C. Fodor and M. Roubens. *Fuzzy Preference Modelling and Multicriteria Decision Support*. Kluwer Academic Pub., Dordrecht, 1994.
9. M.R. Garey and D.S. Johnson. *Computer and Intractability: a Guide to the Theory of NP-Completeness*. Freeman, San Francisco, 1978.
10. M. Gondran and M. Minoux. *Graphs and Algorithms*. Wiley, Chichester, 1984.
11. L. Kitainik. *Fuzzy Decision Procedures with Binary Relations*. Kluwer Academic Pub., Boston, 1993.
12. J. Montero. Arrow's theorem under fuzzy rationality. *Behavioral Science*, 32:267–273 (1987).
13. J. Montero. Social welfare functions in a fuzzy environment. *Kybernetes*, 16:241–245 (1987).

14. J. Montero. Rational aggregation rules. *Fuzzy Sets and Systems* 62:267–276 (1994).
15. J. Montero, J. Tejada and V. Cutello. A general model for deriving preference structures from data. *European Journal of Operational Research*, to appear.
16. S.E. Orlovski. *Calculus of Decomposable Properties, Fuzzy Sets and Decisions*. Allerton Press, New York, 1994.
17. U. Thole, H.J. Zimmermann and P. Zysno. On the suitability of minimum and product operators for the intersection of fuzzy sets. *Fuzzy sets and Systems*, 2:167–180 (1979).
18. R.R. Yager. On ordered weighted averaging aggregation operators in multi-criteria decision making. *IEEE Transactions on Systems, Man and Cybernetics*, 18:183–190 (1988).
19. R.R. Yager. Connectives and quantifiers in fuzzy sets. *Fuzzy sets and Systems*, 40:39–75 (1991).
20. R.R. Yager. Families of owa operators. *Fuzzy sets and Systems*, 59:125–148 (1993).
21. L.A. Zadeh. Similarity relations and fuzzy orderings. *Information Science*, 3:177–200 (1971).

Aggregation in Decision Making with Belief Structures

Maria Teresa Lamata
Departamento de Ciencias de la Computación
e Inteligencia Artificial
Universidad de Granada
18071-Granada, Spain
E-mail: mlamata@goliat.ugr.es

Abstract

We consider the problem of decision making in environments in which there exists some uncertainty about the state of nature. A general approach to the representation of uncertainty using the Dempster-Shafer belief structure is presented.

1 INTRODUCTION

Decision making in environments in which there is some uncertainty with respect to a factor which affects the decision, generically called the state of nature, has been long considered in the literature [6, 9]. The methodologies for solving these problems have been strongly dependent upon the assumption of the type of uncertainty associated with the state of nature. Two assumptions about the knowledge of uncertainty have dominated the literature. The first case is where one assumes a probabilistic knowledge and the second is where one assumes a set of values which includes the actual value, however no probabilistic information is assumed. In the case of probabilistic knowledge considerable use is made of the expected value as a tool for obtaining the optimal alternative.

In the second case, sometimes called decision making under ignorance [9], the introduction of a decision making attitude, optimistic, pessimistic etc, is used to help provide an optimal answer. In [2, 14] the authors suggested that by using the Dempster-Shafer belief structure we can provide a framework which unifies these two assumptions about our knowledge of the uncertainty and in addition provides a structure for representing other more sophisticated types of information about the uncertainty in the environment, this is highlighted in Section 4. Nevertheless, there are times when these actions do not seem reasonable, and that is why instead of assuming a degree of optimism α for deciding, we consider the convenience of creating an index that will be independent the decision-maker. Strat [13] and Nguyen and Walker [8] have also investigated decision making in the Dempster-Shafer environment, with the former proposing a scheme for coding the information that he calls Expected Utility Interval, and by deciding through a degree of optimism.

2 THE STATEMENT OF DECISION PROBLEM

Any decision problem may be expressed by means of the following five elements

$$\{\Omega, D, \mathbf{r}, I, \leq\}$$

where:

1. Ω represents the set of states of nature.

2. D is the set of feasible alternatives to the decision-maker, and from which the decision-maker must choose one. We shall consider Ω and D as finite sets. This allows us to avoid convergence, integrability and measurability problems.

3. $\mathbf{r}: D \times \Omega \to R$ is a function where a real number corresponds to every decision d_i and to every state w_j, i.e.

$$(d_i, \omega_j) \longrightarrow \mathbf{r}(d_i, \omega_j) \equiv r_{ij}$$

with r_{ij} being the reward received by the decision maker when he takes a determined alternative and when a determined nature state is presented.

The function $\mathbf{r}$ can be represented by a matrix called the *matrix of rewards* or *payoffs*:

$D\backslash\Omega$	w_1	w_2	$\cdots$	w_j	$\cdots$	w_m
A_1	r_{11}	r_{12}	$\cdots$	r_{1j}	$\cdots$	r_{1m}
A_2	r_{21}	r_{22}	$\cdots$	r_{2j}	$\cdots$	r_{2m}
$\vdots$						
A_i	r_{i1}	r_{i2}	$\cdots$	r_{ij}	$\cdots$	r_{im}
$\vdots$						
A_n	r_{n1}	r_{n2}	$\cdots$	r_{nj}	$\cdots$	r_{nm}

We do not make any assumption about the construction of r, we only assume, as usual, that the decision maker has a behavior hypothesis.

Under these conditions, the rewards which can be obtained by the decision maker for making a decision will be any value the n-dimensional vector associated with the chosen alternative:

$$A_i \longrightarrow [r_{i1}, r_{i2}, ..., r_{ij}, ..., r_{im}] \equiv \vec{r}_i \in R^m$$

4. A relation of preference "$\leq$" of the decision maker. We shall suppose a coherent decision maker, therefore we shall try to maximize his profits or else minimize his losses.

5. Certain information about the real state of nature. Depending on the decision maker's kind of knowledge about the real state, three environments appear in the literature:

   (a) *Decision making under certainty.* The decision maker knows which state is going to be presented. In particular, that is the case in which $P(s_j) = 1$ or 0 for all $j = 1, \ldots, m$. In this case, the decision maker should choose the alternative which means the maximum payoff / minimum loss to that state of nature.

   (b) *Decision making under risk.* In the decision under risk a probability distribution about the states of nature is assumed to be known. In this case, the standard procedure is to use the expected value:
   1.- For each alternative A_i, calculate its mathematical expectation
   $$E_i = r_{ij} \times p_j$$
   2.- Select as the optimum that decision whose expected value is the greatest, i.e.
   $$E^* = Max_i E_i$$

   (c) *Decision making under ignorance.* The third environment involves the case of ignorance. Since we have no information available about the states of nature, the decision criteria take into account the decision maker's attitude towards risk.

   Among these attitudes the most representative are the following:

   - *Pessimistic attitude.* By means of this strategy, the decision maker has the worst possible point of view. He chooses the worst outcome for every alternative and then he selects the one which achieves the best/worst. This strategy is also known as maximin.
   - *Optimistic attitude.* Under this strategy the decision maker assumes the best result possible selecting the best outcome for each alternative and choosing the alternative which achieves the best/best. This strategy is also known as maximax.
   - *Hurwicz strategy.* With this approach the decision maker selects a value $\alpha \in [0, 1]$ which represents his degree of optimism. For each alternative we weight the maximum and minimum values by α and $(1 - \alpha)$, respectively, i.e.
   $$H_i = \alpha \times Max_i r_{ij} + (1 - \alpha) \times Min_i r_{ij}$$
   Finally, the alternative yielding the maximum of H_i is chosen.

- *Normative approach.* In this approach, the decision maker adds all values relating to every alternative in order to choose that alternative whose sum is the highest.
  In this case, **the decision algorithm** would be as follows:
  1) For each alternative calculate the value of function $V_i = F(r_{i1}, r_{i2}, \ldots, r_{im})$, where F is an aggregation function of the aforementioned functions.
  2) Select an alternative A^* for which the maximum $A^* = Max_i V_i$ is achieved.

3 BELIEF STRUCTURES

In this section we introduce a more general framework for the representation of uncertainty than that of probability. This scheme was suggested by Dempster as multi-valued mappings and later developed by Shafer, it is called the Dempster-Shafer theory of evidence, which is an important and useful tool in the development of expert systems.

A belief structure m on the set Y consists of a collection of non-empty subsets $B_i \subset Y$ and an associated set of weights $m(B_i)$ such that:

1. $m(B_i) > 0$,

2. $\Sigma_i m(B_i) = 1$.

The subsets B_i are called the focal elements of the belief structure.

Assume we have a decision problem in which we have a collection of n alternatives, we denote the set of alternatives as $D = \{A_1, A_2, \ldots, A_n\}$. We further assume that r_{ij} is the payoff to the decision maker if he selects alternative A_i and the state of nature is ω_j. In addition, we assume that our knowledge about the state of nature is captured in terms of a belief structure m on Ω. The focal elements of m are $B_1, B_2, \ldots, B_r$ and associated with each of these there is a probability mass value $m(B_i)$.

We have different ways of distributing the mass of evidence from the focal element amongst the different states that form part of it (it is understood that any of these schemes of behavior is subjective and so that the selection of one or the other will basically depend on the decision-maker's position about the risk):

1. Consists of assigning all the mass of evidence to that state of nature of the focal element for which the associated reward is the greatest. It coincides, as we can see, with an optimistic criterion.

2. Consists of assigning all the mass of evidence to that state of nature belonging to the focal element whose associated reward are the lowest. It coincides, as we can see, with an pessimistic criterion.

3. Consists of distributing the mass of evidence of focal element between all the states but weighted according to the degree of optimism and in increasing order of the payoffs.

Once the decision problem has been established in these terms and the information about the states of nature is known in terms of a B.P.A., our question is: in a determined situation, which is the best decision? We can answer it, in a general way, as follows: we shall choose the decision with the largest associated vector of rewards. This can look like a good answer. In practice, however, this is not easy to do, since it, is necessary to define relations of order in R^m.

The problem is to select the alternative which maximazes the payoff for the decision maker. Supposing that the criteria of dominance do not provide any result, it is the decision maker who must decide, taking into account a subjective rule.

Any reasonable approach to alternative selection should have certain rational properties. The first property is the Pareto Optimality:

Definition. This property requires that given two alternatives, A and B, where A has at least as high a payoff as B for each state of nature, then B should not be more preferred than A.

4 DECISION MAKING UNDER BELIEF STRUCTURES

We know that the procedure for attaining the best decision combines the schemes used in the decision making under ignorance and under risk, both cases being, particular cases of this more general approach that is the decision making under the Dempster-Shafer uncertainty.

Therefore, if we consider that the uncertainty which provides the belief structure is a mixture of uncertainty and ignorance. we shall solve the problem of decision-making in Dempster-Shafer's theory in two stages.

Firstly, by dealing only with the problem of ignorance.

Thus we shall manage to obtain two probability distributions associated with each alternative.

Secondly, by solving the problem of decision-making in the face of risk for each one of the distributions obtained in the first step.

4.1 Decision under ignorance

Our aim shall be, in view of the reward vector and the information available about the states of nature which, in our case, is expressed in terms of evidence, to take the best decision; but this involves having to order the set R^m which is not a body of evidence except in the specific case in which there is an option or alternative which dominates or is preferred to the others.

Therefore, given two alternatives, A_1 and A_2, where A_1 has at least as high a reward as A_2 for each state of nature, then A_2 should not be more preferred than A_1.

$$F(r_{11}, r_{12}, \ldots, r_{1m}) \leq F(r_{21}, r_{22}, \ldots, r_{2m})$$

it being indifferent in the case that $< r_{11}, r_{12}, \ldots, r_{1m} >$ and $< r_{21}, r_{22}, \ldots, r_{2m} >$ are equal.

This property is no easy to attain, especially as m increases, which is why what we have to do is to construct a value function that is highly praised in the literature

$$F : R^m \longrightarrow R$$

and such that the result is a real number. In this context, our best decision shall be to select the alternative for which the result of applying F will be the maximum.

As principal functions describe in the literature we have the following:

Strategy	*AggregationFunction*
Pessimistic	$F = \mathrm{Min}_j r_{ij}$
Optimistic	$F = \mathrm{Max}_j r_{ij}$
Hurwicz	$F = \alpha \times \mathrm{Max}_j r_{ij} + (1 - \alpha) \times \mathrm{Min}_j r_{ij}$
Normative	$F = \Sigma_j r_{ij}$

All these criteria are widely described in the literature and so we shall not refer to them further. Instead of working with this application, F, we shall establish an intermediate step defining an application

$$g : R^m \longrightarrow R^2$$

in such a way that the m-dimensional payoff vector is reduced to a two-dimensional one.

The application

$$g : R^m \longrightarrow R^2$$

is constructed by specifically using the information available about the true state of the nature, which is the same as adopting certain assumptions about behavior. In particular, the assignment of the mass of evidence will vary according to the payoff vector that is considered, since, for each focal element B_k, only the extreme values are contemplated (maximum, minimum) of the constraint of the said vector to the aforementioned focal element.

This way of sharing out the mass of evidence for each focal element means implicitly admitting that the decision maker, in order to choose, only considers the extreme values of the profits and the possibility of obtaining them, without taking the intermediate values into account for this calculation. That is why a specially structure is observed for considering the decision maker's attitude when faced with risk.

For each alternative we obtain two probability distributions, both lying within the true probability distribution.

4.2 Decision under risk

If, for each one of these distributions, we calculate their mathematical expectation, we shall obtain an interval (E_*, E^*) between the two extremes within which would be found the average value corresponding to the real probability distribution. This interval coincides with the one obtained in [2], and which is called the risk interval associated with monotonous expectation of a B.P.A.

All the criteria described in decision making may be applied to this interval, thus:

$$\text{Maximin} \longrightarrow E_*$$
$$\text{Maximax} \longrightarrow E^*$$
$$\text{Hurwicz} \longrightarrow \alpha E_* + (1-\alpha) E^*$$
$$\text{Normative} \longrightarrow 1/2(E_* + E^*)$$

but, if we wanted to decide depending on the risk interval and not just in extremes, we may state several criteria.

Definition. We shall say that the alternative D_i dominates or is preferred to the alternative D_j, written $D_i \succ D_j$, if it is verified that

$$E_{*i} > E_{*j}$$
$$E_i^* > E_j^*$$

Definition. An alternative $d_i \in D$ is efficient (not necessarily dominant) if there is no other alternative $d_i \in D$ such that $D_j \succ D_i$.

We shall call the set of all the alternatives that are not dominated by any other ones a *set of efficient alternatives* $\mathcal{E}$, and it is clear that the later shall be the set on which the decision maker must make his selection, should there be no alternative that dominates the rest.

If we represent each alternative graphically through its risk interval, we obtain a cloud of points located above the main diagonal since $E_* < E^*$, thereby constituting the set of the efficient alternatives, a convex set, in such a way that the closer an alternative is to the diagonal, the smaller the risk interval and so this action shall be one that represents the least uncertainty about the result of the selection, this option is suitable for the decision makers who like minimizing the risk:

$$d_i > d_k \quad \text{if} \quad E_{*di} > E_{*dk}$$

On the other hand, the nearer there presentation of an alternative is to the axis of coordinates the wider the interval and so this alternative implies an optimistic criterion:
$$d_i > d_k \quad \text{if} \quad E^*_{di} > E^*_{dk}$$
In this way of selecting, we can see that a subjective criterion is involved: if we wanted to take a decision with an objective nature, we would have to create an index and decide accordingly.

4.3 Index

So now assume the case of a pessimistic decisionmaker who is asked to select between the following alternatives:
$$d_1 \rightarrow g(r_1) = \{7, 15\}$$
$$d_2 \rightarrow g(r_2) = \{6, 31\}$$

It is obvious that by applying a pessimistic criterion (lexicographic reverse), $d_1 \succ d_2$. But, however, it seems intuitive that any decision maker, however loathe to risk, would not be coherent if he made that decision, since it appears that d_2 is a better option than d_1. That is why we think that, for the set of efficient alternatives, that use of relaxations in the order of R such as "a bit less", "not much more" may lead us to obtaining more coherent fuzzy decision making rules.

Instead of providing these rule, we propose the following index that could include the idea described above
$$I = \left| \frac{E^*_{di} - E^*_{dk}}{E_{*di} - E_{*dk}} \right|$$
with the values that I can adopt being the following:

- If $I > 1$, then the selected alternative is the one for which E^*_d is the maximum.

- If $I = 1$, then the two alternatives are indifferent

- If $I < 1$, then the selected alternative is the one for which E^*_d is the minimum.

An algorithm for the selection of the best alternative according to I is:

1. For each alternative i, do the following:

    (a) Obtain two probability distributions (P_*, P^*) associated with the states of nature through rewards.

    (b) Obtain for (P_*, P^*) the corresponding expected values (E_*, E^*).

2. Apply the Pareto optimality for obtaining the set of efficient alternatives $\{d_i\}$, i=1,2,...,t and $t \leq n$.

3. For the set of efficient alternatives, do the following:

   (a) Choose A_i and A_{i+1} and calculate the index I.
   (b) Select the best alternative.

4. Repeat the process $(t-1)$ times until obtaining one single alternative.

Example. Assume the payoffs matrix is as follows:

	w_1	w_2	w_3	w_4	w_5
A_1	5	8	10	15	25
A_2	12	10	5	12	13
A_3	10	8	4	10	13
A_4	7	18	10	9	15

Assume that our knowledge about the states of nature is given by the following belief structure m:

$Focal elements$	$Weights$
$B_1 = \{w_1, w_3, w_5\}$	0.5
$B_2 = \{w_2, w_4\}$	0.3
$B_3 = \{w_1, w_2, w_3, w_4, w_5\}$	0.2

For A_1, we obtain, as the associated probability distribution, the following:

	w_1	w_2	w_3	w_4	w_5
$m(B_1)$	0.5	0	0	0	0
$m(B_2)$	0	0.3	0	0	0
$m(B_3)$	0.2	0	0	0	0
P_*	0.7	0.3	0	0	0

and in the same way:

	w_1	w_2	w_3	w_4	w_5
$m(B_1)$	0	0	0	0	0.5
$m(B_2)$	0	0	0	0.3	0
$m(B_3)$	0	0	0	0	0.2
P^*	0	0	0	0.3	0.7

with the corresponding expected values being

$$E_{*1} = 0.7 \times 5 + 0.3 \times 8 = 5.9$$

$$E_1^* = 0.3 \times 15 + 0.7 \times 25 = 22$$

Then the alternative A_1 has associated the following as its risk interval

$$[E_{*1}, E_1^*] = [5.9; 22]$$

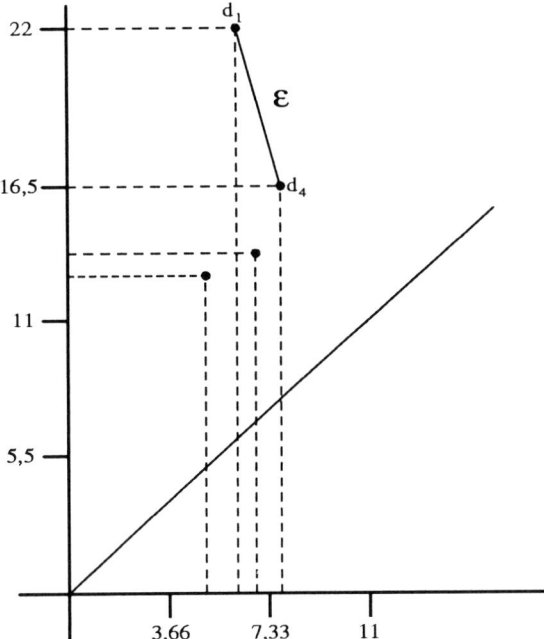

Figure 1:

For A_2, we have

$$P_* = P(w_1) = 0, P(w_2) = 0.3, P(w_3) = 0.7, P(w_4) = P(w_5) = 0$$
$$P^* = P(w_1) = P(w_2) = P(w_3) = 0, P(w_4) = 0.3, P(w_5) = 0.7$$
$$[E_{*2}, E_2^*] = [6.5; 12.7]$$

For A_3, we have

$$P_* = P(w_1) = 0, P(w_2) = 0.3, P(w_3) = 0.7, P(w_4) = P(w_5) = 0$$
$$P^* = P(w_1) = P(w_2) = P(w_3) = 0, P(w_4) = 0.3, P(w_5) = 0.7$$
$$[E_{*3}, E_3^*] = [5.2; 12.1]$$

For A_4, we have

$$P_* = P(w_1) = 0.7, P(w_2) = P(w_3) = 0, P(w_4) = 0.3, P(w_5) = 0$$
$$P^* = P(w_1) = 0, P(w_2) = 0.5, P(w_3) = P(w_4) = 0, P(w_5) = 0.5$$
$$[E_{*4}, E_4^*] = [7.6; 16.5]$$

The situation is depicted in Figure 1.

We obtain therefore $\mathcal{E} = \{d_1, d_4\}$, and the alternatives chosen according to the different criteria are:

- Maximin $\longrightarrow d_4$
- Maximax $\longrightarrow d_1$
- Index $\longrightarrow d_1$

Acknowledgments

This work is partially supported by the DGICYT under project PB95-1181.0.

5 CONCLUDING REMARKS

In this paper we have looked at the problem of decision making in environments in which there exists some uncertainty about the states of nature. We discussed a general approach to the representation of this uncertainty using the Dempster-Shafer belief structure.

We resolve the decision making problem by transforming it into an easier one. To do so, we transform the n-dimensional payoff vector into an interval $[E_*, E^*]$ and we decide about it, using either subjective criteria (lexicographical ones, OWA operators, ...) or subjective ones, for which we define an index depending on the differences between the extreme values.

References

[1] R.Bellman and L.A.Zadeh, (1970) "Decision making in a fuzzy environment". *Management Science*,17, pp. 141-164.

[2] M.J. Bolaños, M.T. Lamata and S. Moral, (1988) Decision making problems in a general environment. *Fuzzy Sets and Systems* 18, 135-144.

[3] A.P. Dempster (1968) A generalization of Bayesian inference, *Journal of the Royal Statistical Society*, 205-247.

[4] P.C. Fishburn (1970) *Utility theory for decision making*. John Wiley: New York.

[5] M.T. Lamata (1986) Problemas de decisión con información general. Ph.D. thesis. Universidad de Granada.

[6] R.D. Luce and H. Raiffa (1967) *Games and Decisions: Introduction and Critical Survey*. Wiley: New York.

[7] M. O'Hagan (1997) Aggregating template rule antecedents in real-time expert systems with fuzzy set logic. *Int. Journal of Man-Machine Studies* (to appear).

[8] H.T. Nguyen and E.A. Walker (1994) On decision making using belief functions, in *Advances in the Dempster-Shafer theory of Evidence.* (M. Fedrizzi, J. Kacprzyk, R.R. Yager, eds). Wiley: New York , pp. 311-330.

[9] S.B. Richmond (1968) *Operations Research for Management Decisions*, Ronald Press: New York.

[10] M. Roubens and Ph. Vincke (1989) *Preference Modeling.* Springer-Verlag: Berlin.

[11] G. Shafer (1976) *A Mathematical Theory of Evidence.* Princeton University Press: Princeton, N.J.

[12] J.Q. Smith (1988) *Decision Analysis: A Bayesian Approach.*

[13] T.M. Strat (1994) Decision analysis using belief functions, in *Advances in the Dempster-Shafer theory of Evidence*: (M. Fedrizzi, J. Kacprzyk, R.R. Yager, eds). Wiley: New York, pp. 275-310.

[14] R.R. Yager (1992) Decision making under Dempster-Shafer uncertainties. *Int. Journal of Intelligent Systems* 20, pp. 233-245.

Multistage Fuzzy Control with a Soft Aggregation of Stage Scores

Janusz Kacprzyk

Systems Research Institute, Polish Academy of Sciences
ul. Newelska 6, 01-447 Warsaw, Poland
E-mail: kacprzyk@ibspan.waw.pl

Abstract

We discuss multistage fuzzy control in Bellman and Zadeh's (1970) setting. Instead of the traditional basic formulation which is to find an optimal sequence of controls best satisfying fuzzy constraints and fuzzy goals at *all* the control stages, we assume a softer requirement that the stage scores (degrees to which the fuzzy constraints and fuzzy goals at a particular control stage are fulfilled) be best satisfied at *most, almost all, much more than a half*, etc. control stages. Zadeh's (1983) fuzzy-logic-based calculus of linguistically quantified statements is employed, and the resulting problem is solved by dynamic programming.

Keywords: multistage fuzzy control, dynamic programming, fuzzy linguistic quantifier.

1 Introduction

Fuzzy (logic) control has been for a couple of years one of much talked about technologies which has been amplified by relevant applications both in specialized equipment as, e.g., cranes or subway trains, and in everyday products as, e.g., electric appliances (washing machines, refrigerators, ...), photographic and video cameras, etc. Needless to say that the latter have been more "visible" to general public, and have attracted attention of the media which has given impetus for further development and financing.

The essence of fuzzy (logic) control is that we do not intend to build a model of the control process itself, as is customary in traditional control, because this may be too difficult or costly, or even such a model may be unknown. We acquire knowledge from, say, experienced process operators or other experts as to *how to control* the process. This knowledge need not be expressed precisely (i.e. using numbers) but may be given as linguistic statements exemplified by "if the pressure is *low*, then increase *slightly* the temperature", with the linguistic terms in italics represented by fuzzy sets. Then, such statements, which stand for control laws for particular situations, are used to infer the control for a current situation which may differ from those in the rules.

It is easy to see that fuzzy (logic) control is clearly a *descriptive* approach since we explictly *describe* how to control the process.

This may be, however, viewed counter-intuitive, and somehow contradicting the long tradition of control. Namely, the essence of control may be generally meant as to *find* a *best* course of action, assuming knowledge of the *behavior* (dynamics) of the system to be controlled, and some *goals* to be attained and *constraints* under which to operate.

Therefore, the essence is not to *describe* how the system is controlled but to *prescribe* how to control the system.

Such a *prescriptive* approach to fuzzy control is even earlier than the descriptive one, as it has appeared as early as in the late 1960's and early 1970's in Chang's (1969a, b) and and Bellman and Zadeh's (1970) papers. It had initially attracted much attention, and even Kacprzyk's (1983a) book discussed it in detail. Then the attention shifted to the descriptive approach to fuzzy control, i.e. to fuzzy (logic) control.

However, it seems that a revival of the prescriptive approach to fuzzy control may happen in near future. First, even if the traditional fuzzy (logic) control is really very effective and has found so many applications, it seems that its potentials from the conceptual point of view are to a large extent exhausted. Even such apparently new approaches as, e.g., the so-called neuro-fuzzy control do not provide conceptually new ideas related to the very essence.

A new generation of fuzzy control should emerge, and we think that some sort of a *prescriptive* approach will be a relevant part of this new generation.

Kacprzyk's (1997) new book provides a detailed description of such a *prescriptive approach* to fuzzy control, and may be viewed as a step toward such a *new generation of fuzzy control* by explicitly incorporating tools for determining *how a system should best be controlled*. In this paper we also assume this general framework.

Basically, the general fuzzy control related framework adopted in this paper is as depicted in Figure 1.

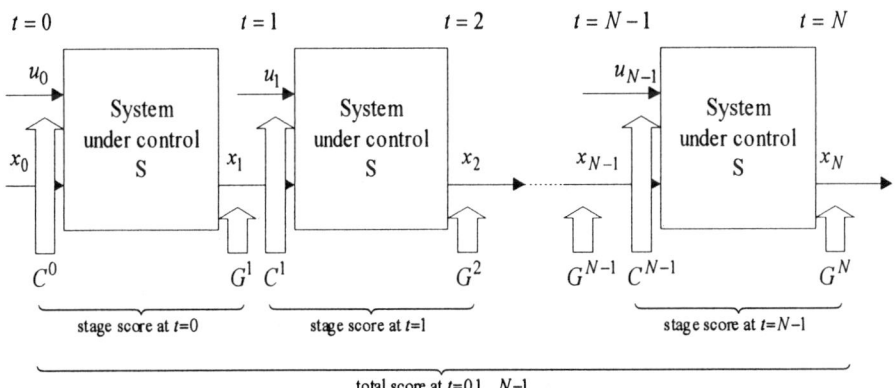

Fig. 1. A general framework of the prescriptive approach to fuzzy control

We start from an initial state at control stage (time) $t = 0$, x_0, apply a control at $t = 0$, u_0, attain a state at time $t = 1$, x_1, apply $u_1, \ldots$. Finally, being at stage $t = N - 1$ in state x_{N-1} we apply control u_{N-1} and attain the final state x_N.

Dynamics of the system under control, S, is assumed known and given by state transitions from state x_t to x_{t+1} under control u_t, the consecutive controls applied, u_t, are subjected to fuzzy constraints, $\mu_{C^t}(u_t)$, and on the states attained, x_{t+1}, fuzzy goals, $\mu_{G^{t+1}}(u_{t+1})$, are imposed, $t = 0, 1, \ldots, N - 1$.

The performance (goodness) of the control process is expressed by some measure of how well, at all the consecutive control stages, the fuzzy constraints on controls and fuzzy goals on states are satisfied. And an optimal sequence of controls at the consecutive control stages, $u_0^*, \ldots, u_{N-1}^*$, is sought (to be determined by an algorithm!).

One may notice that the above general scheme of a prescriptive approach to multistage fuzzy control may be viewed to give rise to the following general *problem classes*, for the following types of the termination time and system under control:

- type of termination time:

    1. fixed and specified in advance,

    2. explicitly specified (as the moment of entering for the first time a termination set of states),

    3. fuzzy, and

    4. infinite;

- type of the system under control:

    1. deterministic,

    2. stochastic, and

    3. fuzzy.

While lokking at the evaluation of the above multistage fuzzy control process we can readily see that it concerns:

- the evaluation of the particular control stage t, $t = 0, 1, \ldots, N - 1$, and

- the evaluation of all the control stages $t = 0, 1, \ldots, N - 1$.

The evaluation of control stage t is basically how well the fuzzy constraint and fuzzy goal are fulfilled. Evidently, this is to a degree, which is assumed in this paper from the unit interval, and is viewed to be a *stage score*.

On the other hand, the evaluation of all the consecutive control stages $t = 0, 1, \ldots, N - 1$ involves some aggregation of the stage scores, and yields a *total score*. Traditionally (cf. Kacprzyk, 1997), we require the fuzzy constraints and fuzzy goals should best be fullfiled at *all* the control stages.

Obviously, such a "hard" requirement can be counter-intuitive in many cases because a weak score at just one control stage may imply a weak evaluation of the whole trajectory, i.e. for *all* control stages, in spite of even quite good scores at virtually all other control stages.

Therefore, a "softening" of the above requirement was proposed by Kacprzyk (1983), Kacprzyk and Iwański (1987) [cf. also Kacprzyk's (1997) book] by requiring that

fuzzy constraints and fuzzy goals should best be fullfiled at Q (e.g., *most, almost all, much more than a half*, etc.) control stages.

Notice that now even a few stage scores may not lower the total score if the rest of stage scores are good. This is certainly intuitively appealing and adequate in many real problems.

In this paper we will deal with such a soft aggregation of stage score in the formulation of multistage fuzzy control. We will employ Zadeh's (1983) fuzzy-logic-based calculus of linguistically quantified statements to derive the problem formulation of Bellman and Zadeh's (1970) type, and then show how dynamic programming can be used for solution.

2 Preliminaries

In this section we will briefly review two issues which will be of crucial importance to our further discussion: Zadeh's (1983) fuzzy-logic-based calculus of linguistically quantified statements, and Bellman and Zadeh's (1970) general approach to decision making under fuzziness.

2.1 A fuzzy-logic-based calculus of linguistically quantified statements

As already indicated in Section 1, a key role in the model considered in this paper is played by fuzzy linguistic quantifiers exemplified by *most, almost all, much more than a half*, etc. However, no tools for handling such linguistic quantifiers are available in conventional mathematics. Fortunately enough, fuzzy logic provides formal means for the representation and handling of such linguistic quantifiers; in fact, such quantifiers are essential to fuzzy logic (cf. Zadeh and Kacprzyk , 1992). In this section we will briefly present the basic fuzzy-logic-based calculus of linguistically quantified statements due to Zadeh (1983). For the use of another approach, due to Yager (1983), in the problem considered, we refer the interested reader to Kacprzyk's (1997) book.

A *linguistically quantified statement* is exemplified by "most experts are convinced" and may be generally written as

$$Qy\text{'s are } F \tag{1}$$

where Q is a linguistic quantifier (e.g., most), $Y = \{y\}$ is a set of objects (e.g., experts) and F is a property (e.g., convinced).

Importance B may also be added to (1) yielding

$$QBy\text{'s are } F \tag{2}$$

exemplified by "most (Q) of the important (B) experts (y's) are convinced (F)".

For our purposes the problem is to find the (degree of) truth of such linguistically quantified statements (1) and (2), denoted truth(Qy's are F) and truth(QBy's are F), respectively, which may be done as shown below.

This is the classical calculus proposed by Zadeh (1983) which is still by far the most widely used. First, a (relative) fuzzy linguistic quantifier is equated with a fuzzy set in $[0, 1]$. For instance, $Q =$ "most" may be given as

$$\mu_{\text{"most"}}(x) = \begin{cases} 1 & \text{for } x > 0.8 \\ 2x - 0.6 & \text{for } 0.3 \leq x \leq 0.8 \\ 0 & \text{for } x < 0.3 \end{cases} \tag{3}$$

which is meant as that if less than 30% of the objects considered possess some property, then it is sure that not most of them possess it, if more than 80% of them possess the property, then it is sure that most of them possess it, and for the cases in-between, this is true (sure) to an extent, from 0 to 1, the more the percentage the higher the truth.

The same argument can be applied for *absolute quantifiers* exemplified by *about 5, much more than 7*, etc. which are however not relevant in our context.

Particularly important are the *non-decreasing fuzzy linguistic quantifiers* defined as

$$x' > x'' \Longrightarrow \mu_Q(x') \geq \mu_Q(x''), \qquad \text{for each } x', x'' \in [0, 1] \tag{4}$$

which reflect an attitude "the more the better"; $Q =$ "most" (3) is clearly non-decreasing.

Property F is defined as a fuzzy set in the set of objects Y, and if $Y = \{y_1, \ldots, y_p\}$, then we assume that truth(y_i is F) $= \mu_F(y_i), i = 1, \ldots, p$.

The (degree of) truth of the linguistically quantified statement (1), that is, truth(Qy's is F), is now calculated in two steps (Zadeh, 1983):

Step 1. Calculate

$$r = \frac{\Sigma\text{count}(F)}{\Sigma\text{count}(Y)} = \frac{1}{p}\sum_{i=1}^{p}\mu_F(y_i) \tag{5}$$

Step 2. Calculate

$$\text{truth}(Qy\text{'s are } F) = \mu_Q(r) \tag{6}$$

In the case with importance, i.e. (2), importance is defined as a fuzzy set $B =$ "important" in Y, such that $\mu_B(y_i) \in [0, 1]$ is a degree of importance of object y_i: from 0 for definitely unimportant to 1 for definitely important, through all intermediate values, such that the higher the more important.

We rewrite first "QBy's are F" as "$Q(B$ and $F)y$'s are B" which clearly implies the following counterparts of the two steps (5) and (6), respectively:

Step 1. Calculate

$$r' = \frac{\Sigma\text{count}(B \text{ and } F)}{\Sigma\text{count}(B)} = \frac{\sum_{i=1}^{p}[\mu_B(y_i) \wedge \mu_F(y_i)]}{\sum_{i=1}^{p}\mu_B(y_i)} \qquad (7)$$

Step2. Calculate

$$\text{truth}(QBy\text{'s are } F) = \mu_Q(r') \qquad (8)$$

where "∧" may be replaced by another suitable operation as, e.g., a *t*-norm.

Example 1. Let $Y = \{X, V, Z\}$, $F = $ "convinced" $ = 0.1/X + 0.6/V + 0.8/Z$, $Q = $ "most" be given by (3), t is "∧," and $B = $ "important" $ = 0.2/X + 0.5/V + 0.6/Z$. Then, $r = 0.5$ and $r' = 0.8$, and

truth("most experts are convinced") = 0.4
truth("most of the important experts are convinced") = 1

□

2.2 Bellman and Zadeh's general approach to decision making under fuzziness

In Bellman and Zadeh's (1970) setting the fuzziness of the environment within which the decision making (or control) process proceeds is modeled by the introduction of fuzzy goals, fuzzy constraints, and a fuzzy decision.

Formally, if $X = \{x\}$ is some set of possible *options* (alternatives, variants, choices, decisions, ...), then:

- the *fuzzy goal* is defined as a fuzzy set G in X, characterized by its membership function $\mu_G : X \longrightarrow [0,1]$ such that $\mu_G(x) \in [0,1]$ specifies the grade of membership of a particular option $x \in X$ in the fuzzy goal G;

- the *fuzzy constraint* is similarly defined as a fuzzy set C in the set of options X, characterized by $\mu_C : X \longrightarrow [0,1]$ such that $\mu_C(x) \in [0,1]$ specifies the grade of membership of a particular option $x \in X$ in the fuzzy constraint C.

The general formulation of decision making in a fuzzy environment is:

"Attain G and satisfy C" (9)

which leads to a fuzzy decision D in X which results from some aggregation G and C corresponding to "and". This aggregation may be different, and the most pupular type of a fuzzy decision is

$$\mu_D(x) = \mu_G(x) \wedge \mu_C(x), \qquad \text{for each } x \in X \qquad (10)$$

where "∧" is the minimum operation, i.e. $a \wedge b = \min(a,b)$.

Such a fuzzy decision will be used throughout this paper. For other types of fuzzy decisions we refer the reader to Kacprzyk's (1997) book.

The *maximizing decision* is defined as an $x^* \in X$ such that

$$\mu_D(x^*) = \max_{x \in X} \mu_D(x) \tag{11}$$

The basic framework presented above can be extended to handle multiple fuzzy constraints and fuzzy goals.

Suppose therefore that we have $n > 1$ fuzzy goals, $G_1, \ldots, G_n$, and $m > 1$ fuzzy constraints, $C_1, \ldots, C_m$, all defined as fuzzy sets in X.

The *fuzzy decision* is then defined as

$$\mu_D(x) = \mu_{G_1}(x) \wedge \ldots \mu_{G_n}(x) \wedge$$
$$\wedge \mu_{C_1}(x) \wedge \ldots \wedge \mu_{C_m}(x), \quad \text{for each } x \in X \tag{12}$$

One can also extend this framework to the case of fuzzy goals and fuzzy constraints that are defined in different spaces. It is clearly the case in virtually all real problems, and is also assumed here.

Suppose that the fuzzy constraint C is defined in $X = \{x\}$, the fuzzy goal G is defined in $Y = \{y\}$, and a function $f : X \longrightarrow Y$, $y = f(x)$, is known. Typically, X and Y may be sets of options and outcomes, notably causes and effects, respectively.

Now, the *induced fuzzy goal* G' in X generated by the given fuzzy goal G in Y is defined as

$$\mu_{G'}(x) = \mu_G[f(x)], \quad \text{for each } x \in X \tag{13}$$

The *fuzzy decision* is now defined analogously as

$$\mu_D(x) = \mu_{G'}(x) \wedge \mu_C(x), \quad \text{for each } x \in X \tag{14}$$

Finally, for $n > 1$ fuzzy goals $G_1, \ldots, G_n$ defined in Y, $m > 1$ fuzzy constraints $C_1, \ldots, C_m$ defined in X, and a function $f : X \longrightarrow Y$, $y = f(x)$, we analogously have

$$\mu_D(x) = \mu_{G'_1}(x) \wedge \cdots \wedge \mu_{G'_n}(x)$$
$$\wedge \mu_{C_1}(x) \wedge \cdots \wedge \mu_{C_n}(x), \quad \text{for each } x \in X \tag{15}$$

In all the above cases the *maximizing decision* is defined as (11), i.e.

$$\mu_D(x^*) = \max_{x \in X} \mu_D(x)$$

In the next section we will show how the above model can be employed to formulate the multistage fuzzy control problem considered in this paper.

3 Multistage fuzzy control in Bellman and Zadeh's setting

As mentioned in Section 2.2, the general Bellman and Zadeh's (1970) approach to decision making in a fuzzy environment is a convenient general framework within which to formulate multistage fuzzy control problems. Decisions are then referred to as *controls*, the discrete time moments at which decisions are to be made – as *control stages*, and the input–output (or cause–effect) relationship – as a *system under control*.

In this paper we will consider the basic, simple case of a deterministic system under control which will best serve the purpose of this paper, i.e. to show how a soft aggregation of stage scores can be employed.

Suppose that the control space is $U = \{u\} = \{c_1, \ldots, c_m\}$, the state space is $X = \{x\} = \{s_1, \ldots, s_n\}$, and both are finite. For simplicity, the control is equated with the input, and the state with the output.

The *control process* proceeds basically as it has already been depicted in Section 1 in Figure 1. In the beginning we are in an initial state $x_0 \in X$. We apply a control $u_0 \in U$ which is subjected to a fuzzy constraint $\mu_{C^0}(u_0)$. We attain a state $x_1 \in X$ via a known input–output (cause–effect) relationship, i.e. a state transition equation of the system under control S; a fuzzy goal $\mu_{G^1}(x_1)$ is imposed on x_1. Next, we apply a control u_1 which is subjected to a fuzzy constraint $\mu_{C^1}(u_1)$, and attain a fuzzy state x_2 on which a fuzzy goal $\mu_{G^2}(x_2)$ is imposed, etc.

The system under control is assumed to be deterministic and its temporal evolution is governed by a *state transition equation*

$$x_{t+1} = f(x_t, u_t), \qquad t = 0, 1, \ldots \qquad (16)$$

where $x_t, x_{t+1} \in X = \{s_1, \ldots, s_n\}$ are the states at control stages t and $t+1$, respectively, and $u_t \in U = \{c_1, \ldots, c_m\}$ is the control at control stage t.

At each control stage t, $t = 0, 1, \ldots$, the control applied $u_t \in U$ is subjected to a *fuzzy constraint* $\mu_{C^t}(u_t)$, and on the state attained $x_{t+1} \in X$ a *fuzzy goal* is imposed

The *initial state* is $x_0 \in X$ and is assumed to be known, and given in advance. The *termination time* (planning, or control, horizon), i.e. the maximum number of control stages, is denoted by $N \in \{1, 2, \ldots\}$, and is assumed to be finite, and fixed and specified in advance, throughout this paper.

The *performance* of the particular control stage t, $t = 0, 1, \ldots, N-1$, is evaluated by the *stage score*

$$v_t = \mu_{C^t}(u_t) \wedge \mu_{G^{N-1}}(x_{N-1}) \qquad (17)$$

beacuse the goal at each stage t is to

satisfy the fuzzy constraint and attain the fuzzy goal.

The *performance* of the whole multistage fuzzy control process is evaluated by the fuzzy decision

$$\mu_D(u_0, \ldots, u_{N-1} \mid x_0) =$$
$$= v_0 \wedge v_1 \wedge \ldots \wedge v_{N-1} =$$
$$= \mu_{C^0}(u_0) \wedge \mu_{G^1}(x_1) \wedge \ldots \wedge \mu_{C^{N-1}}(u_{N-1}) \wedge \mu_{G^N}(x_N) \qquad (18)$$

Very often, without loss of generality, this problem is slightly simplified by assuming that the fuzzy constraints are imposed on all the consecutive control stage, $\mu_{C^0}(u_0), \ldots, \mu_{C^{N-1}}(u_{N-1})$, and the fuzzy goal is imposed only on the final stage, μ_{G^N}. This leads to the fuzzy decision

$$\mu_D(u_0, \ldots, u_{N-1} \mid x_0) =$$
$$= \mu_{C^0}(u_0) \wedge \mu_{C^1}(x_1) \wedge \ldots \wedge \mu_{C^{N-1}}(u_{N-1}) \wedge \mu_{G^N}(x_N) \quad (19)$$

The multistage control problem in a fuzzy environment is now formulated as to find an optimal sequence of controls $u_0^*, \ldots, u_{N-1}^*$ such that

$$\mu_D(u_0^*, \ldots, u_{N-1}^* \mid x_0) = \max_{u_0, \ldots, u_{N-1} \in U} \mu_D(u_0, \ldots, u_{N-1} \mid x_0) \quad (20)$$

As indicated in Section 1, extensions of this basic formulation are possible, but will not be considered here; the reader is referred to Kacprzyk (1997).

Problem (20) can be solved using the following two basic traditional techniques:

− dynamic programming, and

− branch-and-bound,

and also using the following two new ones:

− a neural network, and

− a genetic algorithm.

In this section we will outline the use of the first basic technique, i.e. dynamic programming, as it will be of relevance for our next discussion. For the other solution techniques, we refer the reader to Kacprzyk's (1997) book.

The application of dynamic programming for the solution of problem (20) was proposed in the seminal paper of Bellman and Zadeh (1970).

First, let us slightly rewrite (20) as to find $u_0^*, \ldots, u_{N-1}^*$ such that

$$\mu_D(u_0^*, \ldots, u_{N-1}^* \mid x_0) = \max_{u_0, \ldots, u_{N-1}} [\mu_{C^0}(u_0) \wedge \ldots$$
$$\ldots \wedge \mu_{C^{N-1}}(u_{N-1}) \wedge \mu_{G^N}(f(x_{N-1}, u_{N-1}))] \quad (21)$$

Clearly, its structure makes the application of dynamic programming possible. Namely, the last two right-hand-side terms, i.e.

$$\mu_{C^{N-1}}(u_{N-1}) \wedge \mu_{G^N}(f(x_{N-1}, u_{N-1}))$$

depend only on control u_{N-1} and not on any previous controls, and hence the maximization over $u_0, \ldots, u_{N-1}$ in (20) can be divided into:

− maximization over $u_0, \ldots, u_{N-2}$, and

− maximization over u_{N-1},

which implies

$$\mu_D(u_0^*, \ldots, u_{N-1}^* \mid x_0) =$$
$$= \max_{u_0, \ldots, u_{N-2}} \{\mu_{C^0}(u_0) \wedge \ldots \wedge \mu_{C^{N-2}}(u_{N-2}) \wedge$$
$$\wedge \max_{u_{N-1}}[\mu_{C^{N-1}}(u_{N-1}) \wedge \mu_{G^N}(f(x_{N-1}, u_{N-1}))]\} \quad (22)$$

And further, continuing the same line of reasoning, for u_{N-2}, u_{N-3}, etc. we arrive at the following set of dynamic programming recurrence equations:

$$\begin{cases} \mu_{G^{N-i}}(x_{N-i}) = \max_{u_{N-i}}[\mu_{C^{N-i}}(u_{N-i}) \wedge \mu_{G^{N-i+1}}(x_{N-i+1})] \\ x_{N-i+1} = f(x_{N-i}, u_{N-i}), \quad i = 0, 1, \ldots, N \end{cases} \quad (23)$$

where $\mu_{G^{N-i}}(x_{N-i})$ may be regarded as a fuzzy goal at control stage $t = N-i$ induced by the fuzzy goal at $t = N - i + 1$, $i = 0, 1, \ldots, N$.

The optimal sequence of control sought, $u_0^*, \ldots, u_{N-1}^*$, is given by the successive maximizing values of u_{N-i}, $i = 1, \ldots, N$ in (23). Each of them, u_{N-i}^*, is obviously obtained as a function of x_{N-i}, i.e. as a *policy*.

Example 2. Suppose that $X = \{s_1, s_2, s_3\}$, $U = \{c_1, c_2\}$, $N = 2$, and fuzzy constraints and fuzzy goal are

$$C^0 = 0.7/c_1 + 1/c_2$$
$$C^1 = 1/c_1 + 0.8/c_2$$
$$G^2 = 0.3/s_1 + 1/s_2 + 0.8/s_3$$

and the state transition equation (16) is given as

$$x_{t+1} = \begin{array}{c|cc} & x_t = s_1\ s_2\ s_3 \\ \hline u_t = c_1 & s_1 \quad s_3\ s_1 \\ c_2 & s_2 \quad s_1\ s_3 \end{array}$$

First, using (23) for $i = 1$, we obtain $G^1 = 0.6/s_1 + 0.8/s_2 + 0.6/s_3$, and the corresponding optimal control policy

$$a_1^*(s_1) = c_2 \quad a_1^*(s_2) = c_1 \quad a_1^*(s_3) = c_2$$

Next, (23), for $i = 2$, yields $G^0 = 0.8/s_1 + 0.6/s_2 + 0.6/s_3$ and the corresponding optimal control policy

$$a_0^*(s_1) = c_2 \quad a_0^*(s_2) \in \{c_1, c_2\} \quad a_1^*(s_3) \in \{c_1, c_2\}$$

Therefore, e.g., if we start at $t = 0$ from $x_0 = s_1$, then $u_0^* = a_0^*(s_1) = c_2$ and we obtain $x_1 = s_2$. Next, at $t = 1$, $u_1^* = a_1^*(s_2) = c_1$ and $\mu_D(u_0^*, u_1^* \mid s_1) = \mu_D(c_2, c_1 \mid s_1) = 0.8$. □

3.1 Using fuzzy linguistic quantifiers for a soft aggregation of stage scores in multistage fuzzy control

In this section we will present the use of fuzzy linguistic quantifiers, to be more specific of Zadeh's (1983) calculus of linguistically quantified propositions presented in Section 2.1, to derive a new and interesting class of multistage fuzzy control problems with a "soft" aggregation of stage scores. This was initially proposed by Kacprzyk (1983b) and Kacprzyk and Iwański (1987), and we will further elaborate here on this topic, now from the point of view of stage score aggregation.

We again assume that the system under control is deterministic, and its dynamics is described by the state transition equation (16), i.e.

$$x_{t+1} = f(x_t, u_t), \qquad t = 0, 1, \ldots$$

where $x_t, x_{t+1} \in X\{s_1, \ldots, s_n\}$ are states at control stages t and $t+1$, respectively, and $u_t \in U = \{c_1, \ldots, c_m\}$ is the control at stage t; $t = 0, 1, \ldots, N-1$; $N < \infty$ is fixed and specified in advance.

At each control stage t, a fuzzy constraint $\mu_{C^t}(u_t)$ is imposed on u_t, and the resulting x_{t+1} is subjected to a fuzzy goal $\mu_{G^{t+1}}(x_{t+1})$.

The fuzzy decision [cf. (18)] is therefore

$$\mu_D(u_0, \ldots, u_{N-1} \mid x_0) = \bigwedge_{t=0}^{N-1} [\mu_{C^t}(u_t) \wedge \mu_{G^{t+1}}(x_{t+1})] \qquad (24)$$

and the problem [cf. (20)] is to find an optimal sequence of controls $u_0^*, \ldots, u_{N-1}^*$ such that

$$\mu_D(u_0^*, \ldots, u_{N-1}^* \mid x_0) = \max_{u_0, \ldots, u_{N-1}} \bigwedge_{t=0}^{N-1} [\mu_{C^t}(u_t) \wedge \mu_{G^{t+1}}(x_{t+1})] \qquad (25)$$

If we look at this problem, we can readily notice that its very essence is that at each control stage t, $t = 0, 1, \ldots, N-1$, the fuzzy constraint C^t and the fuzzy goal G^{t+1} are to be satisfied. This may be written as the following statements:

$$\begin{cases} P_1 : \text{"}C^0 \text{ and } G^1 \text{ are satisfied"} \\ \quad \vdots \\ P_N : \text{"}C^{N-1} \text{ and } G^N \text{ are satisfied"} \end{cases} \qquad (26)$$

Evidently, each C^t and G^{t+1} are satisfied to a certain extent, from 0 to 1, which depends on the control applied u_t and the state attained x_{t+1}. This may be represented by the truth of the statement P_t which is

$$\tau(P_t) = \tau(C^t \text{ and } G^{t+1} \text{ are satisfied}) = \mu_{C^t}(u_t) \wedge \mu_{G^{t+1}}(x_{t+1}) \qquad (27)$$

and this is assumed to be the *stage score* v_t [cf. (17)] in our context, i.e. $v_t = \tau(P_t)$, $t = 0, 1, \ldots, N-1$.

Example 3. Let $U = \{c_1, c_2\}$, $X = \{s_1, s_2, s_3\}$, and

$$C^t = 0.5/c_1 + 1/c_2 \qquad G^{t+1} = 0.3/s_1 + 0.7/s_2 + 1/s_3$$

If now $u_t = c_1$ and $x_{t+1} = s_2$, then $\tau(P_{t+1}) = 0.5 \wedge 0.7 = 0.5$. □

The fuzzy decision (24) may now be rewritten as

$$\begin{aligned}
\mu_D(u_0, \ldots, u_{N-1} \mid x_0) &= \\
&= \tau(\text{``}C^0 \text{ and } G^1 \text{ are satisfied''} \text{ and } \ldots \\
&\quad \ldots \text{ and ``}C^{N-1} \text{ and } G^N \text{ are satisfied''}) = \\
&= \tau(P_1 \text{ and } \ldots \text{ and } P_N) = \tau(P_1) \wedge \ldots \wedge \tau(P_N) = \\
&= \bigwedge_{t=0}^{N-1} \tau(P_{t+1}) = \bigwedge_{t=0}^{N-1} [\mu_{C^t}(u_t) \wedge \mu_{G^{t+1}}(x_{t+1})]
\end{aligned} \qquad (28)$$

and the problem (25) may be rewritten as to find $u_0^*, \ldots, u_{N-1}^*$ such that

$$\begin{aligned}
\mu_D(u_0^*, \ldots, u_{N-1}^* \mid x_0) &= \\
&= \max_{u_0, \ldots, u_{N-1}} \tau(P_1 \text{ and } \ldots \text{ and } P_N) = \tau(P_1) \wedge \ldots \wedge \tau(P_N) = \\
&= \max_{u_0, \ldots, u_{N-1}} \bigwedge_{t=0}^{N-1} \tau(P_{t+1}) = \max_{u_0, \ldots, u_{N-1}} \bigwedge_{t=0}^{N-1} [\mu_{C^t}(u_t) \wedge \mu_{G^{t+1}}(x_{t+1})]
\end{aligned} \qquad (29)$$

Thus, for a particular sequence of controls $u_0, \ldots, u_{N-1}$, and a sequence of resulting states $x_0, x_1, \ldots, x_N$, the value of $\mu_D(\cdot \mid \cdot)$ given by (28) may be viewed as the truth value of the statement

the fuzzy constraints and fuzzy goals are satisfied at **all** the control stages,

and the problem (29) is clearly to

find an optimal sequence of controls best satisfying the fuzzy constraints and fuzzy goals at **all** the control stages.

To emphasize this *all*, let us rewrite the fuzzy decision (28) as

$$\begin{aligned}
\mu_D(u_0, \ldots, u_{N-1} \mid x_0, \text{all}) &= \\
= \tau(P_1 \text{ and } \ldots \text{ and } P_N \mid \text{all}) &= \left(\bigwedge_{t=0}^{N-1} \mid \text{all}\right) \tau(P_{t+1}) = \\
&= \left(\bigwedge_{t=0}^{N-1} \mid \text{all}\right) [\mu_{C^t}(u_t) \wedge \mu_{G^{t+1}}(x_{t+1})]
\end{aligned} \qquad (30)$$

and rewrite the problem (29) as to find $u_0^*, \ldots, u_{N-1}^*$ such that

$$\mu_D(u_0^*, \ldots, u_{N-1}^* \mid x_0 \mid \text{all}) =$$

$$= \max_{u_0,\ldots,u_{N-1}} \tau(P_1 \text{ and } \ldots \text{ and } P_N \mid \text{all}) =$$

$$= \max_{u_0,\ldots,u_{N-1}} \left(\bigwedge_{t=0}^{N-1} \mid \text{all} \right) \tau(P_{t+1}) =$$

$$= \max_{u_0,\ldots u_{N-1}} \left(\bigwedge_{t=0}^{N-1} \mid \text{all} \right) [\mu_{C^t}(u_t) \wedge \mu_{G^{t+1}}(x_{t+1})] \qquad (31)$$

where *all* in the above notations is used just to indicate that all the control stages are to be accounted for, and has no other formal meaning.

It is now easy to see that the requirement to satisfy the fuzzy constraints and fuzzy goals at *all* the control stages is very restrictive and often counter-intuitive. It may be fully sufficient to satisfy them at, say, an *overwhelming majority* of the control stages which is a "milder" requirement. Examples may be found in Kacprzyk's (1997) book.

The idea of a milder problem formulation is the essence of Kacprzyk's (1983b), and Kacprzyk and Iwański's (1987) proposals. First, note that *all* may be viewed as a universal quantifier, i.e. *for all*.

Since fuzzy logic provides tools to handle fuzzy linguistic quantifiers like *most, almost all, much more than a half,* etc., then such linguistic quantifiers are used for the derivation of those milder problem formulations of the type

find an optimal sequence of controls that best satisfies the fuzzy constraints and fuzzy goals at Q (e.g., most, almost all, much more than a half, etc.) control stages.

So, analogously to (30), the fuzzy decision for such a milder problem formulation may be written as

$$\mu_D(u_0,\ldots,u_{N-1} \mid x_0, Q) =$$
$$= \tau(P_1 \text{ and } \ldots \text{ and } P_N \mid Q) =$$
$$= \left(\bigwedge_{t=0}^{N-1} \mid Q \right) \tau(P_{t+1}) = \left(\bigwedge_{t=0}^{N-1} \mid Q \right) [\mu_{C^t}(u_t) \wedge \mu_{G^{t+1}}(x_{t+1})] \qquad (32)$$

and the problem (31) may be analogously written as to find $u_0^*, \ldots, u_{N-1}^*$ such that

$$\mu_D(u_0^*,\ldots,u_{N-1}^* \mid x_0, Q) =$$

$$= \max_{u_0,\ldots,u_{N-1}} \tau(P_1 \text{ and } \ldots \text{ and } P_N \mid Q) =$$

$$= \max_{u_0,\ldots,u_{N-1}} \left(\bigwedge_{t=0}^{N-1} \mid Q \right) \tau(P_{t+1}) =$$

$$= \max_{u_0,\ldots u_{N-1}} \left(\bigwedge_{t=0}^{N-1} \mid Q \right) [\mu_{C^t}(u_t) \wedge \mu_{G^{t+1}}(x_{t+1})] \qquad (33)$$

where Q is a fuzzy linguistic quantifier.

For instance, in presumably one of the most natural cases when $Q = $ "most", the fuzzy decision (32) becomes

$$\mu_D(u_0, \ldots, u_{N-1} \mid x_0, \text{most}) =$$
$$= \tau(P_1 \text{ and } \ldots \text{ and } P_N \mid \text{most}) =$$
$$= \left(\bigwedge_{t=0}^{N-1} \mid \text{most} \right) \tau(P_{t+1}) = \left(\bigwedge_{t=0}^{N-1} \mid \text{most} \right) [\mu_{C^t}(u_t) \wedge \mu_{G^{t+1}}(x_{t+1})] \quad (34)$$

and the problem (33) becomes to find $u_0^*, \ldots, u_{N-1}^*$ such that

$$\mu_D(u_0^*, \ldots, u_{N-1}^* \mid x_0, \text{most}) =$$
$$= \max_{u_0, \ldots, u_{N-1}} \tau(P_1 \text{ and } \ldots \text{ and } P_N \mid \text{most}) =$$
$$= \max_{u_0, \ldots, u_{N-1}} \left(\bigwedge_{t=0}^{N-1} \mid \text{most} \right) \tau(P_{t+1}) =$$
$$= \max_{u_0, \ldots u_{N-1}} \left(\bigwedge_{t=0}^{N-1} \mid \text{most} \right) [\mu_{C^t}(u_t) \wedge \mu_{G^{t+1}}(x_{t+1})] \quad (35)$$

Using Zadeh's (1983) fuzzy-logic-based calculus of linguistically quantified propositions presented in Section 2.1, we calculate first (5), i.e.

$$r(u_0, \ldots, u_{N-1} \mid x_0) =$$
$$= \frac{1}{N} \sum_{t=0}^{N-1} \tau(P_{t+1}) = \frac{1}{N} \sum_{t=0}^{N-1} [\mu_{C^t}(u_t) \wedge \mu_{G^{t+1}}(x_{t+1})] \quad (36)$$

and then calculate (6), i.e.

$$\mu_D(u_0, \ldots, u_{N-1} x_0, Q) =$$
$$= \tau(P_1 \text{ and } \ldots \text{ and } P_N \mid Q) = \mu_Q[r(u_0, \ldots, u_{N-1} \mid x_0)] =$$
$$= \mu_Q \left(\frac{1}{N} \sum_{t=0}^{N-1} [\mu_{C^t}(u_t) \wedge \mu_{G^{t+1}}(x_{t+1})] \right) \quad (37)$$

Example 4. Let $U = \{c_1, c_2\}$, $X = \{s_1, s_2, s_3\}$, $N = 3$, and

$$C^0 = 0.5/c_1 + 1/c_2 \quad G^1 = 0.1/s_1 + 0.6/s_2 + 1/s_3$$
$$C^1 = 1/c_1 + 0.7/c_2 \quad G^2 = 0.61/s_1 + 1/s_2 + 0.5/s_3$$
$$C^2 = 1/c_1 + 0.6/c_2 \quad G^3 = 1/s_1 + 0.8/s_2 + 0.3/s_3$$

with the fuzzy linguistic quantifier $Q = $ "most" given as (3), i.e.

$$\mu_{\text{"most"}}(x) = \begin{cases} 1 & \text{for } x \geq 0.8 \\ 2x - 0.6 & \text{for } 0.3 \leq x < 0.8 \\ 0 & \text{for } x < 0.3 \end{cases}$$

Suppose now that we apply the following consecutive controls: $u_0 = c_1$, $u_1 = c_2$ and $u_2 = c_1$, and the resulting states are: $x_1 = s_2$, $x_2 = s_1$ and $x_3 = s_2$. Then

$$r(u_0, u_1, u_2 \mid x_0) =$$
$$= \frac{1}{3}[(\mu_{C^0}(c_1) \wedge \mu_{G^1}(s_2)) + (\mu_{C^1}(c_2) \wedge \mu_{G^2}(s_1)) +$$
$$+ (\mu_{C^2}(c_1) \wedge \mu_{G^3}(s_2))] = \frac{1}{3}(0.5 + 0.6 + 0.8) = 0.63$$

so that

$$\mu_D(u_0, u_1, u_2 \mid x_0) = \mu_{\text{``most''}}[r(u_0, u_1, u_2 \mid x_0)] = \mu_{\text{``most''}}(0.63) = 0.66$$

□

The problem is to find an optimal sequence of controls $u_0^*, \ldots, u_{N-1}^*$ such that

$$\mu_D(u_0^*, \ldots, u_{N-1}^* \mid x_0, Q) =$$
$$= \max_{u_0, \ldots, u_{N-1}} \mu_D(u_0, \ldots, u_{N-1} \mid x_0, Q) =$$
$$= \max_{u_0, \ldots, u_{N-1}} \mu_Q[r(u_0, \ldots, u_{N-1} \mid x_0)] =$$
$$= \max_{u_0, \ldots, u_{N-1}} \mu_Q \left(\frac{1}{N} \sum_{t=0}^{N-1} [\mu_{C^t}(u_t) \wedge \mu_{G^{t+1}}(x_{t+1})] \right) \quad (38)$$

Clearly, it is very difficult, if not impossible, to say something about the solution of this problem for an arbitrary fuzzy linguistic quantifier Q. Fortunately, we may readily confine the class of quantifiers to be considered to the so-called *nondecreasing fuzzy quantifiers* which fulfill [cf. (4)]

$$r_1 > r_2 \Longrightarrow \mu_Q(r_1) \geq \mu_Q(r_2), \qquad \text{for all } r_1, r_2 \in [0, 1] \quad (39)$$

Clearly, only such quantifiers – the essence of which is "the more the better", i.e. in our context "the more (at the more control stages) the fuzzy constraints and the fuzzy goals are satisfied the better" – are relevant in our control context. And, luckily enough, they have some interesting properties which are important for our analysis (cf. Yager, 1983; Kacprzyk, 1983b, 1997; Kacprzyk and Yager, 1984a, b; Kacprzyk and Iwański, 1987).

First, let a *linear quantifier*, denoted Q_L, be defined as

$$\mu_{Q_L}(x) = x, \qquad \text{for each } x \in [0, 1] \quad (40)$$

We now have the following important lemma.

Lemma 1. *For any linear quantifier Q_L, the solution of problem (38), i.e. the optimal sequence of controls $u_0^*, \ldots, u_{N-1}^*$, is given by*

$$\mu_D(u_0^*, \ldots, u_{N-1}^* \mid x_0, Q_L) =$$
$$= \max_{u_0, \ldots, u_{N-1}} r(u_0, \ldots, u_{N-1} \mid x_0) =$$
$$= \max_{u_0, \ldots, u_{N-1}} \frac{1}{N} \sum_{t=0}^{N-1} [\mu_{C^t}(u_t) \wedge \mu_{G^{t+1}}(x_{t+1})] =$$
$$= \frac{1}{N} \sum_{t=0}^{N-1} \{\mu_{C^t}(u_t^*) \wedge \mu_{G^{t+1}}[f(x_t, u_t^*)]\} \tag{41}$$

This lemma implies then an even stronger property that is given by the following proposition (Kacprzyk, 1983b, 1997).

Proposition 2. *If $u_0^*, \ldots, u_{N-1}^*$ is such that (41) holds, i.e. it is a solution of problem (38) for the linear quantifier Q_L, then it is also a solution of problem (38) for any nondecreasing quantifier Q [in the sense of (39)], i.e.*

$$\mu_D(u_0^*, \ldots, u_{N-1}^* \mid x_0, Q) =$$
$$= \max_{u_0, \ldots, u_{N-1}} \mu_Q[r(u_0, \ldots, u_{N-1} \mid x_0)] =$$
$$= \max_{u_0, \ldots, u_{N-1}} \frac{1}{N} \sum_{t=0}^{N-1} [\mu_{C^t}(u_t) \wedge \mu_{G^{t+1}}(x_{t+1})] =$$
$$= \frac{1}{N} \sum_{t=0}^{N-1} \{\mu_{C^t}(u_t^*) \wedge \mu_{G^{t+1}}[f(x_t, u_t^*)]\} \tag{42}$$

Proof. First, for any $u_0, \ldots, u_{N-1}$, we have

$$\mu_D(u_0, \ldots, u_{N-1} \mid x_0, Q) = \mu_Q[r(u_0, \ldots, u_{N-1} \mid x_0)]$$

Suppose now that we have two sequences of controls: $u_0^a, \ldots, u_{N-1}^a$ and $u_0^b, \ldots, u_{N-1}^b$ such that $r^a = r(u_0^a, \ldots, u_{N-1}^a \mid x_0)$ and $r^b = r(u_0^b, \ldots, u_{N-1}^b \mid x_0)$. Then by the assumption of nonincreasingness of Q, we have $\mu_Q(r^a) \geq \mu_Q(r^b)$, and hence

$$\mu_D(u_0^a, \ldots, u_{N-1}^b \mid x_0, Q) = \mu_Q[r(u_0^b, \ldots, u_{N-1}^b \mid x_0)]$$

If now (41) holds, i.e. for each $u_0, \ldots, u_{N-1}$, we have

$$\mu_D(u_0^*, \ldots, u_{N-1}^* \mid x_0, Q_L) =$$
$$= \frac{1}{N} \sum_{t=0}^{N-1} \{\mu_{C^t}(u_t^*) \wedge \mu_{G^{t+1}}[f(x_t, u_t^*)]\} \geq$$
$$\geq \frac{1}{N} \sum_{t=0}^{N-1} \{\mu_{C^t}(u_t) \wedge \mu_{G^{t+1}}[f(x_t, u_t)]\}$$

then, evidently, there holds $r(u_0^*, \ldots, u_{N-1}^* \mid x_0) \geq r(u_0, \ldots, u_{N-1} \mid x_0)$, for any $u_0, \ldots, u_{N-1}$ (and for any x_0, too).

This implies that

$$\mu_D(u_0^*, \ldots, lu_{N-1}^* \mid x_0, Q) \geq \mu_D(u_0, \ldots, u_{N-1} \mid x_0, Q)$$

that is, an $u_0^*, \ldots, u_{N-1}^*$ for which (41) holds is a solution of problem (38) for any nondecreasing quantifier. □

Thus, for any nondecreasing quantifier, the use of the algebraic method results in the following problem to be solved: find an optimal sequence of controls $u_0^*, \ldots, u_{N-1}^*$ such that

$$\mu_D(u_0^*, \ldots, u_{N-1}^* \mid x_0, Q) = \max_{u_0, \ldots, u_{N-1}} \frac{1}{N} \sum_{t=0}^{N-1} [\mu_{C^t}(u_t) \wedge \mu_{G^{t+1}}(x_{t+1})] \quad (43)$$

which may clearly be solved by dynamic programming (cf. Section 3). The set of recurrence equations yielding an optimal solution is then

$$\begin{cases} \mu_{\underline{G}^{N-i}}(x_{N-i}) = \max_{u_{N-i}} \frac{1}{N} [\mu_{C^{N-i}}(u_{N-i}) + \\ \quad + \mu_{G^{N-i+1}}(x_{N-i+1}) + \mu_{\underline{G}^{N-i+1}}(x_{N-i+1})] \\ x_{N-i+1} = f(x_{N-i}, u_{N-i}); \quad i = 1, \ldots, N \end{cases} \quad (44)$$

with $\mu_{\underline{G}^N}(x_N) = 0$, for each $x_N \in X$.

Example 5. Let the problem specifications be as in Example 4, i.e. $U = \{c_1, c_2\}$, $X = \{s_1, s_2, s_3\}$, $N = 3$, and

$$\begin{array}{ll} C^0 = 0.5/c_1 + 1/c_2 & G^1 = 0.1/s_1 + 0.6/s_2 + 1/s_3 \\ C^1 = 1/c_1 + 0.7/c_2 & G^2 = 0.61/s_1 + 1/s_2 + 0.5/s_3 \\ C^2 = 1/c_1 + 0.6/c_2 & G^3 = 1/s_1 + 0.8/s_2 + 0.3/s_3 \end{array}$$

with the fuzzy linguistic quantifier $Q =$ "most" given as (3), i.e.

$$\mu_{\text{"most"}}(x) = \begin{cases} 1 & \text{for } x \geq 0.8 \\ 2x - 0.6 & \text{for } 0.3 \leq x < 0.8 \\ 0 & \text{for } x < 0.3 \end{cases}$$

and the state transitions of the system under control be given by (16), i.e.

$$x_{t+1} = \begin{array}{c|cc} & x_t = s_1 \ s_2 \ s_3 \\ \hline u_t = c_1 & s_1 \quad s_3 \ s_3 \\ c_2 & s_2 \quad s_1 \ s_2 \end{array}$$

By solving the set of recurrence equations (44) subsequently for $i = 1, 2, 3$, we respectively obtain

$$\underline{G}^2 = 0.66/s_1 + 0.53/s_2 + 0.46/s_3$$

and the optimal policy at control stage $t = 2$ is

$$a_2^*(s_1) = c_1 \quad a_2^*(s_2) = c_2 \quad a_2^*(s_3) = c_2$$

Next

$$\underline{G}^1 = 0.75/s_1 + 0.65/s_2 + 0.74/s_3$$

and the optimal policy at control stage $t = 1$ is

$$a_1^*(s_1) = c_1 \quad a_1^*(s_2) \in \{c_1, c_2\} \quad a_1^*(s_3) = c_2$$

Finally, we obtain

$$\underline{G}^1 = 0.75/s_1 + 0.62/s_2 + 0.75/s_3$$

and the optimal policy at control stage $t = 0$ is

$$a_0^*(s_1) \in \{c_1, c_2\} \quad a_0^*(s_2) = c_2 \quad a_0^*(s_3) = c_2$$

□

Thus, on the one hand, the use of the algebraic method is simple since the control problem may be solved by dynamic programming for a large class of (plausible, to say the least) fuzzy linguistic quantifiers. On the other hand, however, the fact that for a large class of quantifiers the solution amounts to solving the same set of recurrence equations may seem to be somewhat counter-intuitive, though one should bear in mind that the solution obtained is just one of the possible solutions, i.e. *a solution*, and not *the solution*.

4 Concluding remarks

In the paper we presented an extended model of multistage fuzzy control in which, through a "softer" aggregation of stage scores (degrees to which the fuzzy constraints and fuzzy goals at the particular control stages are fulfilled), we sought an optimal sequence of controls best fulfilling the fuzzy constraints and fuzzy goals at, say, most, almost all, much more than a half, etc. control stages. This model may be more adequate than the traditional models in many practical problems (cf. Kacprzyk, 1997).

Bibliography

Bellman R.E. and L.A. Zadeh (1970) Decision making in a fuzzy environment. *Management Science* 17: 141–164.

Chang S.S.L. (1969a) Fuzzy dynamic programming and the decision making process. *Proceedings of the Third Princeton Conference on Information Sciences* (Princeton, NJ, USA).

Chang S.S.L. (1969b) Fuzzy dynamic programming and approximate optimization of partially known systems. *Proceedings of the Second Hawaii International Conference on Systems Science* (Honolulu, HI, USA).

Esogbue A.O., M. Fedrizzi and J. Kacprzyk (1988) Fuzzy dynamic programming with stochastic systems. In J. Kacprzyk and M. Fedrizzi (Eds.): *Combining Fuzzy Imprecision with Probabilistic Uncertainty in Decision Making.* Springer-Verlag, Berlin/New York, pp. 266–285.

Kacprzyk J. (1977) Control of a nonfuzzy system in a fuzzy environment with a fuzzy termination time. *Systems Science* 3: 320–334.

Kacprzyk J. (1978) A branch-and-bound algorithm for the multistage control of a nonfuzzy system in a fuzzy environment. *Control and Cybernetics* 7: 51–64.

Kacprzyk J. (1978) Decision-making in a fuzzy environment with fuzzy termination time. *Fuzzy Sets and Systems* 1: 169–179.

Kacprzyk J. (1979) A branch-and-bound algorithm for the multistage control of a fuzzy system in a fuzzy environment. *Kybernetes* 8: 139–147.

Kacprzyk J. (1983a) *Multistage Decision Making under Fuzziness*, Verlag TÜV Rheinland, Cologne.

Kacprzyk J. (1983b) A generalization of fuzzy multistage decision making and control via linguistic quantifiers. *International Journal of Control* 38: 1249–1270.

Kacprzyk J. (1986) Towards 'human-consistent' multistage decision making and control models via fuzzy sets and fuzzy logic. Bellman Memorial Issue (A.O. Esogbue, Ed.), *Fuzzy Sets and Systems* 18: 299–314.

Kacprzyk J. (1992) Fuzzy logic with linguistic quantifiers in decision making and control. *Archives of Control Sciences* 1 (XXXVII): 127–141.

Kacprzyk J. (1994) Fuzzy dynamic programming – basic issues. In M. Delgado, J. Kacprzyk, J.-L. Verdegay and M.A. Vila (Eds.): *Fuzzy Optimization: Recent Advances*, Physica-Verlag, Heidelberg (A Springer-Verlag Company), pp. 321–331.

Kacprzyk J. (1997) *Multistage Fuzzy Control.* Wiley, Chichester.

Kacprzyk J. and A.O. Esogbue (1996) Fuzzy dynamic programming: main developments and applications. *Fuzzy Sets and Systems* 81: 31–46.

Kacprzyk J. and C. Iwański (1987) A generalization of discounted multistage decision making and control through fuzzy linguistic quantifiers: an attempt to introduce commonsense knowledge. *International Journal of Control* 45: 1909–1930.

Kacprzyk J. and R.R. Yager (1984a) Linguistic quantifiers and belief qualification in fuzzy multicriteria and multistage decision making. *Control and Cybernetics* 13: 155–173.

Kacprzyk J. and R.R Yager (1984b) "Softer" optimization and control models via fuzzy linguistic quantifiers. *Information Sciences* 34: 157–178.

Ralescu D.A. (1995) Cardinality, quantifiers, and the aggregation of fuzzy criteria. *Fuzzy Sets and Systems* 69: 355-365.

Yager R.R. (1983) Quantifiers in the formulation of multiple objective decision functions. *Information Sciences* 31: 107–139.

Yager R.R. (1988) On ordered weighted averaging aggregation operators in multi-criteria decision making. *IEEE Transactions on Systems, Man and Cybernetics* SMC-18: 183–190.

Yager R.R. and J. Kacprzyk, Eds. (1997) *The Ordered Weighted Averaging Operators*. Kluwer, Boston.

Zadeh L.A. (1983) A computational approach to fuzzy quantifiers in natural languages. *Computers and Mathematics with Applications* 9: 149–184.

Zadeh L.A. and J. Kacprzyk, Eds. (1992) *Fuzzy Logic for the Management of Uncertainty*, Wiley, New York.

3

FUSION OF COMPLEMENTARY INFORMATION

From Semantic to Syntactic Approaches to Information Combination in Possibilistic Logic

Salem BENFERHAT – Didier DUBOIS – Henri PRADE
Institut de Recherche en Informatique de Toulouse (I.R.I.T.)
Université Paul Sabatier – 118 route de Narbonne
31062 Toulouse – France
Email: {benferha, dubois, prade}@irit.fr

Abstract: This paper proposes syntactic combination rules for merging uncertain propositional knowledge bases provided by different sources of information, in the framework of possibilistic logic. These rules are the counterparts of combination rules which can be applied to the possibility distributions (defined on the set of possible worlds), which represent the semantics of each propositional knowledge base. Combination modes taking into account the levels of conflict, the relative reliability of the sources, or having reinforcement effects are considered.

1 - Introduction

In many situations, relevant information is provided by different sources. Information is also often pervaded with uncertainty. This uncertainty may directly reflect the reliability of the source, or may be attached to the information provided by the source itself. Taking advantage of the different sources of information usually requires to perform some combination operation on the pieces of information, and leads to a data fusion problem. The way this problem is tackled depends on the way the information is represented, which in turn is contingent on the nature of the information. On the one hand, pieces of information pertaining to numerical parameters are usually represented by distribution functions (in the sense of some uncertainty theory). These distributions are directly combined by means of operations, which are in agreement with the uncertainty theory used, and which yield a new distribution on the set of the possible values of the considered parameter. On the other hand, information may be also naturally expressed in logical terms, especially in case of (symbolic) information pertaining to properties, which may be, however, pervaded with uncertainty. In this case, some uncertainty weights are attached to the logical formulas. Although similar issues are raised in the two frameworks, like the handling of conflicting information, the two lines of research in numerical data fusion (e.g., Abidi and Gonzalez, 1992; Flamm and Luisi, 1992) and in symbolic information combination (e.g., Baral et al., 1992; Cholvy, 1992; Dubois et al., 1992; Benferhat et al., 1995) have been investigated independently, and are unequally developed (the second trend being much more recent and the proposals still preliminary).

Possibility theory (Zadeh, 1978; Dubois and Prade, 1988a) offers an uncertainty modelling framework for dealing with possibility distributions representing fuzzy information. This information may pertain to the value of numerical parameters, or to variables ranging in discrete and finite domains. Moreover, possibilistic logic formulas (Dubois et al., 1994a, b) can be handled in a purely syntactic way (as classical logic formulas associated with a weight) in complete agreement with a semantic interpretation in terms of a possibility distribution over the set of possible worlds, which represents the set of possibilistic logic formulas under consideration. The combination of possibility distributions representing ill-known data has been studied by two of the authors in the last past years (Dubois and Prade, 1988b, 1992, 1994), and different combination modes have been proposed depending on the nature of the conflict between the sources and their reliability.

In this paper, we apply these combination modes to the possibility distributions associated with the sets of uncertain propositions provided by each source of information and we look for the syntactic counterparts of these combination on the sets of possibilistic formulas. Section 2 gives the necessary background on possibilistic logic. Section 3 presents the semantical combination modes on possibility distributions, while Section 4 gives their syntactic counterparts applied to the possibilistic propositional logic bases corresponding to each sources.

2 - Possibilistic Logic

2.1 - Possibility Distribution and Possibilistic Entailment

In this paper, we only consider a finite propositional language denoted by $\mathcal{L}$. We denote by $\models$ the classical consequence relation, Greek letters $\alpha, \beta,...$ represent formulas. Let Ω be the finite set of interpretations of the propositional logic language $\mathcal{L}$.

A possibility distribution is a mapping π from Ω to the interval $[0,1]$. π is said to be normal if $\exists \omega \in \Omega$, such that $\pi(\omega)=1$. By convention, π represents some background knowledge about the possible states of the real world; $\pi(\omega)=0$ means that the state ω is impossible, and $\pi(\omega)=1$ means that nothing prevents ω from being the real world. When $\pi(\omega)>\pi(\omega')$, ω is a preferred candidate to ω' for being the real state of the world. π is thus a convenient encoding of a preference relation that can embody concepts such as normality, typicality, plausibility, consistency with available knowledge, etc.

A possibility distribution π induces two mappings grading respectively the possibility and the certainty of a formula ϕ:

– the possibility degree $\Pi(\phi) = \max\{\pi(\omega) \mid \omega \models \phi\}$ which evaluates to what extent ϕ is consistent with the available knowledge expressed by π (Zadeh, 1978). It satisfies the characteristic property:
$$\forall \phi \, \forall \psi \, \Pi(\phi \vee \psi) = \max(\Pi(\phi), \Pi(\psi));$$

– the necessity (or certainty) degree $N(\phi) = \inf\{1 - \pi(\omega) \mid \omega \models \neg \phi\}$ which evaluates to what extent ϕ is entailed by the available knowledge. We have:
$$\forall \phi \, \forall \psi \, N(\phi \wedge \psi) = \min(N(\phi), N(\psi)).$$

The duality between necessity and possibility is expressed by the relation $N(\phi) = 1-\Pi(\neg\phi)$. Possibility and certainty degrees can thus be used for inducing ordering relations on the set of formulas of a language, from the knowledge of a possibility distribution on possible worlds, viewing a formula as the set of possible worlds where the formula is true. The converse is true as well; see Section 2.2.

Given a possibility distribution π on Ω, a notion of preferential entailment $\models_\pi$ can be defined in the spirit of Shoham (1988)'s proposal. First, we need to define the notion of preferential model of a given formula formula ϕ:

Definition 1: An interpretation ω is a π-preferential model of a formula ϕ w.r.t. π, which is denoted by $\omega \models_\pi \phi$, if and only if:

i) $\omega \models \phi$,   ii) $\pi(\omega) > 0$ and iii) $\nexists \omega'$, $\omega' \models \phi$ and $\pi(\omega') > \pi(\omega)$.

A π-preferential model of ϕ is a normal state of affairs where ϕ is true. We distinguish several forms of possibilistic entailments, in the framework of some background knowledge described by a possibility distribution π namely:

Definitions 2: 1. A formula ψ is said to be a *conditional conclusion* of a fact ϕ, with background knowledge π, which is denoted by $\phi \models_\pi \psi$, if and only if $\Pi(\phi) > 0$ and each π-preferential model of ϕ satisfies ψ, i.e., $\forall \omega, \omega \models_\pi \phi \Rightarrow \omega \models \psi$

    2. A formula ψ is said to be a conditional conclusion of a fact ϕ *to a degree a*, with background knowledge π, which is denoted by $\phi \models_\pi (\psi\ a)$, if and only if $\Pi(\phi) > 0$, each π-preferential model of ϕ satisfies ψ, and $1-a \geq \Pi(\phi \wedge \neg\psi)$. This expresses that being certain at least at the degree a that ψ is true in the context where ϕ is true amounts to state that the possibility of the opposite event in this context, i.e., $\phi \wedge \neg\psi$ is less than $1-a$. Clearly, if $\phi \models_\pi (\varphi\ a)$, then $\phi \models_\pi (\psi\ b)$ for $b \leq a$.

    3. In (1) and (2), when $\phi = T$ (namely without additional fact), ψ is said to be an *unconditional* conclusion of π, that is ϕ is normally true, a priori.

Note that for unconditional conclusions of π, it is not necessary to take all the possibility degrees induced by π, but it is enough to consider the most plausible interpretations for performing inferences (because the π-preferential models of T have necesarily the highest degree of possibility in π). So the unconditional conclusions of π are all the propositions that are true in all the normal states of affairs.

2.2 - From Possibilistic Knowledge Bases to Possibility Distributions

A possibilistic knowledge base is a stratified knowledge base, such that a level of certainty a_i is attached to each layer i. Such a knowledge base is made of a finite set of weighted formulas $\Sigma = \{(\phi_i\ a_i), i=1,n\}$ where a_i is understood as a lower bound on the degree of necessity $N(\phi_i)$. A possibilistic knowledge base Σ can be associated with a semantics in terms of possibility distributions, and a preferential entailment can be defined from Σ at the syntactical level (which is sound and complete with respect to this semantics (Dubois et al., 1994a,b)). This is recalled in this section.

A possibility distribution π is *compatible* with $\Sigma = \{(\phi_i\ a_i), i=1,n\}$ if for each $(\phi_i\ a_i) \in \Sigma$, we have

$$1 - \max_\omega \{\pi(\omega)\ /\ \omega \models \neg\phi_i\} \geq a_i\ (\text{i.e., } N(\phi_i) \geq a_i).$$

Of course, in general, there are several possibility distributions compatible with Σ. One way to select one possibility distribution is to use *the minimum specificity principle*. A possibility distribution π is said to be the least specific possibility distribution among the ones which are compatible with Σ, if there is no possibility distribution $\pi' \neq \pi$ compatible with Σ such that $\forall \omega, \pi'(\omega) \geq \pi(\omega)$. The minimum specificity principle allocates the greatest possibility degrees in agreement with the constraints $N(\phi_i) \geq a_i$, and thus does not restrict the levels of plausibility of the possible worlds more than it is necessary w.r.t. the constraints induced by Σ. We denote by π_Σ the least specific possibility distribution which satisfies the set of constraints $N(\phi_i) \geq a_i$, $i=1,n$. This possibility distribution always exists and is defined by (e.g., Dubois et al., 1994a,b)

$$\forall \omega \in \Omega, \pi_\Sigma(\omega) = \min_{i=1,n} \{1-a_i, \omega \models \neg \phi_i\}.$$

A possibilistic knowledge base made of one formula $\{(\phi\ a)\}$ is represented by the possibility distribution:

$$\forall \omega \in \Omega, \pi_{\{(\phi\ a)\}}(\omega) = 1 \quad \text{if } \omega \models \phi$$
$$= 1 - a \quad \text{otherwise.}$$

Thus, π_Σ can be viewed as the result of the conjunctive combination of the $\pi_{\{(\phi_i\ a_i)\}}$'s using the min operator, that is, a fuzzy intersection.

The possibility distribution π_Σ is not necessarily normal, and $\text{Inc}(\Sigma) = 1 - \max_{\omega \in \Omega} \pi_\Sigma(\omega)$ is called the degree of inconsistency of the knowledge base Σ.

Lastly, note that several syntactically different possibilistic knowledge bases may have the same possibility distribution as a semantics counterpart. In such a case, it can be shown that these knowledge bases are equivalent in the following sense: their a-cuts, which are classical knowledge bases, are logically equivalent in the usual sense, where the a-cut of a possibilistic knowledge base Σ is the set of classical formulas whose level of certainty is greater than or equal to a.

2.3 - Possibilistic Inference

The possibilistic logic inference can be performed at the syntactical level by means of a weighted version of the resolution principle:

$$\frac{(\phi \vee \psi\ a)\quad (\neg\phi \vee \delta\ b)}{(\psi \vee \delta\ \min(a,b))}$$

It has been shown that $\text{Inc}(\Sigma)$ corresponds to the greatest lower bound that can be obtained for the empty clause by the repeated use of the above resolution rule and using a refutation strategy. Proving $(\psi\ a)$ from a possibilistic knowledge base Σ comes down to deriving the contradiction $(\perp a)$ from $\Sigma \cup \{(\neg\psi\ 1)\}$ with a weight $a > \text{Inc}(\Sigma)$. It will be denoted by $\Sigma \vdash (\psi\ a)$. This inference method is as efficient as classical logic refutation by resolution, and can be implemented in the form of an A*-like algorithm (Dubois et al., 1987).

This inference method is sound and complete with respect to the possibilistic semantics of the knowledge base. Namely (Dubois et al., 1994b):

$\phi \vDash_{\pi_\Sigma} \psi$ if and only if $\Sigma' \vdash (\psi\ a)$ with $\Sigma' = \Sigma \cup \{(\phi\ 1)\}$, and $a > \text{Inc}(\Sigma \cup \{(\phi\ 1)\})$.

Clearly possibilistic reasoning copes with partial inconsistency. It yields non-trivial conclusions by using a consistent sub-part of Σ, consistent with ϕ, which corresponds to formulas belonging to the layers having sufficiently high levels of certainty. Moreover, possibilistic logic offers a syntactic inference whose complexity is similar to classical logic.

3 - Merging Possibility Distributions

Combination in the possibilistic framework can be performed either at the level of possibility and necessity measures, or directly on possibility distributions.

3.1 - Combining Necessity Measures

The problem of the fusion of m necessity measures is answered by the following result (Dubois and Prade, 1990). The *only* functions f from $[0,1]^n$ to $[0,1]$, satisfying the idempotency constraint $f(a,\ldots,a)=a$, such that the function defined by

$$\forall \varphi, N(\varphi) = f(N_1(\varphi),\ldots,N_n(\varphi))$$

is still a scalar necessity function, are of the form

$$N(\varphi) = \min(g_1(N_1(\varphi)),\ldots,g_n(N_n(\varphi)))$$

where the g_j's are non-decreasing functions from $[0,1]$ to $[0,1]$ such that $\forall j, g_j(1)=1$ and $\exists k, g_k(0)=0$.

Clearly, the idempotency constraint ensures that if all the belief bases Σ_i are consistent and agree on the level of certainty of a proposition, the result of the combination is what each source tells, i.e., $\forall a \in [0,1], N_1(\varphi)=\ldots=N_n(\varphi)=a \Rightarrow N(\varphi)=a$, where N_i (based on π_i) is associated to Σ_i. This is equivalent to the following combination in terms of possibility distributions

$$\pi(\omega) = \max(h_1(\pi_1(\omega)),\ldots,h_n(\pi_n(\omega)))$$

with $\forall j, h_j(0)=0$ and $\exists k, h_k(1)=1$. More precisely we should have $h_j(x)=1-g_j(1-x)$.

This result is in agreement with the fact that in classical logic the intersection of *deductively closed* knowledge bases is itself deductively closed. Indeed the intersection of the closed knowledge bases corresponds to the union of their set of models (assuming $N_j(\varphi) \in \{0,1\}$, and choosing $\forall j, h_j(1)=1$, since $N_j(\varphi)=1 \Leftrightarrow \varphi$ belongs to the deductive closure of Σ_j).

Many other combinations (in particularly the conjunctive ones) may be performed, at the semantic level, on the possibility distributions representing the information provided by each base. Let $\mathcal{C}$ be such a combination different from the above weighted union, i.e., $\pi = \mathcal{C}(\pi_1,\ldots,\pi_n)$. Then due to the above result, there does not exist a function f such that the necessity $N(\varphi)$ induced by π can be expressed in terms of the $N_i(\varphi)$ in a compositional way for all φ. This situation is exemplified by the fact that in classical logic, the union of deductively closed knowledge bases is generally not deductively closed. Indeed, the union of propositional knowledge bases corresponds to the intersection of their sets of models. Even if the Σ_i's are deductively closed, we may have $N_i(\varphi)=0$, $\forall i$ (i.e., $\varphi \notin \Sigma_1 \cup ... \cup \Sigma_n$) while $N(\varphi)=1$ ($\Sigma_1 \cup ... \cup \Sigma_n \vdash \varphi$). Moreover, the union of knowledge bases can be inconsistent, when this intersection is empty. This last phenomenon corresponds to the fact that the intersection of normal possibility distributions may be no longer normal.

In (Dubois and Prade, 1992, 1994), several propositions have been made to address the problem of combining n uncertain pieces of information represented by n possibility distributions $\pi_{i=1,n}$ (encoding the knowledge of n experts or sources of information about some parameters of interest), into a new possibility distribution. All the proposed combination modes are defined at the semantical level, and their syntactic counterpart will be investigated in Section 4. We assume that the experts who provide the π_i's use the same universe of discourse to describe their information, and also use the same scale to evaluate the levels of uncertainty.

Let us now review the different combination modes on possibility distributions.

3.2 - Idempotent Conjunctive and Disjunctive Modes

The basic combination modes in the possibilistic setting are the conjunction (i.e., the minimum) and the disjunction (i.e., the maximum) of possibility distributions. Namely define:

$$\forall \omega, \pi_{cm}(\omega) = \min_{i=1,n} \pi_i(\omega), \qquad \text{(CM)}$$
$$\forall \omega, \pi_{dm}(\omega) = \max_{i=1,n} \pi_i(\omega). \qquad \text{(DM)}$$

If the information provided by a source k is less (resp. more) specific than the information given by all the others then $\pi_{dm}=\pi_k$ (resp. $\pi_{cm}=\pi_k$). The conjunctive aggregation makes sense if all the sources are regarded as equally and fully reliable since all values that are considered as impossible by one source but possible by the others are rejected, while the disjunctive aggregation corresponds to a weaker reliability hypothesis, namely, in the group of sources there is at least one reliable source for sure, but we do not know which one. To clarify this point of view, let us assume that we only have two binary valued possibility distributions π_1 and π_2 such that each π_i partitions the set of classical interpretations into two subsets, namely A_i containing the completely possible interpretations (i.e., $\forall \omega \in A_i$, $\pi_i(\omega)=1$) and $\Omega-A_i$ containing the completely impossible interpretations (i.e., $\forall \omega \in \Omega-A_i$, $\pi_i(\omega)=0$). The result of the combination of π_i using (CM) and (DM) leads respectively to partition Ω into two subsets ($A_1 \cap A_2$, $\Omega-(A_1 \cap A_2)$) and ($A_1 \cap A_2$, $\Omega-(A_1 \cup A_2)$) respectively. The conjunction mode (CM) in this case is natural if $A_1 \cap A_2$ is not empty, while the disjunctive mode makes sense otherwise. Besides, if two sources provide the same information $\pi_1=\pi_2$, the result of the conjunctive, or of the disjunctive combination is still the same distribution; indeed min (resp. max) is the only idempotent conjunction (resp. disjunction) connective.

An important issue with conjunctive combination as defined by (CM) is the fact that the result may be subnormalized, i.e., it may happen that $\nexists \omega$, $\pi_{cm}(\omega)=1$. In that case it expresses a conflict between the sources. Clearly the conjunctive mode makes sense if all the π_i's significantly overlap, i.e., $\exists \omega, \forall i, \pi_i(\omega)=1$, expressing that there is at least a value of ω that all sources consider as completely possible. If $\forall \omega, \pi_{cm}(\omega)$ is significantly smaller than 1 this mode of combination is debatable since in that case at least one of the sources or experts is likely to be wrong, and a disjunctive combination might be more advisable. It is why disjunctive combinations of possibility distributions should be regarded as *cautious* operations, while conjunctive combinations are more *adventurous*.

3.3 - Other t-Norm and t-Conorm-Based Combination Modes

The previous combination modes based on maximum and minimum operators have no reinforcement effect. Namely, if expert 1 assigns possibility $\pi_1(\omega)<1$ to interpretation ω, and expert 2 assigns possibility $\pi_2(\omega)<1$ to this interpretation then overall, in the conjunctive mode, $\pi(\omega)=\pi_1(\omega)$ if $\pi_1(\omega)<\pi_2(\omega)$, regardless of the value of $\pi_2(\omega)$. However since both experts consider ω as rather impossible, and if these opinions are independent, it may sound reasonable to consider ω as less possible than what each of the experts claims. More generally, if a pool of independent experts is divided into two unequal groups that disagree, we may want to favor the opinion of the biggest group. This type of combination cannot be modelled by the minimum operation, nor by any idempotent operation (in particular a similar argumentation holds for the disjunctive combination using maximum). What is needed is a reinforcement effect. A reinforcement effect can be obtained using a triangular norm operation other than min in case of conjunctive, thus adventurous, combination, and a triangular conorm operation other than max for disjunctive, thus cautious, combination.

Definition 3: A triangular norm (for short t-norm) tn is a two place real-valued function whose domain is the unit square $[0,1]\times[0,1]$ and which satisfies the following conditions:
    1. $tn(0,0) = 0$, $tn(a,1) = tn(1,a) = a$ (boundary conditions);
    2. $tn(a,b) \leq tn(c,d)$ whenever $a \leq c$ and $b \leq d$ (monotonicity);
    3. $tn(a,b) = tn(b,a)$ (symmetry);
    4. $tn(a, tn(b,c)) = tn(tn(a,b), c)$ (associativity).

A triangular conorm (for short t-conorm) ct is a two place real-valued function whose domain is the unit square $[0,1] \times [0,1]$ and which satisfies the conditions 2-4 given in the previous definition plus the following boundary conditions:
    5. $ct(1,1)=1$, $ct(a,0)=ct(0,a)=a$.

Any t-conorm ct can be generated from a t-norm through the duality transformation:
    $ct(a,b) = 1 - tn(1-a, 1-b)$
and conversely. Note that t-norms and t-conorms are associative, which makes the definition of the merging n possibility distributions easy. The basic t-norms are the minimum operator, the product operator and the t-norm $\max(0, a+b-1)$ sometimes called "Lukasiewicz t-norm" (since it is directly related to Lukasiewicz many-valued implication). The duality relation respectively yields the following t-conorms: the maximum operator, the "probabilistic sum" $a+b-ab$, and the "bounded sum" $\min(1, a+b)$.

In particular, several authors (e.g., Dubois and Prade, 1992; Boldrin, 1995) have proposed to use Lukasiewicz operator for combining possibility distributions defined by (for two possibility distributions π_1 and π_2):

$$\forall \omega, \pi_{LM}(\omega) = \max(0, \pi_1(\omega) + \pi_2(\omega) - 1). \quad \text{(LM)}$$

Like the product-based combination, the (LM) combination mode is not idempotent; moreover the reinforcement effect with Lukasiewicz t-norm is more drastic since the combination of low degrees (i.e., $\pi_1(\omega) + \pi_2(\omega) < 1$) may result into a null degree of possibility which expresses complete impossibility.

We shall denote by π_{tn} and π_{ct} the possibility distributions resulting from the combination using a t-norm operator tn and a t-conorm operator ct respectively.

3.4 - Normalisation Rules

Disjunctive combination preserves normalisation (for any triangular conorm). When conjunctive combination mode C provides subnormal results ($\nexists \omega, \pi_C(\omega)=1$), we may think of renormalizing π_C. Indeed, let $h(\pi_C) = \max_\omega \{\pi_C(\omega)\}$. It estimates to what extent there exists at least one interpretation which is possible according to each source. Thus, $h(\pi_C)$ is a degree of consistency of the pieces of information provided by the different sources. When $h(\pi_C)<1$ there is a partial inconsistency between the sources. Normalizing a possibility distribution corresponds to taking at least one interpretation and to increase its possibility degree to 1. Of course there are several ways to renormalize π_C into π_{N_C}. The minimal requirements for π_{N_C} are:

1) $\exists \omega, \pi_{N\_C}(\omega) = 1$,
2) if π_C is normal then $\pi_{N\_C} = \pi_C$,
3) $\forall \omega, \omega', \pi_C(\omega) < \pi_C(\omega')$ if and only if $\pi_{N\_C}(\omega) < \pi_{N\_C}(\omega')$.

Note that the third condition entails that only interpretations having possibility degrees equal to $h(\pi_C)$ can receive value 1 in the normalisation process. In the following we consider three noticeable renormalizing procedures defined by the three following equations where it is assumed that $h(\pi_C) > 0$.[1]

- (N-1) $\quad \pi_{N1\_C}(\omega) = \dfrac{\pi_C(\omega)}{h(\pi_C)}$.
- (N-2) $\quad \pi_{N2\_C}(\omega) = 1 \quad\quad$ if $\pi_C(\omega) = h(\pi_C)$
  $\quad\quad\quad\quad\quad = \pi_C(\omega) \quad$ otherwise.
- (N-3) $\quad \pi_{N3\_C}(\omega) = \pi_C(\omega) + (1 - h(\pi_C))$.

All the normalization rules get rid of inconsistency since $h(\pi_{Ni\_C})=1$ for i=1,3, and

---

[1] When $h(\pi_C)=0$ there is a total disagreement between the sources, and conjunctive combination makes no sense in this case.

thus the inconsistency degree of any possibilistic knowledge base associated to $\pi_{N_i\_C}$ for i=1,3 is zero. The normalisation based on the equation (N-1) is the most usual one and is anologuous to the one used in probability theory. As (N-1), normalisation (N-2) is related to a conditioning in possibility theory. Indeed, the conditioning of a possibility distribution π by a formula ϕ held for true is either defined by (Dubois and Prade, 1988a):

- $\pi(\omega|\phi)$ = 1 if $\pi(\omega) = \max\{\pi(\omega) \mid \omega \models \phi\} = \Pi(\phi)$
  = $\pi(\omega)$ if $\omega \models \phi$ and $\pi(\omega) < \Pi(\phi)$
  = 0 if $\omega \not\models \phi$;

in a purely ordinal setting (where the grades are only encoding a linear ordering), or using [0.1] as a ratio scale, by

- $\pi(\omega|\phi)$ = $\dfrac{\pi(\omega)}{\Pi(\phi)}$ if $\omega \models \phi$.
  = 0 if $\omega \not\models \phi$.

Clearly, these two types of conditioning are particular cases of (N-2) and (N-1) respectively (viewing the set of models of ϕ as a binary possibility distribution).

The third normalisation rule (N-3), first introduced by Yager (1987) in the Dempster-Shafer framework, transforms the amount of conflict $1-h(\pi_C)$ into a level of total ignorance. Indeed, any interpretation ω in Ω gets a possibility degree at least equal to $1-h(\pi_C)$.

In general, an associative combination rule does not remain associative under normalization $(N-i)_{i=1,3}$. This is illustrated by the following example: let us consider the combination based on the Lukasiewicz t-norm and the normalisation (N-1). Let us consider the three normal possibility distributions on $\Omega=\{p,\neg p\}$:

$$\pi_1(p)=1, \pi_1(\neg p)=.7,$$
$$\pi_2(p)=.4, \pi_2(\neg p)=1,$$
$$\pi_3(p)=1, \pi_3(\neg p)=.6.$$

Combining π_1 and π_2 first, normalizing the result, and then combining the normalized result with π_3, and finally normalizing this last result leads to the following possibility distribution:

$$\pi(\neg p)=1 \text{ and } \pi(p)=\frac{2}{21}.$$

While if starting the process with π_1 and π_3 and combining the result with π_2 we get:

$$\pi'(p)=1 \text{ and } \pi'(\neg p)=.75,$$

a result which is opposite to the previous one. Clearly from the first combination we get $\neg p$ as a plausible conclusion of the combination while in the second case we get p. However, in the case of rule (N-1), associativity holds if π_C is defined by means of the product t-norm (Dubois and Prade, 1988b). When associativity does not hold, we have first to combine all the possibility distributions before normalizing the result. Combining π_1, π_2 and π_3 using Lukasiewicz t-norm and normalizing by N-1 shows that π' in the above example is the correct result.

For N-2 it is sufficient to consider the following counter-example[2]: $\Omega=\{x,y,z\}$; let $1>e>0$, $\pi_1(x)=1$, $\pi_1(y)=\pi_1(z)=e$; $\pi_2(x)=\pi_2(z)=e$, $\pi_2(y)=1$; $\pi_3(x)=\pi_3(y)=e$, $\pi_3(z)=1$. Indeed $\min(\pi_i,\pi_j)$ assigns equal weights to x, y, z for any i, j and the last possibility distribution to be combined is always the result of the combination. However, the associativity can be recovered in this example if we use the min-combination operation and a lexicographic refinement of N-2. In this refinement, only interpretations with maximal (non-ordered) pairs $(\pi_1(\omega), \pi_2(\omega))$ are put to 1 when combining π_1 and π_2, and the result of combination is given by the min for the other interpretations. Namely for any vector $V=(V_1...V_n)$ of weights, define $\vec{V}$ by reordering the values V_i in a non-decreasing way that is $V_1 \leq V_2 \leq ... \leq V_n$. Then define $\vec{V} <_{lex} \vec{V}'$ if $\exists k \leq n$, such that $\forall i \leq k$, $V_i \leq V'_i$ and $V_k < V'_k$. Then define the leximin complete preordering as follows

$$V <_{leximin} V' \Leftrightarrow \vec{V} <_{lex} \vec{V}'.$$

The improved normalization rule (N-lex) reads

$\pi_{lex-C}(\omega) = 1$ if $(\pi_1(\omega),...,\pi_n(\omega))$ is leximin maximal and $h(\pi_C)>0$
$= \pi_C(\omega)$ otherwise.

Then $\min(\pi_1,\pi_2)$ normalized using (N-lex) leads to π such that $\pi(x)=1$, $\pi(y)=1$, $\pi(z)=e$, and $\min(\pi,\pi_3)$ normalized in the same way gives $\pi'(x)=\pi'(y)=\pi'(z)=1$. But this is not enough for recovering associativity in general as it can be checked in the following example: $\pi_1(x)=\pi_1(y)=1$, $\pi_1(z)=\pi_1(t)=0$; $\pi_2(x)=\pi_2(y)=.1$, $\pi_2(z)=1$; $\pi_2(t)=.3$; $\pi_3(y)=\pi_3(t)=1$, $\pi_3(x)=\pi_3(z)=.4$. Indeed $\min(\pi_1,\pi_2)$, once normalized gives back π_1, and $\pi'=\min(\pi_1,\pi_3)$ is normal, $\pi'(x)=.4$, $\pi'(y)=1$. Combining π_2 with $\min(\pi_1,\pi_3)$ and normalizing using (N-lex) give π such that $\pi(x)=.1$, $\pi(y)=1$.

3.5 - Conflict-Respectful Combination Rule

As explained above the renormalization erases the conflict between the sources expressed by the subnormalization. Even if it is good that the result of the combination focuses on the values on which all the sources partially agree (in the sense that none of them gave to these values a possibility equal to 0), it would be better that the result also takes into account the conflict in some way. Moreover, as pointed out in (Dubois and Prade, 1988b), the normalized rule may be very sensitive to rather small variations of possibility degrees around 0; indeed the rule is not continuous in the vicinity of the total conflict expressed by $h(\pi_C)=0$. A natural idea for taking into account a partial conflict is to discount the result given by (N-i) for i=1,3, by a weight corresponding to the lack of normalization, i.e., $1-h(\pi_C)$, thus partial inconsistency is transformed into partial

---

[2] Pointed out to the authors by S. Moral.

uncertainty. Namely $1-h(\pi_C)$ is viewed as the degree of possibility that both sources are wrong since when $h(\pi_C)=1$ nothing suggests that a source is wrong. This discounting leads to the following modified conjunctive combination rule:

$$\forall \omega, \pi_{CoR-N_i-C}(\omega) = \max(\pi_{N_i-C}(\omega), 1-h(\pi_C)) \quad \text{(CoRM)}$$

The amount of conflict $1-h(\pi_C)$ induces a uniform level of possibility for all values outside the ones emerging in the subnormalized intersection of π_i's, i.e., the result of the combination (N-i-C) is not fully certain. Clearly, for all ω, $\lim_{h(\pi_C) \to 0} \pi_{CoR-N-i-C}(\omega)$ =1, so that the discontinuity effect is coped with, resulting in total ignorance. Lastly, note that the above rule modifies the possibility distribution $\pi_{N-i-C}(\omega)$ only when we use the normalisation defined by (N-1) or (N-2).

Note that (CoRM) is of the form $ct(\pi_{N_i-C}(\omega), 1-h(\pi_C))$ for the t-conorm ct=max. When ct is the bounded sum, $ct(\pi_C, 1-h(\pi_C))$ is normalization rule (N-3). An alternative to (CoRM) could be based on the probabilistic sum; the latter can be justified as the result of discounting the consonant body of evidence induced by π_{N_i-C}, using Shafer (1976)'s method, and leads to $h(\pi_C) \cdot \pi_{N_i-C} + 1 - h(\pi_C)$, which gives back (N-3) when π_{N_i-C} is defined using (N-1).

3.6 - Prioritized Aggregation of Expert Opinions

When combining information conjunctively or disjunctively, we may give priority to some sources over others, and thus discount the information provided by a less prioritary source if this information is not in agreement with the information provided by a more prioritary source. In the following, we only consider min and max-based combinations. First consider the case of two possibility distributions, π_1 and π_2, where π_1 is more reliable than π_2. Then the following combination rule has been proposed (Dubois and Prade, 1988c; Yager, 1991):

$$\forall \omega, \pi_{CPM1>2}(\omega) = \min(\pi_1(\omega), \max(\pi_2(\omega), 1-h(\pi_{cm}))). \quad \text{(CPM)}$$

Note that in case of total conflict, i.e., $h(\pi_{cm})=0$, only the information which has priority, represented by π_1, is retained ($\pi_{CPM1>2} = \pi_1$). More generally, when $1-h(\pi_{cm}) \geq 0.5$ then for any ω, $\pi_{CPM1>2}(\omega)=\min(\pi_1(\omega), 1-h(\pi_{cm}))$.[3] Moreover, if there is no conflict, i.e., $h(\pi_{cm})=1$, the usual conjunctive combination is performed: $\pi_{CPM1>2}=\pi_{cm}$. The CPM rule is easily interpreted as the conjunctive combination of the information supplied by source 1 and the information supplied by source 2, the latter being discounted by a certainty coefficient $h(\pi_{cm})$, such that the degree of possibility that

---

[3] Since $\pi_{CPM1>2}(\omega)$ can be rewritten as $\max(\min(\pi_1(\omega),\pi_2(\omega)), \min(\pi_1(\omega), 1-h(\pi_{cm})))$ and $h(\pi_{cm}) \leq 0.5 \Rightarrow 1-h(\pi_{cm}) \geq 0.5 \geq \min(\pi_1,\pi_2)$.

source 2 is wrong is $1-h(\pi_{cm})$. Of course, $\pi_{CPM1>2}$ can be subnormalized and hence a renormalisation can be used.

The disjunctive counterpart of (CPM) has also been proposed by Dubois and Prade (1988d) and Yager (1991)

$$\forall \omega, \pi_{DPM1>2}(\omega) = \max(\pi_1(\omega), \min(\pi_2(\omega), h(\pi_{cm}))). \quad \text{(DPM)}$$

The effect of this rule is to "truncate" the information supplied by the less prioritary source (by performing $\min(\pi_2, h(\pi_{cm}))$ in the above formula), while it is disjunctively combined with information given by source 1. Again if the two sources disagree ($h(\pi_{cm})=0$) then $\pi_{DPM1>2} = \pi_1$; if $h(\pi_{cm})=1$ then $\pi_{DPM1>2} = \pi_{dm}$.

Note that there are several ways to extend the DMP and CPM rules to the case of n possibility distributions. The simplest method may be the following. Let us assume that the set of n sources can be partionned into classes $n_1, n_2, ..., n_q$ of equally reliable ones, where n_j corresponds to a higher reliability level than n_{j+1}, for $j=1,q$. The combination between results obtained from the n_j's can be performed upon the following principle: the response of group n_2 is used to refine the response of group n_1 using one of the above rules; and then refine the obtained result using the response of group n_3, and so on.

3.7 - Grading the Reliability of the Sources

We assume in this section that a relative level of reliability λ_j is attached to each source s_j. These reliability levels can be used for weighting conjunctive or disjunctive combinations. A normalisation condition is supposed to hold for the λ_j's, namely $\max_j \lambda_j = 1$, i.e., the most reliable source(s) is graded by 1. We only consider min and max-based combination again. The weighted disjunctive combination mode is defined by (e.g., Dubois et al., 1992):

$$\forall \omega, \pi_{DRM}(\omega) = \max_{j=1,n} \min(\pi_j(\omega), \lambda_j) \quad \text{(DRM)}$$

i.e. π_{DRM} is obtained as a weighted union of the possibility distributions associated with each source. Note that π_{DRM} remains normalized as soon as all the π_j's are. If all the λ_j's are equal to 1 then the disjunction mode (DM) applied to the π_j's is recovered. If $\lambda_j=0$ the information provided by the source s_j is not taken into account. For intermediary λ_j, only the possible worlds whose possibility degree is low (which are rather impossible) are taken into account.

The conjunctive counterpart of (DRM) is defined by (e.g., Dubois et al., 1992):

$$\forall \omega, \pi_{CRM}(\omega) = \min_{j=1,n} \max(\pi_j(\omega), 1-\lambda_j) \quad \text{(CRM)}$$

For $\lambda_j=1, \forall j$, the min combination mode is recovered. Clearly π_{CRM} may then be subnormalized even if all the π_j's are normalized.

Note that (CRM) and (DRM) are similar to (CPM) and (DPM). One may think of combining them in order to take into account both the conflict of the sources and the reliability degrees attached to the sources. Assuming that $\lambda_1=1>\lambda_2$ (we give the priority to the most reliable source), for the case of two possibility distributions this leads to:

$$\pi_{CPRM1>2} = \min(\pi_1, \max(\pi_2, 1-\lambda_2, 1-h(\pi_{cm}))); \quad \text{(CPRM)}$$
$$\pi_{DPRM1>2} = \max(\pi_1, \min(\pi_2, \lambda_2, h(\pi_{cm}))). \quad \text{(DPRM)}$$

These types of prioritized combination rules have been proposed by Yager (1991) and studied in detail by Kelman (1996).

4 - Syntactical Combination Modes

In Section 3, several propositions have been made to aggregate possibility distributions. In this section, we are interested in the combination of n possibilistic knowledge bases $\Sigma_{i=1,n}$ provided by n sources. Each possibilistic knowledge base Σ_i is associated with a possibility distribution π_i which is its semantical counterpart (i.e., $\forall \phi, \psi, \Sigma_i \cup \{(\phi,1)\} \vdash (\psi, a)$ if and only if $\phi \models_{\pi_i} (\psi, a)$ with a>Inc($\Sigma_i \cup \{(\phi,1)\}$)). So we need to identify syntactical combination modes C on the Σ_i's which are the counterpart of the combination rules $\tilde{C}$ on the π_i's reviewed in the previous section. More formally, given a semantic combination rule $\tilde{C}$ we look for a syntactic combination C such that:

$$\tilde{C}(\pi_{\Sigma_1}, ..., \pi_{\Sigma_n}) = \pi_{C(\Sigma_1,...\Sigma_n)}.$$

or equivalently,

$\phi \models_{\pi_C} (\psi\ a)$ if and only if $\Sigma_C \cup \{(\phi\ 1)\} \vdash (\psi\ a)$ with a>Inc($\Sigma_i \cup \{(\phi,1)\}$),

where $\pi_C = \tilde{C}(\pi_{\Sigma_1}, ..., \pi_{\Sigma_n})$ and $\Sigma_C = C(\Sigma_1,...\Sigma_n)$.

It is thus clear that the results of the combination by C of Σ_i's which are semantically equivalent, are still semantically equivalent.

4.1 - Discounting and Truncating Possibilistic Knowledge Bases

Before introducing the syntactic counterpart of a combination mode, let us first define two parametered functions which operate on a knowledge base Σ:

- *Truncate(Σ,a)*, which basically consists in removing formulas from the possibilistic knowledge base Σ having certainty degrees strictly less than *a*, and
- *Discount(Σ,a)* which basically consists in decreasing to level *a* the certainty degree of the formulas of Σ whose certainty is higher.

More formally, we have:

Truncate $(\Sigma,a)=\{(\phi\ b) \mid (\phi\ b) \in \Sigma$ and $b \geq a\} \cup \{(\perp a)\}$
Discount $(\Sigma,a)=\{(\phi\ a) \mid (\phi\ b) \in \Sigma$ and $b \geq a\} \cup \{(\phi\ b) \mid (\phi\ b) \in \Sigma$ and $b<a\}$.

It is easy to check that each possibilistic conclusion of Truncate(Σ,a) or of Discount(Σ,a) is also a possibilistic conclusion of Σ. The converse is false.

The following proposition lays bare the semantical counterparts of these two functions.

Proposition 1: Let Σ be a possibilistic knowledge base, and π its associated possibility distribution. Then:
- Truncate (Σ,a) is associated with $\pi'=\min(\pi,1-a)$, and
- Discount (Σ,a) is associated with $\pi'=\max(\pi,1-a)$.

Proof:
• Indeed, let Σ'=Truncate (Σ,a) and π' its associated possibility distribution.
Remember that: $\quad \forall \omega \in \Omega, \pi(\omega)=1-\max\{a_i|\omega \models \neg\phi_i \text{ and } (\phi_i\ a_i)\in\Sigma\}$
and $\quad \pi(\omega)=1$ if ω is a model of all the ϕ_i's in Σ.

Then we have two cases:
if $\omega \models \{\phi \mid (\phi\ b)\in\Sigma \text{ and } b\geq a\}$ then: $\pi(\omega)>1-a$. Hence,
$\quad \pi'(\omega) = 1-a$ (since $(\bot\ a)\in\Sigma'$)
$\quad\quad\quad\ \ = \min(1-a, \pi(\omega))$.
if $\omega \not\models \{(\phi\ a) \mid (\phi\ b) \in \Sigma \text{ and } b\geq a\}$ then $\pi(\omega)\leq 1-a$, hence:
$\quad \pi'(\omega) = \pi(\omega)$ (since $\Sigma'=\Sigma-\{(\phi\ b) \mid (\phi\ b) \in \Sigma \text{ and } b<a\}\cup\{(\bot\ a)\})$
$\quad\quad\quad\ \ = \min(1-a, \pi(\omega))$.

• Now let Σ'=Discount (Σ,a) and π' its associated possibility distribution. Again we can distinguish several cases:
- $\omega \models \Sigma$ then we have $\pi'(\omega)=1=\max(\pi(\omega), 1-a)$ since $\pi(\omega)=1$.
- ω falsifies $\{\phi \mid (\phi\ b)\in\Sigma \text{ and } b\geq a\}$ then
$\quad \pi'(\omega) = 1-a$
$\quad\quad\quad\ \ = \max(\pi(\omega), 1-a)\quad$ since $1-a\geq\pi(\omega)$.
- $\omega \models \{\phi \mid (\phi\ b)\in\Sigma \text{ and } b\geq a\}$ but ω falsifies $\{\phi / (\phi\ b)\in\Sigma \text{ and } b<a\}$
then $\pi(\omega)>1-a$ and hence $\pi'(\omega)= \pi(\omega)= \max(\pi(\omega),1-a)$. ∎

For the sake of simplicity, we only consider the case of two possibility distributions, in the following.

4.2 - Idempotent Conjunctive and Disjunctive Modes

The min-based conjunction mode (CM) leads simply to take the union of Σ_1 and Σ_2 at the syntactic level, namely:

$$\Sigma_{cm} = \Sigma_1 \cup \Sigma_2.$$

It corresponds to the adventurous attitude (i.e., to the fuzzy set intersection at the distribution level, and to a union for the possibilistic knowledge bases). Of course, Σ_{cm} may be inconsistent and the handling of inconsistency is simply achieved by using the possibilistic entailment; see Section 2. We will denote $\text{Inc}(\Sigma_{cm})$ the degree of inconsistency of Σ_{cm}. Note that:

$$\text{Inc}(\Sigma_{cm}) = 1-h(\pi_{cm}).$$

The identification of the syntactic cautious combination mode corresponds to find which possibilistic knowledge base Σ_{dm} is associated with π_{dm}, we first consider the case of two possibilistic knowledge bases $\Sigma_1=\{(\phi\ a)\}$ and $\Sigma_2=\{(\psi\ b)\}$ which exactly contain one formula. Then let us show that in this case Σ_{dm} is defined by:

$$\Sigma_{dm} = \{(\phi\vee\psi\ \min(a,b))\}.$$

Indeed, if a given interpretation ω satisfies ϕ or satisfies ψ then it satisfies $\phi\vee\psi$, hence $\pi_{\Sigma_{dm}}(\omega)=1=\max(\pi_1(\omega),\pi_2(\omega))$ since either $\pi_1(\omega)=1$ or $\pi_2(\omega)=1$. Now, if an interpretation ω neither satisfies ϕ nor ψ, then $\pi_{\Sigma_{dm}}(\omega)=1-\min(a,b)=\max(1-a,1-b)=\max(\pi_1(\omega),\pi_2(\omega))$.

Assume now that the Σ_i's are general sets of possibilistic formulas, namely $\Sigma_1=\{(\phi_i\ a_i)\mid i\in I\}$ and $\Sigma_2=\{(\psi_j\ b_j)\mid j\in J\}$. Then let us show that:

$$\Sigma_{dm} = \{(\phi_i\vee\psi_j\ \min(a_i,b_j))\mid (\phi_i\ a_i)\in\Sigma_1\ \text{and}\ (\psi_j\ b_j)\in\Sigma_2\}.$$

Indeed, if ω satisfies Σ_1 or it satisfies Σ_2 then it satisfies Σ_{dm}, hence $\pi_{\Sigma_{dm}}(\omega)=1=\max(\pi_1(\omega),\pi_2(\omega))$. Now, if ω neither satisfies Σ_1 nor Σ_2, then:

$$\begin{aligned}
\pi_{\Sigma_{dm}}(\omega) &= 1 - \max_{i,j}\{\min(a_i,b_j)\mid \omega\models\neg\phi_i\wedge\neg\psi_j\} \\
&= 1 - \min(\max_i(a_i),\max_j(b_j))\mid \omega\models\neg\phi_i\wedge\neg\psi_j\} \\
&= 1 - \min(\max_i\{a_i\mid \omega\models\neg\phi_i\}, \max_j\{b_j\mid \omega\models\neg\psi_j\}) \\
&= \max(1-\max\{a_i\mid \omega\models\neg\phi_i\}, 1-\max(b_j\mid \omega\models\neg\psi_j\}) \\
&= \pi_{dm}(\omega).
\end{aligned}$$

Note that Σ_{dm} is always consistent (provided that Σ_1 or Σ_2 is consistent).

Now let us see what are the plausible conclusions of Σ_{dm}. We first assume that we have no additional fact, namely we are interested to see what can be unconditionally inferred from Σ_{dm}? In this case, a formula ψ is considered as an unconditional plausible consequence of Σ_{dm} if ψ is inferred both from Σ_1 and Σ_2.

The situation is more complicated when plausibly inferring from a fact ϕ, namely we are interested to compute the set of conditional plausible conclusions of the combination mode. Then in this case, inferring from Σ_{dm} is equivalent to infer from the knowledge base Σ_i which is the most consistent with the fact ϕ, namely:

If $\Pi_1(\phi)>\Pi_2(\phi)$ $\quad(\Leftrightarrow N_1(\neg\phi)<N_2(\neg\phi))$
then $\{(\phi\ 1)\}\cup\Sigma_{dm}\vdash\psi$ if and only if $\{(\phi\ 1)\}\cup\Sigma_1\vdash\psi$.

In case of $\Pi_2(\phi)=\Pi_1(\phi)$
then $\{(\phi\ 1)\}\cup\Sigma_{dm}\vdash\psi$ if and only if $\{(\phi\ 1)\}\cup\Sigma_1\vdash\psi$ and $\{(\phi\ 1)\}\cup\Sigma_2\vdash\psi$.

These results extend states of facts which already appears in classical logic when ϕ is consistent with Σ_1 and ϕ is inconsistent with Σ_2, and respectively when ϕ is consistent with Σ_1 and Σ_2. In the fuzzy case, the fuzzy set of possibilistic consequences of Σ_{dm} can also be viewed as the fuzzy intersection of the fuzzy set of possibilistic consequences of Σ_1 and the fuzzy set of those of Σ_2.

4.3 - Other t-Norm and t-Conorm-Based Combination Modes

In the framework of substructural logics, Boldrin (1995) (see also Boldrin and Sossai, 1995) has proposed an extension of possibilistic logic where a second "and" connective, based on Lukasiewicz t-norm, is introduced for combining information from distinct independent sources. At the syntactic level this new conjunction applied to two one-formula knowledge bases $\Sigma_1=\{(\phi\ a)\}$ and $\Sigma_2=\{(\psi\ b)\}$ results in three possibilistic formulas: $\Sigma_{LM}=\{(\neg\phi\vee\psi\ b), (\phi\vee\neg\psi\ a), (\phi\vee\psi\ \min(1,a+b))\}$ which expresses an adventurous combination mode. Note that this knowledge is equivalent to the following: $\Sigma_{LM}\equiv\{(\psi\ b), (\phi\ a), (\phi\vee\psi\ \min(1,a+b))\}=\Sigma_1\cup\Sigma_2\cup\{(\phi\vee\psi\ \min(1,a+b))\}$, the equivalence being understood as the equality of the associated possibility distributions.

This can be easily checked. Indeed, there are two different cases:

- If $\omega\models\phi$ or $\omega\models\psi$ then this trivially implies that $\omega\models\phi\vee\psi$. Hence:

$$\pi_{\Sigma_{LM}}(\omega) = 1 = \max(0, \pi_1(\omega)+\pi_2(\omega)-1) \quad \text{if } \omega\models\phi\wedge\psi \quad (\text{since } \pi_1(\omega)=\pi_2(\omega)=1)$$
$$= 1-b = \max(0, \pi_1(\omega)+\pi_2(\omega)-1) \quad \text{if } \omega\models\phi\wedge\neg\psi \quad (\text{since } \pi_1(\omega)=1)$$
$$= 1-a = \max(0, \pi_1(\omega)+\pi_2(\omega)-1) \quad \text{if } \omega\models\neg\phi\wedge\psi \quad (\text{since } \pi_2(\omega)=1)$$

- If $\omega\not\models\phi$ and $\omega\not\models\psi$ then this implies that $\omega\not\models\phi\vee\psi$. Hence:

$$\pi_{\Sigma_{LM}}(\omega) = 1-\max(a,b,\min(1, a+b))$$
$$= 1-\min(1,a+b)$$
$$= \max(0, 1-a-b)$$
$$= \max(0, 1-(1-\pi_1(\omega))-(1-\pi_2(\omega))) \quad (\text{since } \pi_1(\omega)=1-a \text{ and } \pi_2(\omega)=1-b)$$
$$= \max(0, \pi_1(\omega)+\pi_2(\omega)-1).$$

The previous remark can be generalized to the case of general possibilistic knowledge bases and to any t-norms and t-conorms. Let $\Sigma_1=\{(\phi_i\ a_i) \mid i\in I\}$ and $\Sigma_2=\{(\psi_j\ b_j) \mid j\in J\}$. Namely, we have the following result.

Proposition 2: Let π_{tn} and π_{ct} be the result of the combination based on the t-norm tn and the t-conorm ct. Then, π_{tn} and π_{ct} are respectively associated with the following knowledge bases:

$$\Sigma_{tn} = \Sigma_1\cup\Sigma_2\cup\{(\phi_i\vee\psi_j\ ct(a_i,b_j)) \mid (\phi_i\ a_i)\in\Sigma_1 \text{ and } (\psi_j\ b_j)\in\Sigma_2\} \quad \text{(adventurous}$$
$$\text{where ct is the t-conorm dual to the t-norm tn.} \quad \text{combination)}$$

$$\Sigma_{ct} = \{(\phi_i\vee\psi_j\ tn(a_i,b_j)) \mid (\phi_i\ a_i)\in\Sigma_1 \text{ and } (\psi_j\ b_j)\in\Sigma_2\} \quad \text{(cautious}$$
$$\text{where tn is the t-norm dual to the t-conorm ct.} \quad \text{combination)}$$

Proof: Let us start with the adventurous combination mode, i.e., with the computation of Σ_{tn}. There are two cases:

- if ω is a model of Σ_1 or of Σ_2 then $\pi_1(\omega)=1$ or $\pi_2(\omega)=1$ and then ω is also a model of $\{(\phi_i\vee\psi_j\ ct(a_i,b_j)) \mid (\phi_i\ a_i)\in\Sigma_1 \text{ and } (\psi_j\ b_j)\in\Sigma_2\}$. Hence in this case:

$\pi_{\Sigma_{tn}}(\omega) = \pi_2(\omega)$ if ω is a model of Σ_1
$= \pi_1(\omega)$ if ω is a model of Σ_2

Hence we have:
$\pi_{\Sigma_{tn}}(\omega) = \pi_{tn}(\omega) = tn(\pi_1(\omega), \pi_2(\omega))$
since by definition of a t-norm $tn(1,a)=a$.

- Now when ω is neither a model of Σ_1 nor of Σ_2, then we have:

$\pi_{\Sigma_{tn}}(\omega) = 1 - \max(\{a_i \mid \omega \models \neg\phi_i\}, \{b_j \mid \omega \models \neg\psi_j\}, \{ct(a_i,b_j) \mid \omega \models \neg\phi_i \wedge \neg\psi_j\})$

due to monotonicity of the t-conorms, $ct(a_i,b_j) \geq a_i$ and $ct(a_i,b_j) \geq b_j$, hence:

$\pi_{\Sigma_{tn}}(\omega) = 1 - ct(\max_i(a_i), \max_j(b_j))$
$= 1 - ct(1-\pi_1(\omega), 1-\pi_2(\omega))$
$= tn(\pi_1(\omega), \pi_2(\omega))$

and hence the result.

In the case of a cautious combination, for computing Σ_{ct}, we also distinguish two cases:

- if ω is a model of Σ_1 or of Σ_2 then $\pi_1(\omega)=1$ or $\pi_2(\omega)=1$ and that ω is also a model of $\{(\phi_i \vee \psi_j, tn(a_i,b_j)) \mid (\phi_i\ a_i) \in \Sigma_1$ and $(\psi_j\ b_j) \in \Sigma_2\}$. Hence in this case:

$\pi_{\Sigma_{ct}}(\omega) = 1 = ct(\pi_1(\omega), \pi_2(\omega))$

since by definition of a t-conorm $ct(1,a)=1$.

- Now when ω is neither a model of Σ_1 nor of Σ_2, then we have:

$\pi_{\Sigma_{ct}}(\omega) = 1 - \max_{i,j}(\{tn(a_i,b_j) \mid \omega \models \neg\phi_i \wedge \neg\psi_j\})$
$= 1 - tn(\max_i(a_i), \max_j(b_j))$
$= ct(1-\max_i(a_i), 1-\max_j(b_j))$
$= ct(\pi_1(\omega), \pi_2(\omega))$. ∎

Let us observe that we recover the result of the previous section letting tn=min and ct=max. Indeed,

$\Sigma_{min} = \Sigma_1 \cup \Sigma_2 \cup \{(\phi_i \vee \psi_j, \max(a_i,b_j)) \mid (\phi_i\ a_i) \in \Sigma_1$ and $(\psi_j\ b_j) \in \Sigma_2\}$
$= \Sigma_1 \cup \Sigma_2$
$\Sigma_{max} = \{(\phi_i \vee \psi_j, \min(a_i,b_j)) \mid (\phi_i\ a_i) \in \Sigma_1$ and $(\psi_j\ b_j) \in \Sigma_2\}$.

4.4 - Normalization

This section discusses the effect of the normalization at the syntactical level. Let $\pi_{\tilde{C}}$ be a sub-normalized possibility distribution obtained by combining π_1 and π_2 using a conjunction operator $\tilde{C}$. Let $\Sigma_{\tilde{C}}$ be the possibilistic knowledge base associated with $\pi_{\tilde{C}}$ and built using the result of the previous section. Let $h(\pi_{\tilde{C}}) = \max_\omega \{\pi_{\tilde{C}}(\omega)\}$. Then we have:

Proposition 3: 1. (N-1) $\pi_{N1\_\tilde{C}}(\omega) = \dfrac{\pi_{\tilde{C}}(\omega)}{h(\pi_{\tilde{C}})}$ is associated with:

$$\Sigma_{N1\_C} = \left\{ \left(\phi_i \ 1 - \frac{1-a_i}{h(\pi_C)} \right) \mid (\phi_i \ a_i) \in \Sigma_C \text{ and } a_i > 1-h(\pi_C) \right\}.$$

2. (N-2) $\pi_{N2\_C}(\omega) = 1$ if $\pi_C(\omega) = h(\pi_C)$
                 $= \pi_C(\omega)$ otherwise
is associated with: $\Sigma_{N2\_C} = \{(\phi_i \ a_i) \mid (\phi_i \ a_i) \in \Sigma_C \text{ and } a_i > 1-h(\pi_C)\}$.

3. (N-3) $\pi_{N3\_C}(\omega) = \pi_C(\omega) + (1-h(\pi_C))$ is associated with:
$$\Sigma_{N3\_C} = \{(\phi_i \ a_i - (1-h(\pi_C))) \mid (\phi_i \ a_i) \in \Sigma_C \text{ and } a_i > 1-h(\pi_C)\}.$$

Proof: Note first that each Σ_{Ni_C} is consistent while Σ_C is inconsistent at level $1-h(\pi_C)$. We need the following notation:
$$I(\omega) = \{a_i \mid (\phi_i \ a_i) \in \Sigma_C \text{ and } \omega \not\models \phi_i \text{ and } a_i > 1-h(\pi_C)\}.$$
Σ_C is semantically equivalent to $\{(\phi_i \ a_i) \mid a_i \in I(\omega)\} \cup \{(\bot \ 1-h(\pi_C))\}$. We always distinguish two cases for each kind of normalisation:

1. • If $\omega \models \Sigma_{N1\_C}$ then $\pi_{\Sigma_{N1\_C}}(\omega) = 1$
                        $= \pi_C(\omega)/h(\pi_C) = \pi_{N1\_C}(\omega)$ since $\pi_C(\omega) = h(\pi_C)$

   • If $\omega \not\models \Sigma_{1N\_C}$ then
$$\pi_{\Sigma_{N1\_C}}(\omega) = 1 - \max\left\{ 1 - \frac{(1-a_i)}{h(\pi_C)} \mid a_i \in I(\omega) \right\}$$
$$= \min\left\{ \frac{(1-a_i)}{h(\pi_C)} \mid a_i \in I(\omega) \right\}$$
$$= \frac{\min\{1-a_i\} \mid a_i \in I(\omega)\}}{h(\pi_C)}$$
$$= \frac{1-\max\{a_i\} \mid a_i \in I(\omega)\}}{h(\pi_C)}$$
$$= \frac{\pi_C(\omega)}{h(\pi_C)}$$
$$= \pi_{N1\_C}(\omega).$$

2. • If $\omega \models \Sigma_{N2\_C}$ then $\pi_{\Sigma_{N2\_C}}(\omega) = 1 = \pi_{N2\_C}(\omega)$ since $\pi_C(\omega) = h(\pi_C)$,
   • If $\omega \not\models \Sigma_{2N\_C}$ then
$$\pi_{\Sigma_{N2\_C}}(\omega) = 1 - \max\{a_i \mid a_i \in I(\omega)\}$$
$$= \pi_C(\omega) = \pi_{N2\_C}(\omega) \text{ since } \pi_C(\omega) < h(\pi_C).$$

3. • If $\omega \models \Sigma_{N3\_C}$ then $\pi_{\Sigma_{N3\_C}}(\omega) = 1$
                      $= \pi_C(\omega) + (1-h(\pi_C)) = \pi_{N3\_C}(\omega)$ since $\pi_C(\omega) = h(\pi_C)$

   • If $\omega \not\models \Sigma_{3N\_C}(\omega)$ then
$$\pi_{\Sigma_{N3\_C}}(\omega) = 1 - \max\{(a_i - (1-h(\pi_C))) \mid a_i \in I(\omega)\}$$
$$= 1 - \max\{(a_i \mid a_i \in I(\omega)\} + (1-h(\pi_C))$$
$$= \pi_C(\omega) + (1-h(\pi_C))$$
$$= \pi_{N3\_C}(\omega). \blacksquare$$

Note that Σ_{N1_C} can be defined from Σ_{N3_C} as follows:

$$\Sigma_{N1\_C} = \left\{ \left(\phi_i \frac{b_i}{h(\pi_C)} \right) \mid (\phi_i \, b_i) \in \Sigma_{N3\_C} \right\}.$$

All the normalization procedures maintain all the formulas of Σ_C whose certainty degrees are higher than the inconsistency degree of Σ_C. The normalization based on (N-2) just forgets the presence of conflicts and does not modify the certainty degrees of the pieces of information encoded by π_C. The two other normalization procedures modify the certainty degrees of the formulas retained in Σ_C. Let us observe that all the normalization modes diminishes the certainty levels of formulas in the wide sense. This diminution is more important with (N3) than with (N1).

4.5 - Conflict-Respectful, Prioritized and Reliability-Based Combination Modes

The corresponding syntactic counterparts of conflict-respectful, prioritized and reliability-based combination modes are immediate using the discussion of sub-sections, we get the corresponding possibilistic knowledge bases:

$$\Sigma_{CoR-Ni-C} = \text{Discount}(\Sigma_{Ni-C}, h(\pi_C))$$
$$C_{CPM1>2}(\Sigma_1, \Sigma_2) = \Sigma_1 \cup \text{Discount}(\Sigma_2, h(\pi_{cm})),$$
$$C_{DPM1>2}(\Sigma_1, \Sigma_2) = C_{dm}(\Sigma_1, \text{Truncate}(\Sigma_2, 1-h(\pi_{cm}))),$$
$$C_{DRM}(\Sigma_i)_{i=1,n} = C_{dm}(\text{Truncate}(\Sigma_i, 1-\lambda_i))_{i=1,n},$$
$$C_{CRM}(\Sigma_i)_{i=1,n} = \bigcup_i \text{Discount}(\Sigma_i, \lambda_i).$$

where C_{dm} denotes the combination operator yielding the possibilistic knowledge base Σ_{dm}, and $C_X(\Sigma_i)_{i=1,n}$ is the syntactic combination operator whose results is associated with π_X for X=CPM1>2, PM1>2, DRM, CRM.

Note that the addition of reliability degrees and priorities aims either to decrease the certainty degrees of highest formulas, or simply to remove formulas which are not sufficiently entrenched.

5 - Concluding Discussions

This paper has proposed a preliminary investigation of syntactic combination modes which can be applied to layered propositional logic knowledge bases, and which are the counterparts of possibilistic combination rules applied to the possibility distributions encoding the semantics of the layered bases. This can be viewed as a logical analog of the fusion of belief networks in agreement with the combination laws of probability distributions, see (Matzkevick and Abramson, 1992, 1993).

This work belongs to a larger research trend which also encompasses belief revision where syntactic encoding of possibilistic revision methods have been devised (e.g., Dubois and Prade, 1996; Williams, 1996). However, in belief revision the input information does not play a symmetrical role with respect to the knowledge we start with, while in this paper all the sources play symmetrical roles. Moreover the syntactic combination methods are exact counterparts of semantic ones. As a consequence they are drastic since they use truncation effects in the presence of inconsistency. All formulas in layers below the inconsistency level are lost. This is called the drowning effect (Benferhat et al., 1993). One might thing of taking into account the syntax of the knowledge bases when combining, and especially when restoring consistency, in order to delete formulas only when necessary.

Lastly, in this paper the weights attached to formulas were understood as uncertainty levels and the combination laws which have been considered are compatible with this view. They might be also interpreted in terms of priority levels, thus expressing preferences rather than uncertainty. In such a case, we would have to look for counterparts of fuzzy set aggregation operations which are meaningful in multiple criteria evaluation when preference profiles can be viewed as fuzzy sets.

Acknowledgements

The authors wish to thank Serafin Moral for helpful comments on an early draft of this paper. This work is partially supported by the European working group FUSION.

References

Abidi M.A., Gonzalez R.C. (eds.) (1992) Data Fusion in Robotics and Machine Intelligence. Academic Press, New York.

Baral C., Kraus S., Minker J., Subrahmanian (1992) Combining knowledge bases consisting in first order theories. Computational Intelligence, 8(1), 45-71.

Benferhat S., Cayrol C., Dubois D., Lang J., Prade H. (1993) Inconsistency management and prioritized syntax-based entailment. Proc. of the 13th Inter. Joint Conf. on Artificial Intelligence (IJCAI'93), Chambéry, France, Aug. 28-Sept. 3, 640-645.

Benferhat S., Dubois D., Prade H. (1995) How to infer from inconsistent beliefs without revising?. Proc. of the 14th Inter. Joint Conf. on Artificial Intelligence (IJCAI'95), Montréal, Canada, Aug. 20-25, 1449-1455.

Boldrin L. (1995) A substructural connective for possibilistic logic In: Symbolic and Quantitative Approaches to Reasoning and Uncertainty (Proc. of Europ. Conf. ECSQARU'95) C. Froidevaux, J. Kohlas, eds.), Springer Verlag, Fribourg, 60-68.

Boldrin L., Sossai C. (1995) An algebraic semantics for possibilistic logic. Proc of the 11th Conf. Uncertainty in Artifucial Intelligence (P. Besnard, S. Hank, eds.) Morgan Kaufmann, San Francisco, CA, 27-35.

Cholvy F. (1992) A logical approach to multi-sources reasoning. In: Applied Logic Conference: Logic at Work, Amsterdam.

Dubois D., Lang J., Prade H. (1987) Theorem proving under uncertainty — A possibility theory-based approach. Proc. of the 10th Inter. Joint Conf. on Artificial Intelligence, Milano, Italy, August, 984-986.

Dubois D., Lang J., Prade H. (1992) Dealing with multi-source information in possibilistic logic. Proc. of the 10th Europ. Conf. on Artificial Intelligence (ECAI'92) Vienna, Austria, Aug. 3-7, 38-42.

Dubois D., Lang J., Prade H. (1994a) Automated reasoning using possibilistic logic: Semantics, belief revision and variable certainty weights. IEEE Trans. on Knowledge and Data Engineering, 6(1), 64-71.

Dubois D., Lang J., Prade H. (1994b) Possibilistic logic. In: Handbook of Logic in Artificial Intelligence and Logic Programming — Vol 3: Nonmonotonic Reasoning and Uncertain Reasoning (Dov M. Gabbay, C.J. Hogger, J.A. Robinson, D. Nute eds.), Oxford Univ. Press, 439-513.

Dubois D., Prade H. (1988a) Possibility Theory — An Approach to Computerized Processing of Uncertainty. Plenum Press, New York.

Dubois D., Prade H. (1988b) Representation and combination of uncertainty with belief functions and possibility. Computational Intelligence, 4, 244-264.

Dubois D., Prade H. (1988c) Default reasoning and possibility theory. Artificial Intelligence, 35, 243-257.

Dubois D., Prade H. (1988d) On the combination of uncertain or imprecise pieces of information in rule-based systems. A discussion in a framework of possibility theory. Int. Journal of Approximate Reasoning, 2, 65-87.

Dubois D., Prade H. (1990) Aggregation of possibility measures. In: Multiperson Decision Making Using Fuzzy Sets and Possibility Theory (J. Kacprzyk, M. Fedrizzi, eds.), Kluwer Academic Publ., 55-63.

Dubois D., Prade H. (1992) Combination of fuzzy information in the framework of possibility theory. In: Data Fusion in Robotics and Machine Intelligence (M.A. Abidi, R.C. Gonzalez, eds.) Academic Press, New York, 481-505.

Dubois D., Prade H. (1994) Possibility theory and data fusion in poorly informed environments. Control Engineering Practice, 2(5), 811-823.

Dubois D., Prade H. (1996) Belief revision with uncertain inputs in the possibilistic setting. Proc. of the 12th Conf. on Uncertainty in Artificial Intelligence (E. Horvitz, F. Jensen, eds.), Portland, Oregon, Aug. 1-4, 1996, 236-243

Flamm J., Luisi T. (Eds.) (1992) Reliability Data and Analysis. Kluwer Academic Publ.

Kelman A. (1996) Modèles flous pour l'agrégation de données et l'aide à la décision. Thèse de Doctorat, Université Paris 6, France.

Matzkevich I., Abramson B. (1992) The topological fusion of Bayes nets. Proc. of the 8th Conf. on Uncertainty in Artificial Intelligence (D. Dubois, M.P. Wellman, B. D'Ambrosio, P. Smets, eds.), Stanford, CA, July 17-19, 191-198.

Matzkevich I., Abramson B. (1993) Some complexity considerations in the combination of belief networks. Proc. of the 9th Conf. on Uncertainty in Artificial Intelligence (D. Heckerman, A. Mamdani, eds.), Washington, DC, July 9-11, 152-158.

Shafer G. (1976) A Mathematical Theory of Evidence. Princeton Univ. Press, Princeton, NJ.

Shoham Y. (1988) Reasoning About Change — Time and Causation from the Standpoint of Artificial Intelligence. The MIT Press, Cambridge, MA.

Williams M.A. (1996) Towards a practical approach to belief revision: Reason-based change. Proc. of the 5th Conf. on Knowledge Representation and Reasoning Principles (KR'96), Cambridge, MA, Nov. 1996.

Yager R.R. (1987) On the Dempster-Shafer framework and new combination rules. Information Sciences, 41, 93-138.

Yager R. R. (1991) Non-monotonic set-theoritic operators. Fuzzy Sets and Systems 42, 173-190.

Zadeh L.A. (1978) Fuzzy sets as a basis for a theory of possibility. Fuzzy Sets and Systems, 1, 3-28.

Aggregation of Imprecise Probabilities*

Serafín Moral[1], José del Sagrado[2]

[1] Dpto. Ciencias de la Computación e Inteligencia Artificial
Universidad de Granada
18071 - Granada - Spain
[2] Dpto. Lenguajes y Computación
Universidad de Almería - 04120 - Almería - Spain

Abstract. Methods to aggregate convex sets of probabilities are proposed. Source reliability is taken into account by transforming the given information and making it less precise. An important property of the aggregation will be that the precision of the result will depend on the initial compatibility of sources. Special attention will be paid to the particular case of probability intervals giving adaptations of aggregation procedures.

1 Introduction

The problem of aggregating probabilities for the same set of events assigned by different experts has recieved a great deal of attention in the literature. See Genest and Zidek [8] and French [7] for surveys of classical statistical methods. In general, it is assumed that there is a finite set of mutually exclusive and exhaustive hypotheses under consideration. It is also considered that each expert expresses his opinion by means of a probability distribution on the set of hypotheses.

In general, the methods for aggregating probabilities calculate a single probability distribution taking into account the expert probabilities and a set of weights expressing the reliability or credibility of the different experts. In some cases, [18] these weights may depend on the different hypotheses, so that one expert weight is different for each individual hypothesis.

Some criteria such as the independence preservation property [14], the strong setwise function property [17], or the marginalization property [17, 8], amongst others, have been proposed in order to determine fusion rules. However, the situation is that no procedure is universally accepted as giving reasonable solutions in every situation, and very often reasonable properties lead to dictatorial aggregation methods (the result is the distribution given by one of the experts).

The use of imprecise probabilities [5, 25] is a generalization of Probability Theory, in which the state of an expert's knowledge is represented by a closed and convex set of probabilities. From a behavioural point of view [25] convex sets arise by assuming that for each event or gamble there is a maximum rate at which you are prepared to bet on it (determining its lower probability) and a minimum rate (determining its upper probability).

---

* This work has been supported by the Commission of the European Communities under ESPRIT IV Working Group 20001: FUSION.

Aggregation of imprecise probabilities has not been studied in-depth in the literature. We can only quote the unpublished research report by Walley [24] and the paper by Cano et al. [3] for the combination of probability intervals, but only in the case of 2 experts. The problem is different from aggregating belief functions which have been dealt with by several authors [10, 1, 28].

Our point of view is that the use of imprecise probabilities can bring a new dimension to the problem of expert opinion pooling, by allowing the experts to give a more imprecise representation of their knowledge state instead of forcing them to give exact probabilities for each event. Among the new possibilities we want to point out the following ones:

- The experts may be consistent, though each one of them has a different knowledge of the problem. That is not possible with a single probability distribution. If each expert gives a convex set of probabilities and the intersection of these sets is not empty, then this intersection is an obvious candidate to be proposed as the result of the pooling operation
- Assigning weights to the experts is not as necessary as in the probabilistic case. An expert may say that the probability of a given event is between 0 and 1 (he knows nothing about it). This seems more natural than forcing the expert to give a number as the probability of this event and, then consider that the expert weight for this event is 0, and so this value of probability has no effect on the final result.

  The precision of the result may depend on the consistency of the experts. This makes sense even in the case of combining single probability distributions. If two different experts give very similar probability distributions then the result should be more precise than in the case in which the same experts express very differing probabilities.
- Weights for imprecise probabilities may represent a greater diversity of situations than probabilistic weights. In general, in the classical procedures there is a weight for each expert and each elementary hypothesis. However, the same expert can be very good in discriminating hypothesis H_i from H_k and very bad in discriminating H_i from H_j. For example, one person can discriminate *red* from *yellow*, but not *red* from *blue*. Imprecise probabilities will allow us to represent expert competence by means of a fuzzy indistinguishability relationship [23], which will be a more appropriate model for this type of situations.

This paper studies the problem of aggregating convex sets of probabilities taking into account the considerations above. It is organized as follows. Section 2 gives a very brief review of classical aggregation methods. Section 3 introduces the necessary concepts about convex sets of probabilities and probability intervals. Section 4 proposes procedures of transforming convex sets of probabilities taking into account source reliability. Section 5 gives methods for aggregating imprecise probabilities. Finally, Section 6 is devoted to the conclusions.

2 Classical Methods of Probability Consensus

In this section we shall study how to obtain mathematical consensus on a set of hypotheses. More details are to be found in [18, 7, 8].

Let $U = \{a_1, a_2, \cdots, a_n\}$ be a finite set of mutually exclusive and exhaustive hypotheses about which we want to obtain m experts' consensus. Every expert will specify a probability distribution on the hypothesis set. The probability distribution, representing the m experts' opinion consensus, is obtained by applying an aggregation function f to m distributions reproducing the experts' opinions. Sometimes, it is interesting that experts assume (or someone assigns to them) credibility degrees that will be represented by vectors with n weights (one for each hypothesis). In this case, to reach consensus the aggregation function will receive as input m weight vectors as well as m probability distributions.

Formally, the consensus of several experts opinions may be computed by the following procedure:

1. Establish a finite set of mutually exclusive and exhaustive hypotheses $U = \{a_1, a_2, \cdots, a_n\}$.
2. Obtain a probability distribution across the hypotheses for each of the m experts, p_i, $i = 1, 2, \cdots, m$, where $p_i(a_j)$ represents the probability expert i assigns to hypothesis j, and a weight vector $W_i = \{w_{i1}, w_{i2}, \cdots, w_{in}\}$, $i = 1, \cdots, m$, where w_{ij}, $0 \leq w_{ij} \leq 1$, represents the credibility assigned to expert i with respect to hypothesis j.
3. Combine the experts' opinions and the credibility they have assigned, using the selected aggregation function f, into a single vector $Q = \{q_1, q_2, \cdots, q_n\}$, such that $q_j = f(w_{1j}, w_{2j}, \cdots, w_{mj}, p_1(a_j), p_2(a_j), \cdots, p_m(a_j))$, $j = 1, \cdots, n$.
4. Generate the consensus distribution p by normalizing Q, that is, $p(a_j) = q_j / \sum_{i=1}^{n} q_i$, $j = 1, 2, \cdots, n$.

The consensus returned by this procedure is a probability distribution across the hypotheses indistinguishable from any of the probability distributions that represent the experts' initial opinions.

Let us note that weights reflect reliability -the credibility degree assigned to each expert- and try to establish a correlation between influence and experience. A more detailed discussion of the aspects implied in weight assignment is to be found in [4, 18].

Combination or aggregation functions also deserve more detailed attention, because they define the aggregation scheme we use to combine experts' opinions. Researchers have gradually developed a great variety of aggregation functions. The most usual combination functions are the following:

- *Linear opinion pool* [18, 8, 22] (also known as weighted average). The consensus probability assigned to a_j, $p(a_j)$, is a weighted sum of all the probabilities that various experts have assigned to it

$$p(a_j) = k * \sum_{i=1}^{m} w_{ij} p_i(a_j) \qquad (1)$$

where k is a normalization constant. According to this scheme, weights assigned to each hypothesis by experts must be normalized and must add up to one ($\forall j, \sum_{i=1}^{m} w_{ij} = 1$), and may not be negative ($\forall i, j, w_{ij} \geq 0$).

- *Logarithmic opinion pool* [18, 8, 7]. Here, the consensus probability assigned to a_j, $p(a_j)$, is a weighted product of all the probabilities that the different experts have assigned to it

$$p(a_j|C) = k * \prod_{i=1}^{m} (p_i(a_j))^{w_{ij}} \qquad (2)$$

where k is a normalization constant and, as in the previous aggregation function, weights must be normalized and must add up to one.

- *Conjugate method* [27]. In this method, the probability assignment expert i attributes to hypothesis a_j is generated by a beta distribution with parameters (α_{ij}, β_i) ($\beta_i \geq \alpha_{ij} > 0$). The consensus of m experts' opinions is obtained by applying Bayes rule in order to combine the m beta distributions on hypothesis a_j which produces a beta distribution (α_j, β), where

$$\alpha_j = \sum_{i=1}^{m} w_{ij} * \alpha_{ij}, \quad \beta = \sum_{j=1}^{n} \alpha_j. \qquad (3)$$

The consensus probability assigned to a_j, $p(a_j)$, is the mean of this distribution:

$$p(a_j) = \frac{\alpha_j}{\beta} \qquad (4)$$

Here, weights need not add up to one, and its originator suggests they should be chosen to satisfy the requirement $\forall j, 1 \leq \sum_{i=1}^{m} w_{ij} \leq m$.

Linear opinion pool differs from logarithmic opinion pool in that the consensus probability assigned to a_j in the former will be zero only if all $p_i(a_j)$ are zero; while in the latter it is suffice that one of the $p_i(a_j)$ be zero. There is another aspect that has to be considered, the influence that the different scales each subject uses to express his/her personal degrees of belief have in the consensus. Whereas linear opinion pool depends on the assumption that all the opinions are expressed using the same probability scale, logarithmic opinion pool is invariant under rescaling of individual degrees of belief.

Many other aggregation methods have been proposed, but the majority of them imply complex analysis of relationships between experts or between a decision maker and an expert, and incorporate psychological principles. The aggregation functions considered here are purely mathematical. The conjugate method presents some problems concerning the determination of the expected sum of the weights and the requirement that all individual probability distributions have to belong to the family of beta distributions. However, linear opinion pool is the simplest method and it seems to perform best in practice giving the best results despite its lack of psychological justification [18].

3 Convex Sets of Probabilities and Probability Intervals

In this section we present some elementary notions about convex sets of probabilities and probability intervals, necessary for the problem of expert opinion pooling. More details are to be found in [25, 26, 2]. We shall consider that if p is a probability distribution on the set of hypotheses, U, then P will denote its associated probability measure.

In this paper, we shall assume that each expert E_k expresses his beliefs as a closed and convex set of probabilities, H_k, with a finite set of extreme points, $H_k = \text{CH}\{p_1^k, \ldots, p_{i_k}^k\}$, where CH stands for the convex hull: H_k is the minimum convex set containing the probability distributions $\{p_1^k, \ldots, p_{i_k}^k\}$.

A convex set of probabilities can be also defined by a set of linear restrictions, $\{L_1, \ldots, L_h\}$, where

$$L_j(x) \equiv \alpha_1^j x_1 + \ldots + \alpha_n^j x_n \leq \beta_j \tag{5}$$

The convex set associated to the set of linear restrictions $\{L_1, \ldots, L_h\}$ is the set of probability distributions p on U such that $(p(a_1), \ldots, p(a_n))$ verifies all the linear restrictions, i.e.,

$$\alpha_1^j p(a_1) + \ldots + \alpha_n^j p(a_n) \leq \beta_j, \quad \forall j = 1, \ldots, h \tag{6}$$

The calculation of a set of linear restrictions from the extreme points can be done by means of a convex hull algorithm [19]. The enumeration of the extreme points from a set of linear restrictions is an interesting problem studied in the field of Computational Geometry. A survey of algorithms is to be found in [16].

Should U have 3 elements, we can use a highly illustrative geometric representation of convex sets of probabilities. Each probability is represented as a point in an equilateral triangle, like the one in Fig. 1. The probability of each single hypothesis, a_i, is the perpendicular distance from the point to each one of the triangle edges.

A system of probability intervals is a pair of mappings:

$$l, u : 2^U \to [0, 1] \tag{7}$$

such that there is a convex set of probability distributions, H, verifying,

$$l(A) = \text{Inf}\{P(A) \mid p \in H\}, \quad \forall A \subseteq U \tag{8}$$

$$u(A) = \text{Sup}\{P(A) \mid p \in H\}, \quad \forall A \subseteq U \tag{9}$$

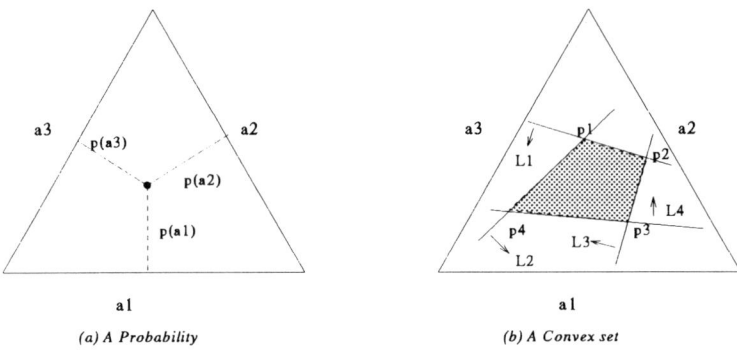

Fig. 1. Geometric Representation

A system of probability intervals defines a convex set of probabilities through the expression:

$$H_{(l,u)} = \{p \mid l(A) \leq P(A) \leq u(A)\} \quad (10)$$

However, this correspondence between convex sets and probability intervals is not one-to-one. Different convex sets can define the same pair of probability intervals by means of expressions (8) and (9). Only one of all the possible ones is $H_{(l,u)}$. We may say that convex sets of probabilities are a more general representation than probability intervals. One convex set H is said to be equivalent to the pair (l, u) if and only if $H = H_{(l,u)}$. Only those convex sets which can be defined by means of linear restrictions which are constant in the non-zero values

$$L_k(x) \equiv x_{i_1} + \ldots + x_{i_k} \leq \beta_k \quad (11)$$

are equivalent to a pair of probability intervals. In Fig. 2, the case (a) represents a general convex set, non equivalent to a system of probability intervals. Case (b) in which the relevant restrictions are parallel to the triangle edges, is equivalent to a system of probability intervals.

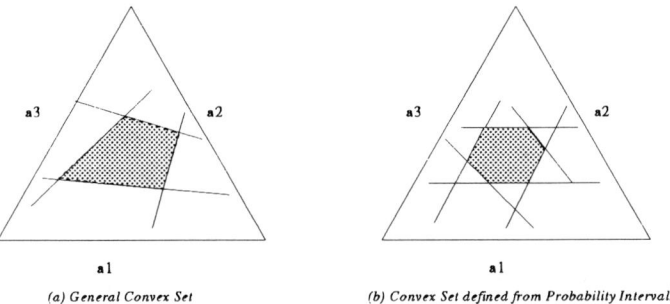

Fig. 2. Convex Sets and Probability Intervals

It is not necessary to store all the values $(l(A), u(A))$ for every $A \subseteq U$. First, we have that $l(A) = 1 - u(\overline{A})$. So, the upper or lower values are enough.

Now, in the case of using upper bounds, each value $u(A)$ determines the linear restriction:

$$L_{u(A)}(x) \equiv \sum_{a_i \in A} x_i \leq u(A) \qquad (12)$$

Then, we only have to store the values $u(A)$ such that $L_{u(A)}$ is non-redundant, or relevant, in the system of linear inequations $\{L_{u(A)} | A \subseteq U\} \cup \{\sum_{a_i \in U} x_i = 1\}$. See [9] for algorithms to determine redundant restrictions. If $L_{u(A)}$ is redundant, then the value $u(A)$ may be calculated by linear programming from the non-redundant restrictions.

A particular case of probability bounds are those used in the Theory of Evidence [20, 21]. A pair (Bel, Pl) of belief-plausibility intervals is a pair of mappings:

$$Bel, Pl : 2^U \to [0, 1] \qquad (13)$$

such that there is a mapping $m : 2^U \to [0, 1]$ (the mass assignment) verifying,

$$m(\emptyset) = 0 \qquad \sum_{A \subseteq U} m(A) = 1 \qquad (14)$$

$$Bel(B) = \sum_{A \subseteq B} m(A) \qquad Pl(B) = \sum_{A \cap B \neq \emptyset} m(A) \qquad (15)$$

A pair of belief-plausibility measures is always a pair of probability intervals, in the sense that they verify the definition above.

Usually, a pair of belief plausibility measures is given by its mass assignment, and this is specified only for the focal sets (those for which the mass is different from 0).

A pair of proper and reachable elementary probability intervals (elementary intervals in short) is a pair of mappings [2]:

$$p_*, p^* : U \to [0, 1] \qquad (16)$$

verifying $p_*(a_i) \leq p^*(a_i), \quad \forall a_i \in U$ and

$$\sum_{j \neq i} p_*(a_j) + p^*(a_i) \leq 1, \quad \forall a_i \in U \qquad (17)$$

$$\sum_{j \neq i} p^*(a_j) + p_*(a_i) \geq 1, \quad \forall a_i \in U \qquad (18)$$

A system of elementary probability intervals corresponds to a system of intervals in which only the intervals for the singleton events, $\{a_i\}$ are relevant. A system of elementary intervals (p_*, p^*) always defines a system of intervals for all the events through the expressions [2]:

$$l(A) = \text{Max } \{\sum_{a \in A} p_*(a), 1 - \sum_{a \notin A} p^*(a)\} \qquad (19)$$

$$u(A) = \text{Min } \{\sum_{a \in A} p^*(a), 1 - \sum_{a \notin A} p_*(a)\} \qquad (20)$$

In general, for two pairs of probability intervals, (l_1, u_1) and (l_2, u_2), we shall say that (l_1, u_1) is included in (l_2, u_2) ($(l_1, u_1) \subseteq (l_2, u_2)$) if and only if the convex set associated to (l_1, u_1) is included in the convex set associated to (l_2, u_2), which is equivalent to

$$l_1(A) \geq l_2(A), \quad u_1(A) \leq u_2(A), \qquad \forall A \subseteq U \qquad (21)$$

Analogous definitions of inclusion are considered for belief-plausibility intervals and elementary intervals.

Given convex sets of probabilities, H_1 and H_2, and $\alpha \in [0, 1]$, the convex combination of H_1 and H_2 is the convex set given by,

$$H = \alpha H_1 + (1 - \alpha)H_2 = \{\alpha p_1 + (1 - \alpha)p_2 | p_1 \in H_1, p_2 \in H_2\} \qquad (22)$$

If H_1 has $\{p_1^1, \ldots, p_{i_1}^1\}$ as extreme points, and H_2 the set $\{p_1^2, \ldots, p_{i_2}^2\}$, then it is very easy to show that $H = \alpha H_1 + (1 - \alpha)H_2$ is generated by points $\{\alpha p_i^1 + (1 - \alpha)p_j^2 \mid i = 1, \ldots, i_1, \; j = 1, \ldots, i_2\}$.

If H_1 and H_2 are given by linear restrictions $\{L_1^1, \ldots, L_{h_1}^1\}$ and $\{L_1^2, \ldots, L_{h_2}^2\}$, respectively and H is not degenerated (H does not verify any equality restriction apart from $\sum_{i=1}^n x_i = 1$), then the calculation of a set of restrictions defining H is not very complicated. The procedure is as follows:

- Initially, the set of restrictions of H is empty.
  For each restriction of H_1,

$$L_i^1 \equiv \alpha_1^{i,1} x_1 + \ldots + \alpha_n^{i,1} x_n \geq \beta_i^1 \qquad (23)$$

minimize $\alpha_1^{i,1} x_1 + \ldots + \alpha_n^{i,1} x_n$ under linear restrictions defining H_2. Let the minimum value be γ_i^1. Add the restricition

$$\alpha_1^{i,1} x_1 + \ldots + \alpha_n^{i,1} x_n \geq \alpha \beta_i^1 + (1 - \alpha)\gamma_i^1 \qquad (24)$$

to the set of restrictions of H.
- For each restriction of H_2,

$$L_j^2 \equiv \alpha_1^{j,2} x_1 + \ldots + \alpha_n^{j,2} x_n \geq \beta_j^2 \qquad (25)$$

minimize $\alpha_1^{j,2} x_1 + \ldots + \alpha_n^{j,2} x_n$ under linear restrictions defining H_1. Let the minimum value be γ_j^2. Add the restricition

$$\alpha_1^{j,2} x_1 + \ldots + \alpha_n^{j,2} x_n \geq \alpha \gamma_j^2 + (1 - \alpha)\beta_j^2 \qquad (26)$$

to the set of restrictions of H.

The degenerated case is much more complicated and out of the scope of this paper. Procedures in [15] may be adapted to do this operation. Besides, this method can always be used as an approximate calculation of the convex combination. In the degenerated case, the result is a convex set H' containing the true convex set H (it is more imprecise, but no probability distribution is lost).

In the case of probability intervals, computations are quite simple. If H_1 and H_2 are associated to elementary probability intervals (p_{*1}, p_1^*) and (p_{*2}, p_2^*), respectively, then $H = \alpha H_1 + (1-\alpha) H_2$ is associated with the probability interval, (p_*, p^*) where:

$$p_*(a_i) = \alpha p_{*1}(a_i) + (1-\alpha) p_{*2}(a_i); \quad p^*(a_i) = \alpha p_1^*(a_i) + (1-\alpha) p_2^*(a_i) \quad (27)$$

Analogously for probability intervals, if H_1 and H_2 are associated with (l_1, u_1), (l_2, u_2) respectively, then H is associated with (l, u) given by

$$l(A) = \alpha l_1(A) + (1-\alpha) l_2(A), \quad u(A) = \alpha u_1(A) + (1-\alpha) u_2(A), \quad \forall A \subseteq U \quad (28)$$

For mass assignments, we have a similar result. If H_1 and H_2 are associated to m_1 and m_2, respectively, then $H = \alpha H_1 + (1-\alpha) H_2$ is associated to m given by: $m(A) = \alpha m_1(A) + (1-\alpha) m_2(A), \quad \forall A \subseteq U$.

4 Discounting Experts Opinions

In classical methods, the weights measuring experts' realibility are used in the aggregation function. However, we shall take the approach of Dempster-Shafer theory of evidence, acccording to which [21] the weights are used to transform expert opinions. If an expert is completely reliable, then his opinion is not changed. If an expert is not very reliable, then his opinion is transformed into a much less precise one.

In general, we have a convex set of probabilities, H, and a measure of reliability, R, which may be a weight in $[0, 1]$, a vector of weights, an equivalence relationship in U, or a fuzzy indistinguishability relationship [23]. The problem will be to calculate the result of discounting H by R: $H' = D(H, R)$.

Our basis will be the discounting of a mass assignment, m, by a factor $c \in [0, 1]$. The result is the mass assignment given by [21]:

$$m'(A) = (1-c).m(A), \quad \forall A \neq U; \quad m'(U) = (1-c).m(U) + c \quad (29)$$

If (Bel, Pl) is the pair of belief-plausibility measures associated to m, then the pair associated to m' is given by the following espressions:

$$Bel'(A) = (1-c).Bel(A), \quad A \neq U; \quad Bel'(U) = 1, \quad (30)$$

$$Pl'(A) = (1-c).Pl(A) + c, \quad A \neq \emptyset; \quad Pl'(\emptyset) = 0, \tag{31}$$

If a source has a weight of $w \in [0,1]$, then our approach will be to discount it with a factor $c = 1 - w$. Then the result is combined with the rest of sources as it were completely reliable.

In the following we propose definitions to perform this discounting and procedures to calculate it. Particular cases of single probabilities, probability intervals and belief-plausibility intervals are considered.

4.1 Discounting a Probability Distribution

In this case the convex set contains only one probability distribution $H = \{p\}$.

Discounting by a constant weight In the case of discounting a probability by a constant weight $w \in [0,1]$, then as a probability distribution is a particular case of mass assignment, we can apply formulae (29) for $c = 1 - w$. The result is the convex set $D(p, w)$ associated to the mass assignment given in its focal elements by,

$$m'(\{a_i\}) = w.p(a_i), \quad \forall a_i \in U; \quad m'(U) = (1-w) \tag{32}$$

In terms of intervals, the resulting belief and plausibilities are given by:

$$Bel'(A) = w. \sum_{a_i \in A} p(a_i), \quad A \neq U; \quad Bel'(U) = 1, \tag{33}$$

$$Pl'(A) = w. \sum_{a_i \in A} p(a_i) + (1-w), \quad A \neq \emptyset; \quad Pl'(\emptyset) = 0, \tag{34}$$

Geometrically, what it is obtained by discounting a probability distribution is an equilateral triangle like the one in Fig. 3.

Discounting by a Vector of Weights Now let us consider the case in which R is a vector of weights $W = (w_1, \ldots, w_n)$, where each w_i represents the source reliability in relation to hypothesis a_i. An immediate generalization is to define $D(p, W)$ as the convex set associated to mass, m', where

$$m'(\{a_i\}) = w_i p(a_i), \quad i = 1, \ldots, n; \quad m'(U) = \sum_{a_i \in U}(1 - w_i)p(a_i) \tag{35}$$

The associated beliefs and plausibilities are:

$$Bel'(A) = \sum_{a_i \in A} w_i p(a_i), \quad A \neq U; \quad Bel'(U) = 1 \tag{36}$$

$$Pl'(A) = \sum_{a_i \in A} w_i p(a_i) + \sum_{a_i \in U}(1 - w_i)p(a_i), \quad A \neq \emptyset; \quad Pl'(\emptyset) = 0 \tag{37}$$

The result of discounting a probability, p, by vector $W = (0.1, 0.5, 0.8)$ is shown in Fig. 4.

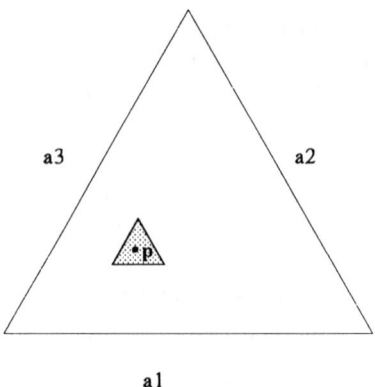

Fig. 3. Discounting a Probability

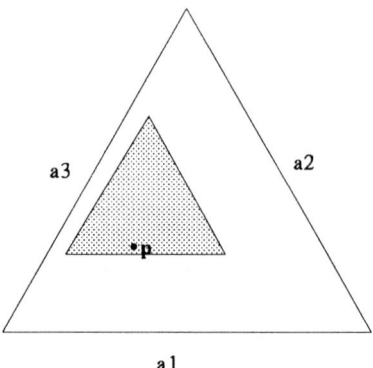

Fig. 4. Discounting a Probability with Non-Uniform Weights

Discounting by an Equivalence Relationship Now let us consider that we want to discount by an equivalence relationship R defined in U. The idea is that if a_i and a_j are equivalent, then the source cannot discriminate between a_i and a_j, and it makes no sense for it to allocate a probability to a_i or a_j. All the mass should go to a set containing $\{a_i, a_j\}$. Imagine that we want to know the colour and shape, (C, S), of a given object and that a source is not able of discriminating between different colours. In this case, all the pairs, (C_1, S), $C_2, S)$, with the same shape are equivalent, and there is no basis for specifying probabilities as $p(C_1, S) = 0.2$. The only thing justified is assigning a mass to an equivalence class $\{(C_i, S) | C_i$ is a possible colour $\}$ or a superset, i.e., to specify probabilities to the shapes.

Following the ideas above, the discounting of a probability distribution p by an equivalence relationship, R, moves the probabilities assigned to a single hypothesis, a_i, to its equivalence class, $[a_i]$. More precisely, the discounting of p by R is the convex set given by the mass assignment,

$$m'([a_i]) = \sum_{a_j \in [a_i]} p(a_j) \qquad (38)$$

Fig. 5 shows the result of discounting a probability p by an equivalence relationship saying that $a_1 R a_2$. The result is the straight line, H', in which the distance to the a_3 edge remains constant.

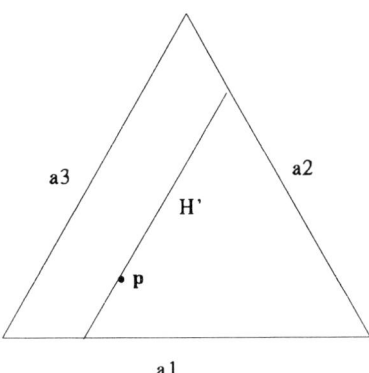

Fig. 5. Discounting a Probability with an Equivalence Relationship

Discounting by a Fuzzy Indistinguishability Relationship In general, an equivalence relationship is a very sharp representation of the reliability of a source. Usually, it is more realistic to assign a number in the interval $[0, 1]$ to the indistinguishability of each pair of hypothesis. Indistinguishability relationships with respect to a t-norm, T, have been defined by Trillas and Valverde [23]. In this paper, we consider the Min t-norm, and then a fuzzy indistinguishability relationship is a mapping, $R : U \times U \to [0, 1]$, verifying

i) *Reflexivity.-* $R(a_i, a_i) = 1, \quad \forall a_i \in U$
ii) *Symmetry.-* $R(a_i, a_j) = R(a_j, a_i), \quad \forall a_i, a_j \in U$
iii) *Transitivity.-* $R(a_i, a_j) \geq \text{Min} \{R(a_i, a_k), R(a_k, a_j)\}, \quad \forall a_i, a_j, a_k \in U$

This type of indistinguishability relationship verifies that for every $x \in [0, 1]$, the crisp relationship,

$$R_x(a_i, a_j) = \begin{cases} 1 \text{ if } R_x(a_i, a_j) \geq x \\ 0 \text{ otherwise} \end{cases} \qquad (39)$$

is an equivalence relationship.

Let m'_x be the result of discounting p by R_x, then the discounting of p by R will be the mass asignment m', given by

$$m'(A) = \int_0^1 m'_x(A).dx \qquad (40)$$

Assume as an example that $U = \{a_1, a_2, a_3\}$ and that p is given by

$$p(a_1) = 0.2, \qquad p(a_2) = 0.5, \qquad p(a_3) = 0.3$$

and that R is given by, $R(a_1, a_2) = 0.7, R(a_1, a_3), R(a_2, a_3) = 0.2$. Then m'_x in its focal elements is equal to:

a) For $x \in [0, 0.2]$, $m'_x(U) = 1$
b) For $x \in (0.2, 0.7]$, $m'_x(\{a_1, a_3\}) = 0.7, m'_x(\{a_3\}) = 0.3$
c) For $x \in (0.7, 1]$, $m'_x(\{a_1\}) = 0.2, m'_x(\{a_2\}) = 0.5, m'_x(\{a_3\}) = 0.3$

So m' calculated according to formula (40) is given by,

$$m'(\{a_1\}) = 0.06, m'(\{a_2\}) = 0.15, m'(\{a_3\}) = 0.24$$

$$m'(\{a_1, a_2\}) = 0.35, m'(U) = 0.2$$

4.2 Discounting a Convex Set of Probabilities

Assume now that each expert expresses his beliefs by means of a convex set of probabilities H. An immediate generalization of the methods to discount a single probability can be obtained by discounting all the probabilities in H. So the discounting of a convex set H by a reliability measure R is given by

$$D(H, R) = \bigcup_{p \in H} D(p, R) \qquad (41)$$

In general, in all the methods we have studied, it is enough to discount the extreme points and then take the convex hull. In the following, we consider how the discounting can be done when the convex set is given by linear restrictions or probability intervals, without calculating the extreme points.

Discounting by a Constant Weight Let H be a convex set defined by linear restrictions $\mathcal{L} = \{L_1, \ldots, L_h\}$ and assume that H is not degenerated (there is not an equality linear restriction verified by all the points in H, apart from the fact that the addition of the probabilities is 1). The degenerated case is much more complicated and it is not considered here. Anyway this method may be applied in the degenerate case as an approximation: The result is a convex set containing the original one.

We want to calculate a set of linear restrictions $\mathcal{L}'$ defining $D(H, w)$. The procedure is as follows:

1. Transform each linear restriction $L_k, k = 1,\ldots,n$ into a restriction with the following form

$$L_k \equiv \alpha_1^k x_1 + \ldots + \alpha_n^k x_n \geq \beta_k, \quad \alpha_i^k \geq 0 \quad (42)$$

and where at least one of the parameters α_i^k be equal to 0. This can be done by taking into account that all the points in the convex set verify equality $x_1 + \ldots + x_n = 1$.

2. Calculate Inf x_j subject to $\mathcal{L}$ and $\sum_{x_i \in U} x_i = 1$, for each $j = 1,\ldots,n$. Call the result b_j. Add to $\mathcal{L}$ the restrictions:

$$L_{h+j} \equiv x_j \geq b_j, \quad j = 1,\ldots,n \quad (43)$$

3. Transform each restriction $L_k \equiv \alpha_1^k x_1 + \ldots + \alpha_n^k x_n \geq \beta_k$ into the restriction

$$L_k' \equiv \alpha_1^k x_1 + \ldots + \alpha_n^k x_n \geq \beta_k . w \quad (44)$$

$\mathcal{L}'$ is given by the set of transformed restrictions L_k', where $L_k \in \mathcal{L}$.

As an example, consider the convex set if Fig. 6 (a) which is given by linear restrictions, $x_2 - x_3 \geq -0.2, x_2 - x_3 \leq 0.2, x_1 \geq 0.2, x_1 \leq 0.6$. Assume that we want to discount it with $w = 0.8$.

After step 1, the transformed restrictions are $x_1 + 2x_2 \geq 0.8, x_1 + 2x_3 \geq 0.8, x_1 \geq 0.2, x_2 + x_3 \geq 0.4$.

The minimum values are $b_1 = 0.2, b_2 = 0.1, b_3 = 0.1$, and the added restrictions in step 2 are $x_1 \geq 0.2, x_2 \geq 0.1$, and $x_3 \geq 0.1$.

After step 3, the restrictions in $\mathcal{L}'$ are $x_1 + 2x_2 \geq 0.64, x_1 + 2x_3 \geq 0.64, x_1 \geq 0.16, x_2 + x_3 \geq 0.32, x_2 \geq 0.08, x_3 \geq 0.08$. The discounted convex set is in Fig. 6 (b)

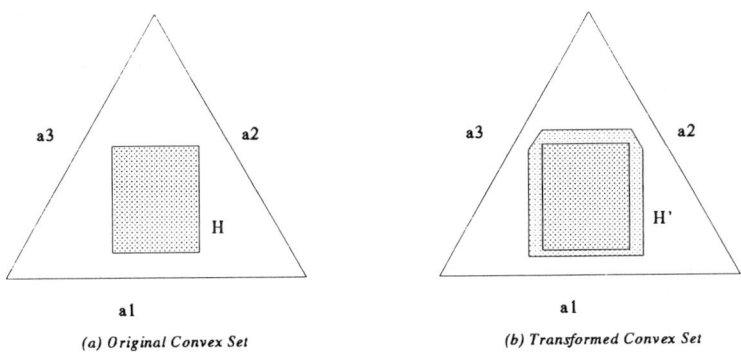

(a) Original Convex Set

(b) Transformed Convex Set

Fig. 6. Discounting a Convex Set of Probabilities

In the case of elementary probability intervals, (p_*, p^*), the discounting by weight w: $(p_*', p^{*'})$ can be calculated in a very simple way:

$$p_*'(a_i) = w.p_*(a_i), \quad p^{*'}(a_i) = w.p^*(a_i) + (1-w), \quad \forall a_i \in U \quad (45)$$

For general probability intervals, the result of discounting (l, u) is the pair (l', u') given by:

$$l'(A) = w.l(A), \quad A \neq U; \quad l'(U) = 1 \quad (46)$$

$$u'(A) = w.l(A) + (1-w), \quad A \neq \emptyset; \quad u'(\emptyset) = 0 \quad (47)$$

For the particular case of Belief-Plausibility intervals the result is the same as the previously defined discounting for $c = 1 - w$, which is given by expressions (30) and (31).

Discounting by a Vector of Weights Now we want to discount using a vector $W = (w_1, \ldots, w_n)$.

For the case of transforming a non-degenerated convex set given by linear restrictions $\mathcal{L} = \{L_1, \ldots, L_h\}$, the procedure is completely analogous to the case of a constant weight. The only difference is in step 3, which now becomes:

3'. Transform each restriction

$$L_k \equiv \alpha_1^k x_1 + \ldots + \alpha_n^k x_n \geq \beta_k, \quad \alpha_i^k \geq 0 \quad (48)$$

verifying that $\alpha_i^k = 0$ for $w_i = 0$, into the restriction

$$L_k' \equiv \frac{\alpha_1^k}{w_1} x_1 + \ldots + \frac{\alpha_n^k}{w_n} x_n \geq \beta_k, \quad \alpha_i^k \geq 0 \quad (49)$$

If a restriction has $\alpha_i^k > 0$ for $w_i = 0$, then it is not considered.

For probability intervals (and elementary intervals too), the problem is that if H is associated to (l, u), then $H' = D(H, W)$ cannot always be defined by probability intervals. As an example, consider the convex set in Fig. 7 (a), which is defined by probability intervals $p^*(a_1) = 0.7, p_*(a_1) = 0.5$. Assume now the vector of weights $(0, 0, 0.5)$. The result of the transformation is convex set H' in Fig. 7 (b), which cannot be defined by probability intervals: the edges are not parallel to the triangle edges.

Discounting by an Equivalence Relationship For a general convex set defined by restrictions $\mathcal{L} = \{L_1, \ldots, L_h\}$, the problem of calculating a set of restrictions $\mathcal{L}'$ defining convex set $H' = D(H, R)$ where R is an equivalence relationship is quite difficult. The most direct method we know is based on algorithms to remove quantifiers in convex sets [15]. In the following we explain how these algorithmos can be applied.

Let U/R be the set of equivalence classes of U under R, and $\mathbf{P}(U), \mathbf{P}(U/R)$ the sets of probability distributions on U and U/R, respectively. Let f be the mapping $f : \mathbf{P}(U) \to \mathbf{P}(U/R)$, given by,

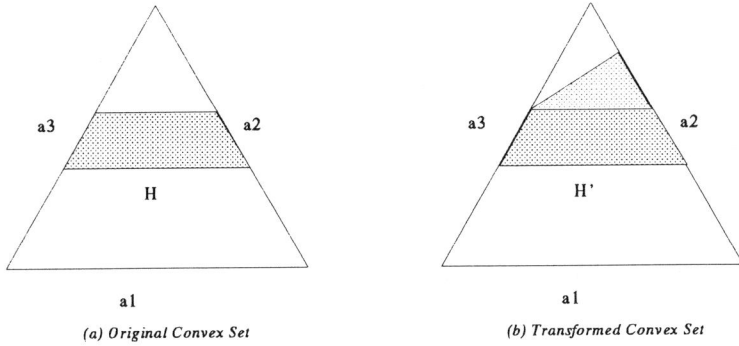

Fig. 7. Discounting Probability Intervals by $W = (0, 0, 0.5)$

$$f(p)([a_j]) = \sum_{a_i \in [a_j]} p(a_i) \qquad (50)$$

The image of H under f: $f(H)$, is a convex set and the algorithms in [15] can be directly applied to calculate a set of restrictions $\mathcal{L}^* = \{L_1^*, \ldots, L_r^*\}$ defining $f(H)$. Finally, the set $\mathcal{L}'$ may be calculated by transforming each restriction $L_k^* \in \mathcal{L}^*$ given by

$$L_k^* \equiv \sum_{[a_i] \in U/R} \alpha_{[a_i]} x_{[a_i]} \geq \beta_k \qquad (51)$$

into the restriction

$$L_k' = \sum_{a_i \in U} \alpha_i x_i \geq \beta_k \qquad (52)$$

where $\alpha_i = \alpha_{[a_i]}$. $\mathcal{L}'$ is given by all the tranformed restrictions $\{L_1', \ldots, L_r'\}$.

In the case of elementary intervals (p_*, p^*), the result is a convex set which, in general, cannot be defined by elementary intervals, but by a pair of general intervals, (l', u'), given by

$$l'(A) = \text{Max} \left\{ \sum_{[a_j] \subseteq A} \left(\sum_{a_i \in [a_j]} p_*(a_i) \right), 1 - \sum_{a_j \in \overline{[A]}} p^*(a_j) \right\} \qquad (53)$$

$$u'(A) = \text{Min} \left\{ \sum_{a_j \in [A]} p^*(a_j), 1 - \sum_{a_j \notin [A]} p_*(a_j) \right\} \qquad (54)$$

where $[A] = \bigcup_{a_i \in A} [a_i]$.

For belief functions, the result of discounting m is the mass given by:

$$m'(B) = \sum_{[A]=B} m(A) \qquad (55)$$

and in terms of belief-plausibility measures:

$$Bel'(A) = Bel(\{a_j \mid [a_j] \subseteq A\}); \quad Pl'(A) = Pl([A]) \tag{56}$$

For general probability intervals, the result is very similar:

$$l'(A) = l(\{a_j \mid [a_j] \subseteq A\}); \quad u'(A) = u([A]) \tag{57}$$

Discounting by an Indistinguishability relationship Let $T = \{R(a_i, a_j) \mid a_i, a_j \in U\}$ be the set of possible values of R. Since U is finite, this set is finite. Assume that $T = \{t_1, \ldots, t_s\}$, where $t_1 \leq \ldots \leq t_s = 1$.

Let R_x be the equivalence relationship defined by (39), then according to expression (40) the discounting of a convex set H by R is the convex set

$$H' = \sum_{j=2}^{s}(t_j - t_{j-1})\mathrm{D}(H, R_{t_j}) + t_1.\mathrm{D}(H, R_{t_1}) \tag{58}$$

Once it is known how to calculate the discounting of a convex set of probabilities by and equivalence relationship, R_{t_j}, the problem of discounting by a fuzzy relationship is reduced to the calculation of a convex combination of convex sets:

$$H' = \sum_{j=1}^{s} r_j H'_j, \quad \sum_{j=1}^{s} r_j = 1 \tag{59}$$

$H'_j = \mathrm{D}(H, R_{t_j}), r_j = t_j - t_{j-1} \ (j \geq 2), r_1 = t_1$.

Procedures for carrying out this convex combination have been considered in Section 3.

5 Aggregation of Convex Sets of Probabilities

In this section we consider the problem of aggregating m convex sets of probabilities $H_1, \ldots, H_m$, where each H_i is assumed to be non-empty. It is assumed that all the sources have the same importance. If we had information about the sources reliabilities, this would have been used to discount initial convex sets in a previous step. The result of the aggregation will be denoted as $A(H_1, \ldots, H_m)$. Some basic properties are:

P1. *Symmetry.-* $A(H_1, \ldots, H_m) = A(H_{\sigma(1)}, \ldots, H_{\sigma(m)})$, where σ is a permutation on $\{1, \ldots, m\}$.
P2. If we have $H_1, \ldots, H_m$ and H'_1 is such that for every $p \in H_1, q \in H_j (j \geq 2)$ there is a $p' \in H'_1$ such that p' is a convex combination of p and q, then

$$A(H'_1, H_2, \ldots, H_m) \subseteq A(H_1, H_2, \ldots, H_m) \tag{60}$$

P3. If $H_1 \cap \ldots \cap H_m \neq \emptyset$, then $H_1 \cap \ldots \cap H_m \subseteq A(H_1, H_2, \ldots, H_m)$

P1 states that the order in which convex sets are considered is not relevant. P2 states that if H_1' is closer (agrees better) to every convex set in $\{H_2, \ldots, H_m\}$, then the aggregation of $\{H_1', H_2, \ldots, H_m\}$ should be more precise than the aggregation of $\{H_1, H_2, \ldots, H_m\}$. P3 states that if the experts agree on some probabilities, then these probabilities should be in the pooling result. The possibility is open for including more probabilities, which in some cases is not unreasonable. For example, consider the convex sets in Fig. 8, their intersection is probability p, but there are a lot of probabilities in H_1 which are very close to probabilities in H_2. This fact can be taken into account by considering more probabilities in the result apart from p: distributions taken from the region between H_1 and H_2.

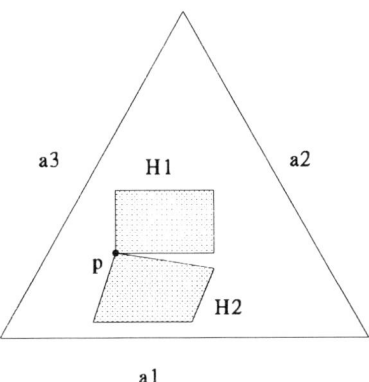

Fig. 8. Including more probabilities in the aggregation

Additionally, we may consider the following properties:

P4. *Irrelevance of Ignorance.-* $A(H_1, \ldots, H_m) = A(H_1, \ldots, H_m, \mathbf{P}(U))$
P5. *Strong P3.-* If $H_1 \cap \ldots \cap H_m \neq \emptyset$, then $A(H_1, H_2, \ldots, H_m) = H_1 \cap \ldots \cap H_m$.
P6. $A(H_1, \ldots, H_m) \subseteq \text{CH}(H_1 \cup \ldots \cup H_m)$

These properties are not verified in all the procedures we are going to propose. Perhaps P4 looks quite natural, but it is difficult to make it compatible with the idea that the number of people expressing an opinion should be taken into account ($A(H_1, H_1, H_2)$ different from $A(H_1, H_2)$) and the situations in which P5 is not verified. Anyway the effect of the ignorance will always be small, only slightly increasing the imprecision of the final result. Almost all the proposed procedures will verify P6, except one of the variations of the main method.

Below, we first consider the case of combining single probability distributions with an imprecise result, and then the general problem with convex sets and probability intervals.

5.1 Combining Single Probabilities

The method will depend on a parameter $c \in [0,1]$. The result of aggregating probabilities $\{p_1, \ldots, p_m\}$ will be denoted by $H_c = A_c(p_1, \ldots, p_m)$. This set H_c is calculated as follows:

1. Calculate $p = \frac{1}{m}\sum_{i=1}^{m} p_i$, the mid point of the probabilities.
2. Calculate $p_c^i = c.p + (1-c)p_i, \quad \forall i = 1, \ldots, m$
3. Make H_c equal to the convex set determined by points $p_c^i, \quad i = 1, \ldots, m$.

Fig. 9 shows the aggregation of 4 probabilities for $c = 0.5$.

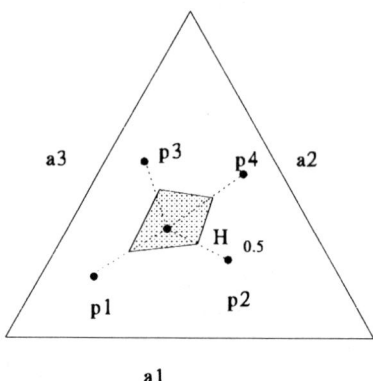

Fig. 9. Aggregation of single probabilities. $c = 0.5$

Parameter c really determines the behaviour of the procedure. For $c = 0$ it coincides with the classical probabilistic linear aggregation with uniform weights. The result is always precise and depends on the number of experts expressing an opinion: if the number of experts giving a probability p_i increases then the final result is closer to this probability, p_i. For $c = 1$, $A_1(p_1, \ldots, p_m)$ is the convex set determined by points $\{p_1, \ldots, p_m\}$. In this case the number of experts expressing a given distribution is irrelevant, but the imprecision of the result will reflect the contradiction of the sources. Intermediate values of c seem to be more appropriate.

This procedure expands the mid point in the direction of given probabilities (See Fig. 10 (a) with $c = 0.5$). However, it could be argued that the mid point should be expanded in every direction. An alternative method is to calculate the minimum value w such that $D(p, w)$ contains all the initial probabilities, and then consider $H'_c = D(p, w.c)$. See Fig. 10 (b) for this alternative expansion. The resulting aggregation does not verify P6 and will not be considered further in this paper.

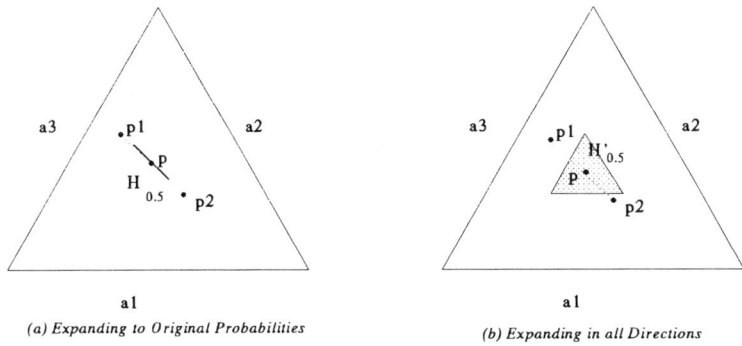

Fig. 10. Different ways of expanding the middle point

5.2 Aggregating General Convex Sets

We are going to generalize the method to combine single probabilities to the case of general convex sets. The result will be denoted as $H_c = A_c(H_1, \ldots, H_m)$, First we start with 3 preliminary steps. The first and the second will be considered as optional. The first will guarantee that P4 is verified and the second will be a guarantee for P5. The third is enough for condition P2.

S1 If there is H_i such that $H_j \subseteq H_i$, $\forall j = 1, \ldots, m$, then remove H_i. That is, make

$$A_c(H_1, \ldots, H_m) = A_c(H_1, \ldots, H_{i-1}, H_{i+1}, \ldots, H_m) \tag{61}$$

This simplification is repeated while there is a convex set including all the others.

S2 Calculate the maximal subsets $T \subseteq \{H_1, \ldots, H_k\}$ such that $\bigcap_{H_i \in T} H_i \neq \emptyset$. Let $\mathcal{T}$ be the family of all these subsets. Transform the original problem into the problem of aggregating the sets $\bigcap_{H_i \in T} H_i$, where $T \in \mathcal{T}$. Assign to each $\bigcap_{H_i \in T} H_i$ a degree of importance according to:
  1. Calculate for each H_i ($i = 1, \ldots, m$) the value n_i equal to the number of subsets $T \in \mathcal{T}$, such that $H_i \in T$.
  2. Assign a degree of importance to $\bigcap_{H_i \in T} H_i$ equal to

$$I = \sum_{H_i \in T} \frac{1}{n_i} \tag{62}$$

In the case of Fig. 11, the aggregation of H_1, H_2, and H_3 is transformed into the aggregation of $H_1 \cap H_2$ and H_3, but $H_1 \cap H_2$ will have an importance of 2 and H_3 an importance of 1. These degrees of importance have a different nature to the degrees of reliability and will be used in a different way.

S3 For each convex set H_i, calculate the minimum convex set H_i' such that for every $p \in H_i$ and $q \in \bigcup_{j \neq i} H_j$ there exists $p' \in H_i'$, such that q' is a convex combination of p and q.

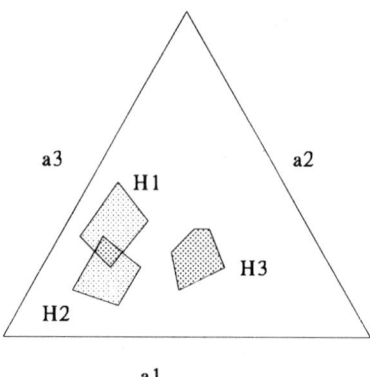

Fig. 11. Transformation in S2

Change the set $\{H_1, \ldots, H_m\}$ by the set $\{H'_1, \ldots, H'_m\}$. Fig. 12 shows how 3 convex sets are reduced in this step (darker parts).

Algorithms to carry out this reduction are not very simple. The basic computation is to determine, for a convex set and a point, (an extreme point of another convex set) whether it is interior or exterior and, in the latter case, which are the points of the convex set that can be seen directly from the point. This is done as part of the Beneath-Beyond algorithm to calculate the convex hull of a given set of points [6, 19].

An alternative way of performing this reduction is to change H_i for the convex set generated for all the probabilities at minimum distance with any other convex set H_j. This alternative will not be studied further in this paper.

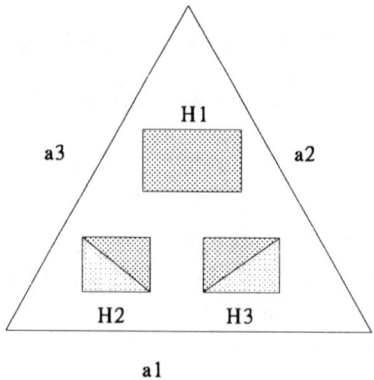

Fig. 12. Reduction in S3

Next, we generalize the probabilistic method to calculate $A_c(H_1, \ldots, H_m)$ with I_i being the value of importance of H_i. Calculation is as follows:

1. Calculate normalized degrees of importance:
$$I_i^n = \frac{I_i}{\sum_{j=1}^m I_j} \tag{63}$$

2. Calculate $H = \sum_{i=1}^m I_i^n . H_i$.
3. Calculate $H_c^i = c.H + (1-c)H_i, \quad i = 1, \ldots, m$.
4. Make H_c equal to the convex hull of $\bigcup_{i=1}^m H_c^i$

Fig. 13 shows how this method works after the preliminary steps: First it calculates a convex set, H, intermediate to the convex sets H_1, H_2, H_3 and then it expands it in the direction of the original convex sets (in this case with a factor $c = 0.25$).

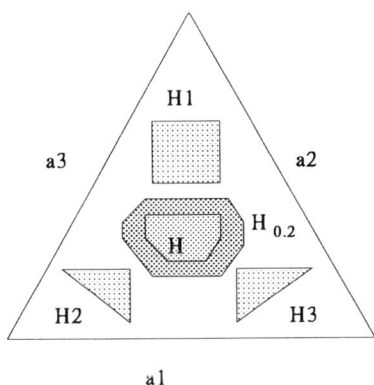

Fig. 13. Aggregation of three convex sets

Now we consider the case in which each convex set, H_i, is associated to a pair of elementary intervals, (p_{*i}, p_i^*). Then calculations are easier and can be done as follows:

S1 It is immediate. The verification of $H_i \subseteq H_j$ can be done by testing,
$$p_{*i}(a_l) \geq p_{j*}(a_l); \quad p_i^*(a_l) \geq p_j^*(a_l), \quad \forall a_l \in U \tag{64}$$

S2 The intersection of two pairs of probability intervals (p_{*i}, p_i^*) and (p_{*j}, p_j^*) is done in two steps [2]. First the pair (t_*, t^*) is calculated:
$$t_* = \text{Max}\{p_{*i}, p_{*j}\}, \quad t^* = \text{Min}\{p_i^*, p_j^*\} \tag{65}$$

The intersection is not empty if and only if $\sum_{a_i \in U} t_*(a_i) \leq 1 \leq \sum_{a_i \in U} t^*(a_i)$. Should it not be empty, then the intersection is given by the pair of elementary intervals:

$$p_*(a_i) = \text{Max } \{t_*(a_i), 1- \sum_{a_j \neq a_i} t^*(a_j)\}, \quad p^*(a_i) = \text{Min } \{t^*(a_i), 1- \sum_{a_j \neq a_i} t_*(a_j)\}, \tag{66}$$

S3 The reduction of convex sets given by probability intervals raises a problem: the reduction is not always a convex set associated to probability intervals. In Fig. 14, we may see an example of two convex sets H_1 and $H_2 = \{p\}$ which can be defined by probability intervals, but the reduction of H_1 (darker part) does not have its edges parallel to the triangle edges.

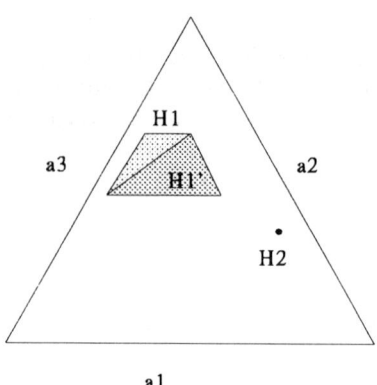

Fig. 14. Reduction of elementary intervals

As an alternative to the general procedure, in the following we propose an approximate method which always gives elementary intervals as the result. Assume that $(p_{*1}, p_1^*), \ldots, (p_{*m}, p_m^*)$ are the intervals given. The reduction of the interval (p_{*i}, p_i^*) is the interval $(p'_{*i}, p_i^{*'})$ calculated as follows:

- Determine all the lower bounds, $p_{*i}(a_k)$ for which

$$p_{*i}(a_k) > \text{Min } \{p_{*j}(a_k) \mid j = 1, \ldots, m\} \tag{67}$$

Call these lower bounds fixed (they can not be changed).
- Determine all the upper bounds, $p_i^*(a_k)$ for which

$$p_i^*(a_k) < \text{Max } \{p_j^*(a_k) \mid j = 1, \ldots, m\} \tag{68}$$

Call these upper bounds fixed (they can not be changed).
- If there is a fixed upper bound, $p_i^*(a_l)$, $l \neq k$, then $p'_{*i}(a_k) = p_{*i}(a_k)$.
If there is no such upper bound, then $p'_{*i}(a_k) = \text{Max } \{p_{*i}(a_k), t\}$, where

$$t = \text{Min } (\{p_{*j}(a_k) \mid j \neq i\} \cup \{1 - \sum_{l \neq k,r} p_i^*(a_l) - p_{*i}(a_r) \mid p_{*i}(a_r) \text{ is fixed }\}) \tag{69}$$

- If there is a fixed lower bound, $p_{*i}(a_l)$, $l \neq k$, then $p_i^{*'}(a_k) = p_i^*(a_k)$.
  If there is no such lower bound, then $p_*^{*'}(a_k) = \text{Min}\{p_i^*(a_k), t\}$, where

$$t = \text{Max}\left(\{p_j^*(a_k) \mid j \neq i\} \cup \{1 - \sum_{l \neq k, r} p_{*i}(a_l) - p_i^*(a_r) \mid p_i^*(a_r) \text{ is fixed }\}\right) \quad (70)$$

As an example of how this procedure works, see Fig. 15, in which the general convex set reduction is given by the darkest parts and in this approximation the parts of intermediate darkness are added.

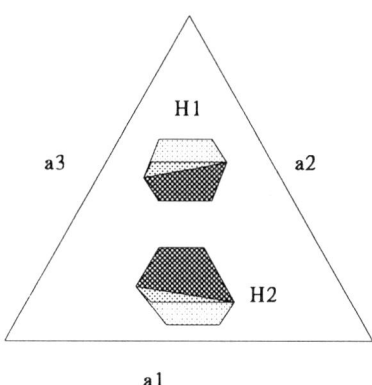

Fig. 15. Approximated reduction of elementary intervals

That is all as regards the preliminary steps. For the combination itself, the aggregation is based on the calculation of convex combinations and this is easy in terms of elementary probability intervals. Only point 5 requires the calculation of the convex hull of $\bigcup_{i=1}^{m} H_c^i$, which is not immediate in terms of elementary intervals. Assume that each convex set, H_c^i, is associated with intervals (p_{c*}^i, p_c^{*i}). Then the difficulty is that, in general, the convex hull of $\bigcup_{i=1}^{m} H_c^i$ cannot be defined in terms of probability intervals [2]. However the minimum convex set which can be defined by elementary probability intervals and contains the sets $H_c^i, (i = 1, \ldots, m)$ is very easy to calculate. It is associated to probability intervals (p_{c*}, p_c^*) where

$$p_{c*} = \text{Min}\{p_{c*}^i \mid i = 1, \ldots, m\}, \quad p_c^* = \text{Max}\{p_c^{*i} \mid i = 1, \ldots, m\} \quad (71)$$

This pair will be the aggregation result when we want to work with elementary probability intervals.

Finally, let us consider the case in which each convex set, H_i, is associated to a pair of probability intervals (l_i, u_i). We shall only consider the steps which are not trivial in terms of probability intervals.

Preliminary step 2 involves the combination of two probability intervals (l_i, u_i) and (l_j, u_j). This can be done by determining the linear restrictions associated to upper bounds u_i and u_j: $\mathcal{L}_{ij} = \mathcal{L}_i \cup \mathcal{L}_j = \{L_{u_i(A)} \mid A \subseteq U\} \cup \{L_{u_j(A)} \mid A \subseteq U\}$, where $L_{u_i(A)}, L_{u_j(A)}$ are calculated according to expression (12). The combination is empty if the set of restrictions $\mathcal{L}_{ij} \cup \{\sum_{i=1}^{n} x_i = 1\}$ is inconsistent (which can be checked by using conventional linear programming techniques). Should it be consistent, the intersection of the associated convex sets is given by the pair (l, u), where $l(A)$ $(u(A))$ is the minimum (maximum) of $\sum_{a_i \in A} x_i$ under $\mathcal{L}_{ij} \cup \{\sum_{i=1}^{n} x_i = 1\}$.

Preliminary step 3 presents the same difficulty as in the case of elementary intervals: the result of the reduction is a convex set which cannot always be expressed in terms of probability intervals. Here we give an approximate method for the case of general intervals. Assume that we want to reduce (l_i, u_i) in the aggregation of $(l_1, u_1), \ldots, (l_m, u_m)$. The result (l'_i, u'_i) is calculated taking the following points as basis:

- Determine all the upper bounds, $u_i(A)$, for which $u_i(A) < \text{Min } \{u_j(A) \mid j = 1, \ldots, m\}$. These upper bounds are called fixed and cannot be changed.
- For each fixed upper bound $u_i(A)$ and for each $B \subseteq U$, calculate the value $M_{u_i(A)}(B)$, which is equal to the maximum of $\sum_{a_i \in B} x_i$ under restrictions $\mathcal{L}_i, \sum_{a_i \in A} x_i = u_i(A)$, and $\sum_{i=1}^{n} x_i = 1$.
- If $u_i(B)$ is fixed then $u'_i(B) = u_i(B)$.
- If $u_i(B)$ is not fixed, then

$$u'_i(B) = \text{Min } \{u_i(A), \text{Max } (\{u_j(B) \mid j \neq i\} \cup \{M_{u_i(A)}(B) \mid u_i(A) \text{ is fixed }\})\} \tag{72}$$

- Calculate the lower bound by duality: $l'_i(B) = 1 - u'_i(\overline{B})$.

That is all for the preliminary steps, for the final aggregation the problem is also similar to the case of elementary intervals. If each convex set H_c^i is associated to intervals (l_i^c, u_i^c), then the convex hull of the union $\bigcup_{i=1}^{m} H_c^i$ is not associated to a system of probability intervals. The solution will be to calculate the smallest convex set containing convex sets H_c^i and associated to a pair of probability intervals. This convex set is associated to intervals (l^c, u^c), which can be easily calculated [2] as

$$l^c(A) = \text{Min } \{l_i^c(A) \mid i = 1, \ldots, m\}, \quad u^c(A) = \text{Max } \{u_i^c(A) \mid i = 1, \ldots, m\} \tag{73}$$

6 Conclusions

In this paper we have studied the problem of aggregating imprecise probabilities. This imprecision has allowed a richer handling of the reliability of sources, considering more realistic situations than in the classical probabilistic case. The proposed methods also take into account the inconsistency of the sources to determine the precision of the final result: more inconsistent sources give rise to a less precise result.

Special attention has been paid to the calculations involved in the methods, showing how these operations can be carried out in the particular case of probability intervals. Some difficulties remain as how to carry out some computations in degenerated convex sets.

We consider this paper to be a starting point, showing the new possibilities offered by the use of convex sets. The most relevant work in the future will be to study the meaning of the parameters and to give normative procedures to determine them. The comparison of different alternatives in practical experiments will also be of great interest.

Acknowledgments

We are very grateful to Peter Walley for very interesting and useful discussions on this topic. Some of the ideas were born by his suggestions to the authors. For example, the preliminary step 2, in aggregating general convex sets.

References

1. D.A. Bell, J.W. Guan, S.K. Lee: Generalized union and project operations for pooling uncertain and imprecise information, *Data & Knowledge Engineering*, 18, 89-117, 1996.
2. L.M. Campos, J.F. Huete, S. Moral: Probability intervals: a tool for uncertain reasoning, *International Journal of Uncertainty, Fuzziness and Kowledge-Based Systems*, 2, 167-196, 1994.
3. J.E. Cano, S. Moral, J.F. Verdegay-López: Partial inconsistency of probability envelopes, *Fuzzy Sets and Systems*, 52, 201-216, 1992.
4. R.M. Cooke, T. Bedford: *Expert opinion*, Cambridge Program for Industry, 1996.
5. A.P. Dempster: Upper and lower probabilities induced by a multivalued mapping. *Ann. Math. Statist.* 38, 325-339, 1967.
6. H. Edelsbruner: *Algorithms in Combinatorial Geometry*, Springer-Verlag, Berlin, 1987.
7. S. French: Group consensus probability distribution: A critical survey, *Bayesian Statistics 2*, J.M. Bernardo et al. (Eds.) Elsevier, 183-202, 1985.
8. C. Genest, J.V. Zidek: Combining probability distributions: a critique and an annotated bibliography. *Statistical Science* 1, 114-146, 1986.
9. J.L. Imbert, P. Van Hentenryck: Redundancy elimination with a lexicographic solved form. *Annals of Mathematics and Artificial Intelligence*, 17, 85-106, 1996
10. R. Hummel, M. Landy: Evidence as opinions of experts. In: *Uncertainty in Artificial Intelligence 2*, J.F. Lemmer and L.N. Kanal, eds., North-Holland, 43-53, 1988.
11. R. Hummel, L. Manevitz: Combining bodies of dependent information. *Proc. Tenth Int. Conf. Artificial Intelligence,* Milan, 1015-1017, 1987.
12. R. Hummel, L. Manevitz: Combination calculi for uncertain reasoning: Representing uncertainty using distributions. *Annals of Mathematics and Artificial Intelligence*, 1996.

13. R. Hummel, L. Manevitz: A statistical approach to the representation of uncertainty in beliefs using spread of opinions. *IEEE Transactions on Systems, Man, and Cybernetics - Part A: Systems and Humans*, 26, 378-384, 1996.
14. R. Laddaga: Lehrer and the consensus proposal. *Synthese* 36, 473-477, 1977.
15. C. Lassez, J.L. Lassez: Quantifier elimination for conjunctions of linear constraints via a convex hull algorithm. In: *Symbolic and Numerical Computation for Artificial Intelligence*, B.R. Donald et al. (eds.) Academic Press, London, 103-119, 1992.
16. T.H. Mattheis, D.S. Rubin: A survey and comparison of methods for finding all vertices of convex polyhedral sets. Technical Report N. 77-14, Department of Operations Research and Systems Analysis, University of Carolina at Chapel Hill, 1977.
17. K.J. McConway: Marginalization and linear opinion pools. *J. Amer. Statist. Assoc.* 76, 410-414, 1981.
18. K.C. Ng, B. Abramson: Consensus diagnosis: a simulation study, *IEEE Transactions on Systems, Man and Cybernetics*, Vol. 22, 1992.
19. F.P. Preparata, M.I. Shamos: *Computational Geometry. An Introduction.* Springer Verlag, New York, 1985.
20. G. Shafer: *A Mathematical Theory of Evidence*, Princeton University Press, Princeton, 1976.
21. Ph. Smets: Belief functions, *Non-Standard Logics for Automated Reasoning*, Ph. Smets, E.H. Mamdani, D. Dubois, H. Prade (eds.) Academic Press, London, 253-286, 1988.
22. M. Stone: The opinion pool: *Ann. Math. Stat.*, 32, 1961, 1339-1342.
23. E. Trillas, L. Valverde: On implication and indistinguishability in the setting of fuzzy logic, *Management Decision Support Systems using Fuzzy Sets and Possibility Theory*, J. Kacprzyk and R. Yager, eds., TÜV Rheinland, Köln, 198-212, 1985.
24. P. Walley: The elicitation and aggregation of beliefs. Statistics Research Report. University of Warwick, 1982.
25. P. Walley: *Statistical Reasoning with Imprecise Probabilities.* Chapman and Hall, London, 1991.
26. P. Walley: Measures of uncertainty in expert systems, *Artificial Intelligence*, 31, 1-58, 1996.
27. R.L. Winkler: The consensus of subjective probability distributions. *Management Science*, 15, B61-B75, 1968.
28. R.R. Yager: On the aggregation of prioritized belief structures. *IEEE Transactions of Systems, Man, and Cybernetics*, 26, 708-717, 1996.

Fusion of Image Information under Imprecision

Isabelle Bloch, Henri Maître

Ecole Nationale Supérieure des Télécommunications, département Images
CNRS URA 820
46 rue Barrault, 75013 Paris, France
Tel: +33 1 45 81 75 85 - Fax: +33 1 45 81 37 94 - E-mail : bloch@ima.enst.fr

Abstract. We present in this paper a review of image fusion techniques paying a special attention to the management of uncertainty and imprecision. We investigate the three main numerical methods based on probabilistic reasoning, fuzzy set theory and Dempster-Shafer evidence theory. We show how it is possible to introduce imprecision at the three basic levels of modelling, combining and deciding. We underline the main advantages of each method at these three levels. We also explore some promising fields where innovative works will most certainly take place in the coming years. They concern the management of spatial information within the framework of fusion and require the development of new tools or the extension of yet established ones.

1 Introduction

Data (or information) fusion appeared in the domain of image processing a few years ago, as a necessary stage for several applications like medical imaging, aerial and satellite imaging, quality control, robot vision, vehicle or robot guidance. The need for information fusion in these domains originates from the increasing number of information sources, mostly images issued from different sensors, but other types of information as well (maps, atlases, expert knowledge, etc.). Imprecision and fuzziness being inherent to these pieces of information to merge, data fusion methods have to take this specificity of images into account, in order to improve the decision.

The aim of this paper is to underline the key points of image fusion and the ways to explore the implications of fusion in image interpretation, without providing a deep study of all these points. The aim of this paper is thus to provide an overview of data fusion in image processing under fuzziness, by highlighting how imprecision in the image information is modelled and introduced in the fusion process. Methods for considering spatial imprecision, which are specific to image information fusion, will be presented here. We mainly address all along this paper the problem of numerical fusion.

This paper is organized as follows. In Section 2, we try to classify problems in image fusion according to several criteria and points of view. In Section 3, we describe several aspects that are specific to image fusion, and highlight where imprecision occurs. In Section 4, we shortly mention how probabilistic approaches are used in the domain of image fusion. In our opinion, fuzzy methods are very appropriate to deal with imprecision in image fusion, and this is addressed in Sections 5 and 6. In Section 7, we deal with spatial imprecision, which is one of the main aspects of image fusion.

2 Fusion in image processing

In this Section, we make a guided tour of image information fusion, providing some basic distinction between different classifications of the topic. Then we shortly describe the main fusion schemes and the main steps of the fusion process. Then we take the point of view of the level of representation used for combining information.

2.1 Data fusion / data association

The term data fusion being rather vague, we restrict it to a more precise domain, and distinguish data association and data fusion.

Data association is the action which allows to find from many different signals issued from two or more sources, those which are emitted by the same object (source or target). According to Bar-Shalom and Fortmann [5], data association is the most difficult stage in multi-target tracking. It consists in detecting and associating some noise corrupted measurements, the origins of which are uncertain due to many factors, including random false alarms in the detection, clutter, interfering targets, decoys ot other counter-measures. Although uncertainty has to be understood here in the most general sense, possibly including imprecision and fuzziness, most of the methods developed for data association and tracking are mainly dedicated to the processing of probabilistic uncertainty. The main models used in this field are either deterministic (involving classical hypothesis testing) or probabilistic (relying mainly on the Bayesian framework) [5], [74], [53]. The most common method [5] relies on Kalman filter under Gaussian hypothesis.

This domain will not be further detailed in this paper, since it faces problems rather different from those we have in multi-image fusion. We will restrict our investigations to the domain of data fusion in the following sense: data fusion consists in combining several pieces of information issued from different sources about the same phenomenon, in order to take a better decision on this phenomenon. In image fusion, information is typically provided by several images of a scene. They may come from different sensors, obtained using different imaging techniques or with different acquisition parameters or at different times. Decision can take various forms, ranging from hypothesis evaluation, classification to object recognition, and to scene interpretation and understanding. This definition of data fusion excludes all domains where the pieces of information are not really combined in order to improve decision, like registration, color composition of multi-sources images, visualization, etc. We also assume here the most general case of heterogeneous information, i.e. where no metrics can be defined among sources, therefore excluding the classical data analysis methods [35] based on vectorial representations of data in a convenient metric space.

2.2 Fusion schemes

A general image fusion problem can be stated in the following terms: given l images I_j representing heterogeneous data on the observed phenomenon, take a decision D_i on an element x, where x can be a pixel or any other object extracted from the images (see Section 2.3), and D_i belongs to a decision space $D = \{D_1, ..., D_n\}$ (or set of hypotheses). As opposed to symbolic fusion methods (see e.g. [86], [62], [36]), in numerical fusion methods, the information relating x to each possible decision D_i according to each image I_j is represented as a number M_i^j having different properties and different meanings depending on the mathematical fusion framework. The global fusion scheme consists in taking a decision on x as a function of all $M_i^j, 1 \leq i \leq n, 1 \leq j \leq l$. Typically, M_i^j evaluates how much source j supports decision i. In this scheme, all information is taken into account, but is complex and hardly tractable. Therefore, two degraded systems are often defined, that will be be refered as centralized and distributed schemes in the following.

In the distributed scheme, decision is taken locally on each image I_j separately, from $M_i^j, 1 \leq i \leq n$. Then the local decisions are merged in a global one. Many distributed systems have been developed for real-time and military applications, since partial decision are taken as soon as information is available. This scheme is also called decision fusion [91], [25]. Methods in distributed decision often assume 2 possible decisions only D_0 or D_1 (or sometimes 3, the third one representing no decision). In this case, the problem amounts to find an application f from $\{D_0, D_1\}^l$ into $\{D_0, D_1\}$ [73]. The main methods rely on probabilistic models, empirical estimations based on samples, nearest neighbour rules, etc. In this scheme, imprecision can be taken into account for each sensor or each information source separately. Models are then designed adaptively for each type of information, and the addition of a new source of information is easily done. However, no process in such schemes leads to a cooperation between sensors for reducing imprecision or ambiguity provided by one source by using information provided by another one, except at the decision level, which is quite rough for this task. Moreover, conflicts between local decisions are difficult to solve, since all numerical information on image imperfections is lost[1].

In the centralized scheme, the measures related to each possible decision i and provided by all images are combined in a global (still numerical) evaluation of this decision, taking the form, for each i: $M_i = F[M_i^1, M_i^2, ..., M_i^l]$, where F is a fusion operator. Then a decision is taken from the set of $M_i, 1 \leq i \leq n$. In this scheme, no intermediate decision is taken and the final (binary) decision is issued at the end of the processing chain. Therefore we avoid to take decision at intermediate steps with partial information only, and thus we diminish contradictions and conflicts, which usually require a difficult control or arbitration step. In centralized schemes, imprecision and more generally imperfection of image information is combined at a numerical level, and therefore fusion between

---

[1] One possible solution consists in keeping several possible individual decisions for each source, and in considering all of them for the fusion.

sources can be better exploited to reduce these imperfections before the final decision is taken.

What follows in this paper relies mainly on the centralized scheme, which is for us the most efficient when real time decision is not mandatory[2]. The main steps of image fusion can then be described as [14], [56]: (i) modelling image information and its imprecision, (ii) estimation of the $M_i^j, 1 \leq i \leq n$ according to the chosen mathematical framework, (iii) combination, i.e. choice of an appropriate fusion operator F [11], (iv) decision.

2.3 Fusion levels

In image fusion, like in most image processing domains, several levels of information representation are distinguished. At the lowest level, information carried by each pixel is considered. Fusion schemes at this level aim at taking a decision for each pixel, e.g. attributing each pixel to a class of interest. All raw image information is then used, but such schemes are lacking in structural information and spatial coherence [21]. A precise registration between all images to be combined is generally considered as unavoidable at this level. At an intermediate level, the fusion is performed on features extracted from each image using some image processing tools. These features are typically segments, contours, or regions. Precise registration is then crucial. At the highest level, objects extracted from images are combined. This involves a preliminary recognition step and is therefore related to decision fusion. Structural information can be easily introduced at this level [64], and therefore makes registration less mandatory. However, the richness of the basic numerical information is lost, and conflicts may be difficult to solve, unless some numerical evaluation is attached to the objects and structures extracted from each image. Applications where segmentation results are fused (e.g. [23], [20]) can be considered as intermediate level fusion if the segmented regions or contours have no semantic interpretation (i.e. no real decision is taken on each image), or as high level fusion otherwise.

Imprecision and fuzziness can occur at each level of representation. This will be detailed in the next Section.

3 Imprecision in images

In this Section, we summarize some aspects of image information that have to be taken into account in image fusion. Some of them are very specific to image processing, and have to be introduced in fusion schemes and mathematical frameworks, which are generally issued from other domains. In particular, we focus on imprecision and fuzziness in images.

---

[2] A scheme that can be seen as intermediary between centralized and distributed decision consists in choosing the necessary information among the $M_i^j, 1 \leq i \leq n, 1 \leq j \leq l$ in order to solve some specific decision problems. Such hybrid schemes often involve expert knowledge and are mainly used in rule-based and multi-agent systems.

First, the information to handle in image fusion is often **heterogeneous**. In several applications, several imaging techniques have to be used together to answer a specific question. They provide different aspects and different points of view on the problem by exploiting different physical properties. For instance when planning some surgical operation in medical imaging, necessary images can be as heterogeneous as anatomical images (provided by MRI or CT), angiographic images (MRA, spiral CT, etc.), functional images (PET, functional MRI). These images are not informative about the features they are not dedicated to. Similar examples can be found in other domains, in particular in satellite or aerial imaging. An additional cause of heterogeneity comes from the fact that image information needs to be combined with external information to make sense. This can be information on acquisition conditions, on the observed phenomenon, or more generally expert knowledge related to the problem at hand. This knowledge can be expressed either in "iconic" terms, under the form of atlases or maps, or in "propositional" terms, under the form of linguistic expressions or rules. This is an aspect that is quite specific to image processing. Indeed, unlike in other domains where data fusion is used, especially in artificial intelligence, symbolic information is not only the information injected in the process by the expert in a propositional way. It may also be derived from graphical documents since such representations carry a rather abstract interpretation of the objects contained in the scene. Such information is considered as symbolic, as opposed to the numerical information constituted by location, intensity, etc. that may be found in images.

Duality between numerical and symbolic information emerges at this level. Let us take the example of a map and an aerial image of the same area. The numerical information carried by the image provides a quite accurate description of the scene, but the interpretation attached to it is hard to derive. For instance, it is generally difficult to assess the type of a building, although its drawing on the image is accurate. On the contrary, the map carries symbolic information as a semantic meaning of the objects represented on the map but its shape is often sketchy. Here lies also an example of duality between imprecision and uncertainty (imprecision on location and uncertainty on the nature of the objects). This example illustrates how imprecision may be quite different depending on the level of representation (pixel in an image, object in a map for instance).

The situation is even more complex if we include in this scheme image processing algorithms used to extract information about objects, and the attributes describing the objects. These attributes can be numerical or symbolic, and they contribute to the nature of the objects. Therefore, the information we have to deal with is embedded in a complex net, where images, objects, algorithms, measures, interpretations are closely related to each other. Imprecision can be attached to any of these elements. One of the problems to be solved is therefore to understand and modell the influence, on the nature of one element, of the imprecision associated with the others. In particular, in order to define accurately these imprecisions and their relationships, we found it difficult to provide some reasonable constructive intermediate between generic definitions and naive instances.

Numerical-symbolic fusion and it specificities when applied to image information under imprecision is a very complex problem, that still did not find a definite solution and that remains a very active field of research.

However, a first step towards understanding and management of imprecisions and their relationships can be performed by studying the different **causes of imprecision**. Indeed, imprecision in image information may be due to several factors, ranging from the observed phenomenon to the algorithms artefacts. In biology, a soft transition between tissues (e.g. healthy and pathological tissues) is surely a cause of imprecision inherent to the nature of the observed objects. Also, if tissues have similar characteristics, images that represent these characteristics will poorly discriminate these tissues. This will result in an uncertainty on the belonging of a pixel to one or the other tissue. Another cause of imprecision is due to the discrete nature of numerical images, resulting in a delocalization of information contained in a small volume at only one point. The partial volume effect (the presence of several tissues in one pixel or voxel) belongs also to this type of spatial imprecision. Other image imperfections can be caused by numerical reconstruction algorithms in computed imaging. One example is the Gibbs effect that may appear in MRI around sharp transitions. At the processing level, imprecision is often induced by the chosen algorithms (e.g. filtering, contour detection, registration between images, etc.), another place where the duality between imprecision and uncertainty can be found [81].

Another aspect that is quite specific to image processing, as compared to other application domains of data fusion, is the **complexity of the information**. This is partly due to the previously mentioned characteristics but also to the increasing number of acquisition techniques and to the huge data sets that have to be dealt with. Typically, one MRI brain image contains $256 \times 256 \times 128$ voxels, one satellite image contains 6000×6000 pixels, and several images of this size have to be combined in a fusion process. The large data volumes, and the statistical measures that are therefore made possible, may explain the use of statistical approaches in most image fusion schemes. The complexity of the fusion process also comes from the simultaneous redundancy and complementarity between images, closely related to the heterogeneity aspects. One of the main tasks of image fusion is to exploit redundancy, in order to increase the global information, and complementarity, to improve certainty and precision. The decision is thus improved by the fusion in terms of both quantity and quality.

The main information in image processing, which is specific to this domain, is the **spatial information**. The previous examples have already shown how it appears at all levels, and the types of imprecision attached to it. When using a fusion method, often issued from another domain, we thus have to incorporate spatial information in the process. This point will be addressed in Section 7.

One of the main problems in image fusion (like in other fields of image processing) concerns the **validation** of methods. If the truth generally more or less exists, it is often difficult or impossible to access[3].

---

[3] In that sense, aerial or satellite imaging is probably a better field of investigation

To conlude this Section, let us briefly illustrate another reason why image fusion is so different from other application domains of data fusion. Of course, the nature of information is an important point, and the spatial aspects as well. But, to our opinion, with respect to multi-criteria aggregation and optimization, the main difference relies in the fact that in that field, the aim is to try to find a solution that satisfies as well as possible some flexible constraints. In image fusion, the sensors do provide (more or less explicitely) a degree of satisfaction (of membership to classes for instance) and the decision amounts rather to choose the best one. With respect to voting problems, the difference is that in those problems, no truth exists, and that probably subjectivity has more to be taken into account, but according to some "ethic" rule. Subjectivity is also part of combination of expert opinions. In that domain, information is moreover usually more sparse, which makes learning perhaps more difficult than in image processing. However, this last problem is probably the closest one to our.

4 Probabilistic approaches for dealing with imprecision in image fusion

The most used framework in image fusion is undoubtly the probabilistic framework, and in particular Bayesian approaches.

Modelling and estimation In the Bayesian framework, the M_i^j's represent conditional probabilities: $p(x \in D_i | I_j)$. These probabilities are computed from characteristics extracted from the images, typically grey-levels or texture indices at low level, or other features and object properties at higher level. They represent mainly the probabilistic uncertainty attached to image information. One of the advantages of Bayesian fusion relies on the large experience on learning that allows the user to perform estimation of these conditional probabilities.

Uncertainty may be expressed in three different ways depending on our a priori knowledge of the system:

- as extreme values which limit the possible excursion of parameters,
- as confidence intervals where the parameter is supposed to have a uniform probability,
- as an exact distribution law (often reduced to a normal distribution).

Another advantage of probability originates from the notion of entropy, of conditional entropy and of mutual information [45], [50], [57]. They can be efficiently used in order to assess the complementarity and redundancy between images at any level of representation. Let us just mention one example, where information extracted at several levels and different spatial scales are fused using mutual information between scales in order to provide an interpretation of aerial images [94]. Although similar entropy concepts have been defined for the types

---

than medical imaging. At least, it allows an easier evaluation.

of information represented by fuzzy sets or belief functions (see e.g. [55], [17], [87]), they are still not used in image fusion.

Because of the very strict dependencies between the many propositions of a probabilistic modelling, too complex systems cannot be developed without care in this framework. Arbitrary probabilities, inconsistent propositions, and logical dead locks are the most probable pittfalls of very large probabilistic systems. To avoid these limitations Bayesian networks have been proposed [68].

Combination In the Bayesian framework, the combination of image information as well as of prior probabilities is performed through the Bayes theorem. In image processing, due to the difficulty to learn joint distributions[4], independence between sources is often assumed. However, statistical tests show that this is seldom the case, and that better results can be obtained without independence assumption [21].

Bayesian methods are probably the most widely used in probabilistic image fusion. They lead mainly to conjunctive data fusion. Other probabilistic techniques have been proposed, that are able to model different kinds of fusion logic (e.g. [70]).

Decision The most used decision rule in Bayesian decision is the maximum a posteriori. However, many other criteria have been developed by probabilists and statisticians, including maximum posterior marginal, maximum likelihood, maximum entropy, minimum expected risk, etc. Each of them is answering a specific problem which better fits the user's demand or the user's constraints. Quality of decision can be estimated using confidence levels, and rules leading to ambiguity and distance rejection have been proposed. Such rules are widely used in image pattern recognition, either as tables or criteria to validate the decision. For instance using Markov random fields, probabilistic approaches may allow not only to decide on the local best solution, but also on the global one for the complete images to be processed.

5 Fuzzy approaches for dealing with imprecision in image fusion

5.1 Modelling and information representation

In the framework of fuzzy sets and possibility theory [97], [98], [31], the M_i^j's represent membership degrees to a fuzzy set or possibility distributions. One possible model consists in setting, for each element x, $M_i^j(x) = \mu_i^j(x)$, where $\mu_i^j(x)$ denotes the membership degree of x to the class i according to image j. A global evaluation of the membership of x to class i is then provided by the combination of $\mu_i^j(x), 1 \leq j \leq l$. Another model consists in interpreting $M_i^j(x)$ as a possibility

---

[4] When combining n images with 256 grey levels, probability tables of dimension 256^n are needed.

degree that x belongs to class j, and therefore setting $M_i^j = \pi_i^j$, where π_i^j is a possibility distribution according to image j^5. Such models explicitely represent imprecision in the information provided by the images, as well as possible ambiguity between classes or decisions (for instance, it is possible to evaluate the possibility and necessity of any subset A of D: $\Pi(A) = \sup_{d \in A} \pi(d)$).

Fuzzy sets have powerful features when used specifically for image information processing [48], in particular for three points:

- the ability of fuzzy sets to represent spatial information in images along with its imprecision,
- the adequacy of operations recently generalized to fuzzy sets in order to manage spatial information,
- and the flexibility of information fusion using fuzzy combination operators.

The two first points will be detailed in this Section, and the third point in the next one.

Some examples of applications using fuzzy sets have already shown excellent results, even if applications of fuzzy logic and fuzzy sets to image processing are quite recent, compared with other fields like control theory. Fuzzy sets techniques give rise now to a large development in image processing. This may be explained by several factors, and in particular by the ability of fuzzy sets to represent and to manage imprecise spatial information, which is essential for image fusion and decision. Another factor is that this development can take advantage of the degree of maturity of both image processing and fuzzy logic fields.

Fuzzy sets have several advantages for representing imprecision inherent to image processing. First, they are able to represent several types of imprecision in images, as, for instance, imprecision in spatial location of objects, or imprecision in membership of an object to a class. In the case of partial volume effect, which occurs frequently in medical imaging, a consistent representation is given with fuzzy sets (membership degrees of a voxel to tissues or classes directly represent partial membership to the different tissues mixed up in this voxel, leading to a modelling very close to reality). Second, image information can be represented at different levels with fuzzy sets (local, regional, or global), as well as under different forms (numerical, or symbolic). For instance, classification based only on grey levels involves very local information (at the pixel level); introducing spatial coherence in the classification [18] or relationships between features like parallelism [47] involves regional information; and introducing relationships between objects or regions for scene interpretation involves more global information [72]. Third, the fuzzy set framework allows for the representation of very heterogeneous information, and is able to deal with information extracted directly from the images, as well as with information derived from some external knowledge, like expert knowledge for instance. This is exploited in particular in model-based pattern recognition (see e.g. [72], [27]), where fuzzy information extracted from

---
[5] Approaches based on fuzzy sets consider classes as intrinsically fuzzy, while approaches based on possibility theory consider them as crisp.

images is compared and matched to a model representing knowledge expressed in fuzzy terms.

The representation of image information using fuzzy sets can be seen from two different points of view: one leads to a symbolic approach and the other to a numerical one. The first one already gave rise to large developments, whereas the second one is worth to be amplified and extended to other problems than the traditional classification problem. In the first approach, knowledge, reasoning, rules are represented using concepts of fuzzy logic, but the basic spatial information in image is represented in a traditional way [27]. This approach is used for reasoning about image information by means of fuzzy rules and fuzzy external knowledge, for instance for scene interpretation tasks by fusing several sources of information. However, the specificity of image information, i.e. spatial information, is not explicitely taken into account using this approach. In the second approach, fuzzy sets are used to represent directly spatial structures in images (regions, classes, contours). The modelling consists in assigning, to each image point, a membership degree to a class of interest. This approach was adopted mostly for fuzzy classification or clustering for pattern recognition ([8] and the widely used fuzzy C-means algorithm, [69]), and for measures on fuzzy sets (set operations, topological or geometrical characteristics, issued from the works of Zadeh [97] and Rosenfeld [76]). Both approaches have advantages the combination of which can be very fruitful: the first, symbolic, approach can take advantage of the granularity concept that allows to represent symbolic information by linguistic variables for instance taking only a few values. For example, considering distance between objects, it can be evaluated using a few fuzzy sets representing "far of", "near to" (e.g. [27]), the same for color description [7]. The second, numerical, approach uses fuzzy sets through their spatial interpretation, and is therefore very close to the basic information in image processing. Again considering distances with this approach, the problem amounts to defining a distance between two fuzzy sets, where each fuzzy set defined on the image represents an object or a region. Distance is an example of useful relationships in image processing which gains from being expressed with fuzzy sets. Since the introduction of fuzzy sets, several measures or operators have been defined on fuzzy sets. They are able to take into account several aspects of the information, that are widely used in image processing, like set theoretical, geometrical, topological, morphological characteristics [75], [76], [67], [49], [100], [15]. They have helped in several applications in particular for image segmentation (e.g. [67], [26]). We distinguished two kinds of spatial operations in [10]. Those well defined if the objects are crisp, like adjacency, inclusion or other set relationships may benefit from fuzzifying these concepts when dealing with imprecise spatial image information, since they are highly sensitive to errors or imprecision in segmentation. Other relationships are inherently vague concepts, like relative position. Fuzzy definitions of such relationships are more consistent than crisp ones.

5.2 Combination

For the combination step in the fusion process, the advantages of fuzzy sets and possibilities rely in the variety of combination operators, which may deal with heterogeneous information [32], [95], [34]. We proposed a classification of these operators with respect to their behaviour (in terms of conjunctive, disjunctive, compromise [32]), the possible control of this behaviour, their properties and their decisiveness, which proved to be useful for several applications in image processing [11]. It is of particular interest to note that, unlike other data fusion theories (like Bayesian or Dempster-Shafer combination), fuzzy sets provide a great flexibility in the choice of the operator, that can be adapted to any situation at hand. Indeed, image fusion has often to deal with situations where an image is reliable only for some classes, or does not provide any information about some class, or is not able to discriminate between two classes while another does. In this context, some operators are particularly powerful, like operators that behave differently depending if the values to be combined are of the same order of magnitude or not, if they are small or high, and operators that depend on some global knowledge about source reliability about classes, or conflict between images (global or related to one particular class) [34]. Again, the combination process can be done at several levels of information representation, from pixel level to higher level. Whatever the level of representation, such a process corresponds to the numerical approach described above. A noticeable advantage of this approach is that it is able to combine heterogeneous information, like it is usually the case in multi-image fusion, and to avoid to define a more or less arbitrary and questionable metric between pieces of information issued from these images. Another scheme relies on the symbolic approach and consists in combining information described from images but not directly attached to pixels or regions. This is typically used for combining rules where the rule elements involve fuzzy attributes, fuzzy measures or fuzzy relationships. Some authors use a neural architecture for such schemes, that allows for learning and several combination operations (see e.g. [60]).

5.3 Decision

Decision is usually taken from the maximum of membership values after the combination step. Constraints can be added to this decision, typically for checking for the reliability of the decision (is the obtained value high enough?) or for the discrimination power of the fusion (is the difference between the two highest values high enough?). In image fusion, very few works addressed this decision problem, unlike in probabilistic approaches, and very few tools allow to take refined decisions.

5.4 Applications of fuzzy image fusion

The concept of data fusion (combination of several images as well as combination of several measures extracted from an image) is, in our opinion, a central point in the application of fuzzy sets to image processing, and it is found in almost all applications. For preprocessing, it is typically possible to make use of fuzzy fusion operators or fuzzy rules for combining local or regional measures of homogeneity and measures of contrast in order to smooth homogeneous regions while preserving or even sharpening edges (see e.g. [78]). Fuzzy classification is up to now the widest developed application of fuzzy sets for image processing. Most applications make use of the variety of information that can be extracted from one or several images, and therefore, such applications are related to data fusion. At pixel or intermediate level, applications can be found for color image classification [42], for medical imaging [27], [18], satellite imaging [21], [60], etc. Low level information, represented as membership degrees to a class, can be combined with some more global information like uncertainty due to feature acquisition, and credibility of a set of features to provide reliable information concerning each class. This has been done using possibility integrals e.g. in [96], [90]. When several levels of representation have to be combined, the most widely developed approach consists in using fuzzy rules (see e.g. [41] for medical image classification, [46] for combining numerical information with model-based and knowledge-based information, [60] for multi-spectral satellite image classification). Model-based recognition is a more recent application of fuzzy sets to image processing, that is likely to lead to a large development in a near future. Indeed, it allows to combine low and intermediate level spatial information with a more structural information, for instance provided by spatial relationships like described in the previous section. The problem amounts to the (fuzzy) matching between image information and models. Structural descriptions like graphs and fuzzy graphs are often used for this task [28], [72]. The matching requires the definition of a similarity measure combining the different features of interest using different levels of representation. Let us mention applications for recognition of anatomical structures in medical imaging [27], recognition of facial expressions [71], character recognition [58], partial shape matching [66], space shuttle simulator panel [72], flexible interrogation of digital angiography database [44], all of them involving aggregation of fuzzy information.

6 Fuzzy and evidential approaches for dealing with imprecision in image fusion

Dempster-Shafer evidence theory (DS) has been already widely used in satellite image processing (see e.g. [52], [93], [24], [99]). On the contrary, when considering medical applications, it seems that DS has been applied mainly for topics derived from artificial intelligence (DS actually issued from this domain, for reasoning under uncertainty), and thus addresses the medical diagnosis, where propositional representations of evidences and knowledge are set in the DS framework [82], [39], [4]. Evidences issued from the image signal (of "iconic" type rather than propositional) are seldom considered. To our knowledge, only a few papers report on image fusion by DS in medical imaging [51], [22] (both dealing with brain MRI), [89] (applied for MRI left ventricle). However, DS offering a number of advantages for image fusion deserves to be applied, also in medical imaging [12].

6.1 Modelling and information representation

DS, as possibility theory, allows to represent both imprecision and uncertainty, using plausibility and belief functions derived from a mass function defined on 2^D rather than on D only [80], [40], [83]. This is one of the main advantages of the DS approach. Indeed, it leads to a very flexible and rich modelling, able to fit a very large class of situations, occurring in particular in image fusion. A few examples of situations where DS theory may be successfully used are:

- in ideal cases where all information relevant to the problem is known, i.e. cases where Bayesian fusion applies;
- when a source provides information concerning only a few of several classes: for instance brain PET images under some conditions allow for the detection of the brain surface but not of the head surface;
- when a source differentiates two classes and another does not: DS allows to deal with hesitation or ambiguity between these two classes;
- in the case of partial volume effects (often at the border of classes): it can also be taken into account by assigning masses to the union of the two classes mixed in the considered area;
- in cases where global source reliability has to be taken into account: this may be done by weakening all masses and reinforcing $m(D)$, using a discounting process;
- in cases where knowledge of source reliability is available only for some classes: it can be taken into account by modifying accordingly the masses assigned to these classes and by introducing ignorance (for instance, PET images are not very reliable concerning anatomical information whereas MRI images are more reliable);
- in cases where a priori information has to be introduced: even if it is not represented in a probabilistic manner, it can be taken into account if it induces a way to assign masses, in particular to compound hypotheses; for instance if we know that a source is not able to distinguish between two classes, then it is not worth trying to estimate masses for these two classes separately, but the estimation has to be made on the union of the classes and mass is then assigned to the corresponding compound hypothesis.

To our opinion, one of the main differences between DS and fuzzy sets lies at the level of flexibility. The situations described above are easily taken into account in the DS framework at the modelling level, while in the fuzzy set framework this flexibility relies on the wide range of available combination operators.

6.2 Estimation

The definition of mass functions remains a largely unsolved problem, which did not yet find a general answer. In image processing, they may be derived at three different levels. At the highest, most abstract level, information representation is used in a way similar to that in artificial intelligence and masses are assigned to propositions, often provided by experts [65], [4], [39]. Up to now, this kind of information is usually not derived from measures on the images. At an intermediate level, masses are computed from attributes, and may involve simple geometrical models [22], [93], [24], [1]. This is well adapted to model-based pattern recognition but it is difficult to use for image fusion classification of complex structures without a model. At the pixel level, mass assignment is inspired from statistical pattern recognition. The most widely used approach is as follows: masses on simple hypotheses are computed from probabilities or from the distance to a class centre [52], [88], [89], [51], [54], [99], [2], [43]. Then a global ignorance $m(D)$ is introduced as a discounting factor, often as a constant on all pixels (e.g. [51], [99]). In most cases no other compound hypothesis is considered, and this drastically under-exploits the power of DS. One way to overcome this problem consists in using the decomposability property on the masses and the Barnett approach (only masses on singletons and on their complements are considered) [6], [51], [22], [39]. Although this approach may be interesting for pattern recognition purposes where each class has to be tested against all the others, it may however be questionable when a source provides information on a compound hypothesis which is neither a singleton nor the complement of a singleton. In [52], the use of mixed pixels in a probabilistic like approach is suggested. In [89], nested focal elements are used, by sorting probabilities. In our opinion, these approaches are quite promising but are also restrictive since they do not allow to take into account all occurring situations. The way we assigned masses in the example presented in [12] is based on a reasoning approach where knowledge about the information provided by each image is used to choose the focal elements. For instance, we just consider a disjunction of two classes for images that are not able to discriminate between these classes. Partial volume effect is modelled in a similar way. However, in case of large numbers of classes, this process would become too tedious, and unsupervised methods are needed, for instance like the one proposed in [61] for SAR imaging or in [63] for fusion of several classifiers.

6.3 Combination

In the DS framework, masses are combined by the orthogonal rule of Dempster [80]. For m_j being the mass function associated with source j ($j = 1...l$), this rule is written, for all non-empty subset A of D:

$$(m_1 \oplus m_2 \oplus ... \oplus m_l)(A) = \frac{\sum_{B_1 \cap ... \cap B_l = A} m_1(B_1)m_2(B_2)...m_l(B_l)}{1-k},$$

and $(m_1 \oplus m_2 \oplus ... \oplus m_l)(\emptyset) = 0.$ (1)

The combination is defined if $k < 1$, where:

$$k = \sum_{B_1 \cap ... \cap B_l = \emptyset} m_1(B_1)m_2(B_2)...m_l(B_l). \qquad (2)$$

Similar equations can be derived for directly combining belief or plausibility functions. To some extent, k can be interpreted as a measure of conflict between the sources and is directly taken into account in the combination as a normalization factor. It represents the mass which would be assigned to the empty set if masses were not normalized. It is very important to take this value into account for evaluating the quality of the combination: when it is high (in case of strong conflict: $k \approx 1$), the combination may not make sense and may lead to questionable decisions (moreover, the combination rule is not continuous if k is very close to 1 [33]). Several authors suggest not to normalize the combination result (see e.g. [84]).

This fusion operator has a conjunctive behaviour[6]. We claim that this means that all imprecision on the data has to be introduced explicitly at the modelling level, in particular in the choice of the focal elements. For instance, ambiguity between two classes in one image has to be modelled using a disjunction of hypotheses, so that conflict with other images can be limited and ambiguity can be possibly solved during the combination.

6.4 Decision

After the combination, the final decision is usually taken in favour of a simple hypothesis using one of several rules [29]: for instance, the maximum of plausibility (generally over simple hypotheses), the maximum of belief, the maximum of belief without overlapping of belief intervals, i.e. in favour of $d \in D$ such that $Bel(d) \geq \max_{d' \in D, d' \neq d} Pls(d')$ (a very strict condition), the pignistic decision rule [85], or rules using expected utility [88].

In [12], the decisions have been taken according to two rules: a classical one, where a decision is always taken in favour of a simple hypothesis, and a second, original one, where it can also be decided in favour of a compound hypothesis. This second rule has several advantages: it is robust with respect to evidence weighting, as seen when testing different weights on mass functions and on partial volume, and it fits reality by highlighting for instance regions with partial volume effect. Moreover it is adapted to the expert's way of reasoning.

7 Taking spatial imprecision into account in image fusion

As stated before, spatial information is an essential aspect of image information, which suffers from several types of imprecision (delocalization at each pixel, imprecision in location or in registration, imprecision in delineation of objects, in descriptions and interpretations that can be extracted from the images, etc.). In this Section, we investigate how spatial information can be integrated in a fusion process, and propose some solutions to take into account the imprecision attached to it using mathematical morphology.

---

[6] Other combination rules have been proposed in the DS framework, in particular disjunctive rules [83].

One of the main stakes is to guarantee that the resulting decision is spatially consistent. As a simple example, let us mention the example of multi-source classification, where isolated pixels of a class embedded in another class have to be avoided. Such knowledge on the spatial coherence of the expected result is often related to the application at hand. Here, we suggest to consider spatial context as an additional source of information. It is not directly image information although it is issued from the very image nature, but is of a different type, often related to expert knowledge. In our opinion, this consideration should allow to develop fusion schemes where spatial context is introduced (possibly at the same level than other pieces of information) in the fusion process like any other information.

7.1 Introducing spatial information in image fusion

There are several ways to introduce spatial information in a fusion process depending on the level of representation. At pixel level, spatial context consists of a few pixels in the neighbourhood of the considered point. At feature level (e.g. segments, contours, regions), spatial context involves relationships between features, and contains more structural information, like parallelism, similarities in directions, continuity, etc. (see e.g. [92]). At object level, spatial context involves topological and geometrical relationships between objects.

Spatial context can also be introduced at all steps of the fusion process. At the first modelling and estimation step, spatial context can help in validating the measures describing the belonging of an element to a class. This has been done especially at pixel level, for instance using statistical approaches, using measures performed not only at a point, but also in a neighbourhood of this point. At the next combination step, our interpretation of spatial context as additional source of information is particularly useful, since it can be combined with image information using the fusion operator. However, this aspect has not been widely developed until now, except in probabilistic methods. For the other methods, it is still an open field for research. At the final decision step, the most common way to introduce spatial context relies on some rules. For instance in [18], local spatial information is taken into account in cases where the decision is not reliable or not discriminating. In such schemes, spatial context can be seen as an additional information for taking a decision on rejected points. For instance in fuzzy fusion, a decision at an element x can be rejected if:

$$\max_i \mu_i(x) < \text{decision-threshold, or:}$$

$$\max_i \mu_i(x) - \max_i 2\mu_i(x) < \text{discrimination-threshold}$$

which express that the decision is not enough supported (distance rejection) or not discriminating enough (ambiguity rejection) (with max 2 representing the second highest value). In such cases, one common rule consists in assigning the considered point to the most represented class in a neighbourhood $\mathcal{V}(x)$ of x, for instance:

$$x \in C_i \text{ if } |\{y \in \mathcal{V}(x), y \in C_i\}| \geq |\mathcal{V}(x)|/2$$

Similar examples can be found in Dempster-Shafer fusion. Rejection rules can be stated for instance as:

$$\max_A Bel(A) < \text{decision-threshold},$$

and similar reclassification can be performed (see e.g. [61]). Until know, such rules are quite heuristic.

Finally, spatial information can be introduced in different mathematical frameworks, and therefore under different forms. This is described in the following.

7.2 Probabilistic methods

The oldest probabilistic method for assuring spatial coherence is found in relaxation schemes [77]. The principle consists in updating probabilities of a point to belong to a class D_i using the probability of neighbourhood points to belong to classes that are compatible (in a topological sense) with D_i. If we denote by $p^{(k)}(x \in D_i)$ this probability at iteration k, and by $r_{xy}(D_i, D_j)$ a degree of compatibility expressing if $x \in D_i$ is consistent with $y \in D_j$, relaxation can be expressed as:

$$p^{(k+1)}(x \in D_i) = \alpha p^{(k)}(x \in D_i)[1 + \sum_{y \in \mathcal{V}(x)} d_{xy} \sum_j r_{xy}(D_i, D_j) p^{(k)}(y \in D_j)]$$

Another approach has been proposed in the framework of Markov random fields [37]. Bayesian decision is taken from the posterior probability:

$$p(x \in D_i | I_1, I_2, ..., I_l) \propto p(I_1, I_2, ..., I_l | x \in D_i) p(x \in D_i),$$

and in cases of independant sources:

$$p(x \in D_i | I_1, I_2, ..., I_l) \propto \Pi_{j=1}^n p(I_j | x \in D_i) p(x \in D_i).$$

In image fusion, Markov random fields are often used by stating that the prior probability $p(x \in D_i)$ depends only on a neighbourhood $\mathcal{V}(x)$ of x. Therefore, the fusion appears as a conjunctive fusion of l sources and a spatial regularization term.

This is probably currently the most common approach, and it has been applied at several levels. At local level, a lot of examples can be found (see e.g. [30], [3]). At a more structural level, this approach has been applied on graphs (neighbourhood is then defined in terms of graphs and no more at pixel level), at feature level (see e.g. [92] for road detection on SAR images, [38] for segmentation of brain MR images using adjacency graphs) or at object level (see e.g. [64] for urban area interpretation from an aerial image using a map, [59] for recognition of cortical sulci using structural anatomical models).

Although similar approaches could be theoretically developed with fuzzy sets or belief functions, this has not been widely addressed in the literature to our knowledge.

A promising research field consists in combining imprecision represented as fuzzy sets with uncertainty represented as probabilities. For instance, in [19] and [79], fuzzy sets are used together with a probabilistic Markovian approach, allowing to take into account fuzzy classes and mixed pixels as well as contextual spatial information.

7.3 Fuzzy and evidential methods: using fuzzy mathematical morphology

Spatial information could be introduced at combination level in fuzzy fusion by defining a membership of x to each class D_i according to the neighbourhood $\mu_i^\mathcal{V}(x)$ of x^7, which is considered as an additional source of information. The combination then becomes:

$$\mu_i(x) = F[\mu_i^1(x), \mu_i^2(x), ..., \mu_i^n(x), \mu_i^\mathcal{V}(x)].$$

The problem of the choice of the operator will be even more crucial than in the standard scheme. Here again, it can rely on considerations about desired behaviour, properties, etc. This scheme can be applied at low level (pixel) or at higher level by introducing in $\mu_i^\mathcal{V}(x)$ also spatial fuzzy relationships (distances, relative position, adjacency, etc.) at feature or object level.

We think that such fusion schemes deserve to be investigated.

At modelling and estimation step, we suggest to introduce imprecision on spatial information using fuzzy mathematical morphology [15]. A general principle for defining fuzzy mathematical morphology relies on the translation of set equations defining morphological operations on binary sets into their functional (or fuzzy) equivalents [15] using t-norms and t-conorms. We obtain then for the basic operators excellent properties with respect to mathematical morphology and with respect to fuzzy sets. Moreover, these definitions inherit the properties of t-norms and t-conorms in terms of data fusion, reasoning under uncertainty, and decision making.

A first application concerns modelling of spatial imprecision due to misregistration. We suggested in [16] to introduce imprecision in registration between two modalities using fuzzy dilation. In this way we reduce contradictions or conflicts between data and allow the use of a conjunctive fusion operator, reducing imprecision during the combination process. An example of this method concerns 3D reconstruction of blood vessels based on fusion of digital angiography and endovascular echography data, without any geometrical a priori of the vessel model [16]. For this application, fuzzy mathematical morphology proved to be very useful for introducing the imprecision on the geometrical parameters in an efficient way: the different positions of a point in the 3D space, along with their possibility degrees (depending on the imprecision on the geometrical parameters) are represented through a fuzzy structuring element, and included in the reconstruction through a fuzzy dilation. This application shows also how

---

[7] The neighbourhood can typically be defined according to the spatial extent of the information we want to take into account around a point.

a fuzzy structuring element can be built directly from the data, without any arbitrary choice.

Another example of the use of fuzzy morphology can be given in the framework of Dempster-Shafer theory. One of the difficulties in using DS for image fusion consists in estimating mass functions on compound hypotheses, to provide a proper representation of imprecision in the information provided by each image (spatial imprecision, imprecision in grey-level characteristics, ability of an image source to discriminate between some classes or not, etc.). It happens that fuzzy mathematical morphology provides useful tools for introducing imprecision in the mass functions while estimating compound hypotheses. The basic idea is that fuzzy erosion and dilation (resp. opening and closing) are dual with respect to complementation, and therefore can be interpreted as belief and plausibility functions. Starting from an initial estimate of simple hypotheses, we have proposed in [13] to derive expressions for belief and plausibility by computing fuzzy erosion and dilation (or opening and closing), from which new mass functions are deduced, both on simple and compound hypotheses, while taking into account the imprecision modelled as a fuzzy structuring element. We proved that the obtained belief functions satisfy all expected properties. Appropriate choices of the fuzzy structuring element allow to propagate information in a controlled way. For instance, if the plausibility that a point belongs to a class is high, then the effect of dilation will be to increase the plausibility that its neighbours belong to the same class. Several schemes are possible for applying this approach. As our method is particularly simple for 2-class problems, a first scheme consists in deriving, from each image, estimates of each class against all the others. A second one relies on successive refinements of the space of discernement. A third scheme consists in applying directly an extension of the method to more than two classes.

Finally, fuzzy mathematical morphology is also involved in the definition of several spatial relationships, both at low level and at structural level (e.g. [9], [10]). Such relationships are more and more involved in image fusion problems, particularly at higher level, in order to guaranty a global coherence to the interpretation of the scene.

8 Conclusion

We have seen the large variety of techniques used in image fusion under imprecision. This proliferation is due to the many different tasks which contribute to the decision from the management of different sources of information. Probabilistic approaches still remain the most widely used methods, because of the exceptional achievement of the tools developed under its umbrella. They benefit from a very long experience and may drive the user from the very elementary modelling stage up to decision. At this level (decision making) probability theory reveals as the most efficient tool, the modelling one being on the contrary rather questionable. Fuzzy set theory proposes a quite intuitive modelling well adapted to ill-posed problems. Its decision stage, as far as image processing applications are concerned, is unfortunately still rather coarse and most of the user's knowledge has to be put at the fusion stage, to model with adapted operators its knowledge on the problem. DS evidence theory has the most powerful modelling tools with capacities to handle uncertainty and imprecision as well. It is rather rigid in its way to combine information and allows up to now only few different strategies at the decision level. Manipulating spatial information has not yet been so much investigated in most of the fusion theories. Nevertheless it will certainly deserve a special attention in the very near future with two possible different tracks: one where extent of information propagation is controlled only with the image content (i.e. Markov random fields techniques), the other one where geometrical and topological constraints are the mere guidelines of propagation (i.e. morphological techniques). The future will tell which one is the most successful.

References

1. K .M. Andress and A. C. Kak. Evidence Accumulation and Flow Control in a Hierarchical Spatial Reasoning System. *AI Magazine*, pages 75–94, 1988.
2. A. Appriou. Formulation et traitement de l'incertain en analyse multi-senseurs. In *Quatorzième Colloque GRETSI*, pages 951–954, Juan les Pins, 1993.
3. L. Aurdal, X. Descombes, H. Maître, I. Bloch, C. Adamsbaum, and G. Kalifa. Fully Automated Analysis of Adrenoleukodystrophy from Dual Echo MR Images: Automatic Segmentation and Quantification. In *Computer Assisted Radiology CAR'95*, pages 35–40, Berlin, Germany, June 1995.
4. J. F. Baldwin. Inference for Information Systems Containing Probabilistic and Fuzzy Uncertainties. In L. Zadeh and J. Kacprzyk, editors, *Fuzzy Logic and the Management of Uncertainty*, pages 353–375. J. Wiley, New York, 1992.
5. Y. Bar-Shalom and T. E. Fortmann. *Tracking and Data Association*. Academic Press, San Diego, 1988.
6. J. A. Barnett. Computational Methods for a Mathematical Theory of Evidence. In *Proc. of 7th IJCAI*, pages 868–875, Vancouver, 1981.
7. E. Benoit and L. Foulloy. Capteurs flous multicomposantes : applications à la reconnaissance des couleurs. In *Les Applications des Ensembles Flous*, pages 167–176, Nîmes, France, October 1993.
8. J. C. Bezdek. *Pattern Recognition with Fuzzy Objective Function Algorithms*. Plenum, New-York, 1981.
9. I. Bloch. Distances in Fuzzy Sets for Image Processing derived from Fuzzy Mathematical Morphology. In *Information Processing and Management of Uncertainty in Knowledge-Based Systems*, pages 1307–1312, Granada, Spain, July 1996.
10. I. Bloch. Fuzzy Spatial Relationships: A Few Tools for Model-based Pattern Recognition in Aerial Images. In *SPIE/EUROPTO Conference on Image and Signal Processing for Remote Sensing*, volume 2955, pages 141–152, Taormina, Italy, September 1996.
11. I. Bloch. Information Combination Operators for Data Fusion: A Comparative Review with Classification. *IEEE Trans. on Systems, Man, and Cybernetics*, 26(1):52–67, 1996.

12. I. Bloch. Some Aspects of Dempster-Shafer Evidence Theory for Classification of Multi-Modality Medical Images Taking Partial Volume Effect into Account. *Pattern Recognition Letters*, 17(8):905–919, 1996.
13. I. Bloch. Using Fuzzy Mathematical Morphology in the Dempster-Shafer Framework for Image Fusion under Imprecision. In *IFSA '97*, Prague, 1997.
14. I. Bloch and H. Maître. Fusion de données en traitement d'images : modèles d'information et décisions. *Traitement du Signal*, 11(6):435–446, 1994.
15. I. Bloch and H. Maître. Fuzzy Mathematical Morphologies: A Comparative Study. *Pattern Recognition*, 28(9):1341–1387, 1995.
16. I. Bloch, C. Pellot, F. Sureda, and A. Herment. 3D Reconstruction of Blood Vessels by Multi-Modality Data Fusion using Fuzzy and Markovian Modelling. In *CVRMed'95*, pages 392–398, Nice, France, April 1995.
17. B. Bouchon-Meunier and R. R. Yager. Entropy of Similarity Relations in Questionnaires and Decision Trees. In *Second IEEE Int. Conf. on Fuzzy Systems*, pages 1225–1230, San Francisco, California, March 1993.
18. N. Boujemaa, G. Stamon, J. Lemoine, and E. Petit. Fuzzy Ventricular Endocardium Detection with Gradual Focusing Decision. In *14th IEEE EMBS Conference*, pages 1893–1894, Paris, France, 1992.
19. H. Caillol, A. Hillion, and W. Pieczynski. Fuzzy Random Fields and Unsupervised Image Segmentation. *IEEE Trans. on Geoscience and Remote Sensing*, 31(4):801–810, 1993.
20. B. Charroux. *Image Analysis: Interpretation-Guided Cooperation between Segmentation Operators (in French)*. PhD thesis, University Paris XI, January 1996.
21. S. Chauvin. *Evaluation of Decision Theories Applied to Data Fusion in Satellite Imaging (in French)*. PhD thesis, ENST and Nantes University, December 1995.
22. S. Y. Chen, W. C. Lin, and C. T. Chen. Evidential Reasoning based on Dempster-Shafer Theory and its Application to Medical Image Analysis. In *SPIE*, volume 2032, pages 35–46, 1993.
23. C. C. Chu and J. K. Aggarwal. The Integration of Image Segmentation Maps using Region and Edge Information. *IEEE Trans. on Pattern Analysis and Machine Intelligence*, 15(12):1241–1252, 1993.
24. P. Cucka and A. Rosenfeld. Evidence-based Pattern Matching Relaxation. Technical Report CAR-TR-623, Center of Automation Research, University of Maryland, May 1992.
25. B. V. Dasarathy. Fusion Strategies for Enhancing Decision Reliability in Multi-Sensor Environments. *Optical Engineering*, 35(3):603–616, March 1996.
26. S. Dellepiane, F. Fontana, and G. Vernazza. A Robust Non-Iterative Method for Image Labelling using Context. In *IEEE Int. Conf. on Image Processing*, volume II, pages 207–211, Austin, Texas, November 1994.
27. S. Dellepiane, G. Venturi, and G. Vernazza. Model Generation and Model Matching of Real Images by a Fuzzy Approach. *Pattern Recognition*, 25(2):115–137, 1992.
28. C. Demko, P. Loonis, and E. H. Zahzah. Isomorphism of Fuzzy Structures: a New Method for Image Classification. In *9th Scandinavian Conference on Image Analysis*, pages 297–304, Uppsala, Sweden, June 1995.
29. T. Denœux. A k-nearest Neighbor Classification Rule based on Dempster-Shafer Theory. *IEEE Trans. on Systems, Man and Cybernetics*, 25(5):804–813, 1995.
30. X. Descombes, M. Moctezuma, H. Maître, and J.-P. Rudant. Coastline Detection by a Markovian Segmentation in SAR Images. *Signal processing*, 55(1):123–132, November 1996.

31. D. Dubois and H. Prade. *Fuzzy Sets and Systems: Theory and Applications*. Academic Press, New-York, 1980.
32. D. Dubois and H. Prade. A Review of Fuzzy Set Aggregation Connectives. *Information Sciences*, 36:85–121, 1985.
33. D. Dubois and H. Prade. Representation and Combination of Uncertainty with Belief Functions and Possibility Measures. *Compu. Intell.*, 4:244–264, 1988.
34. D. Dubois and H. Prade. Combination of Information in the Framework of Possibility Theory. In M. Al Abidi et al., editor, *Data Fusion in Robotics and Machine Intelligence*. Academic Press, 1992.
35. R. Duda and P. Hart. *Pattern Classification and Scene Analysis*. Wiley, New-York, 1973.
36. S. J. Gee and A. M. Newman. RADIUS: Automating Image Analysis Through Model-Supported Exploitation. In *Image Understanding Workshop*, pages 185–196, Washington D.C., 1993.
37. S. Geman and D. Geman. Stochastic Relaxation, Gibbs Distribution and the Bayesian Restoration of Images. *IEEE trans. on Pattern Analysis and Machine Intelligence*, PAMI-6:721–741, 1984.
38. T. Géraud, J.-F. Mangin, I. Bloch, and H. Maître. Segmenting Internal Structures in 3D MR Images of the Brain by Markovian Relaxation on a Watershed Based Adjacency Graph. In *ICIP-95*, pages 548–551, Washington DC, October 1995.
39. J. Gordon and E. H. Shortliffe. A Method for Managing Evidential Reasoning in a Hierarchical Hypothesis Space. *Artificial Intelligence*, 26:323–357, 1985.
40. J. Guan and D. A. Bell. *Evidence Theory and its Applications*. North-Holland, Amsterdam, 1991.
41. L. O. Hall, T. L. Machrzak, and M. S. Silbiger. Obtaining Fuzzy Classification Rules in Segmentation. In *IPMU*, pages 619–624, Paris, France, 1994.
42. T. L. Huntsberger, C. Rangarajan, and S. Jayaramamurthy. Representation of Uncertainty in Computer Vision using Fuzzy Sets. *IEEE Trans. on Computers*, C-35(2):145–156, 1986.
43. H. H. S. Ip and J. M. C. Ng. Human Face Recognition using Dempster-Shafer Theory. In *ICIP*, volume II, pages 292–295, Austin, Texas, 1994.
44. M. C. Jaulent and A. Yang. Application of Fuzzy Pattern Matching to the Flexible Interrogation of a Digital Angiographies Database. In *IPMU*, pages 904–909, Paris, France, 1994.
45. E. T. Jaynes. Information Theory and Statistical Mechanics. *Physical Review*, 106(4):620–630, 1957.
46. A. Kandel, M. Schneider, and G. Langholz. Autonomous Fuzzy Intelligent Systems for Image Processing. In *IPMU*, pages 613–618, Paris, France, 1994.
47. H. B. Kang and E. L. Walker. Characterizing and Controlling Approximation in Hierarchical Perceptual Grouping. *Fuzzy Sets and Sytems*, 65:187–223, 1994.
48. R. Krishnapuram and J. M. Keller. Fuzzy Set Theoretic Approach to Computer Vision: an Overview. In *IEEE Int. Conf. on Fuzzy Systems*, pages 135–142, San Diego, CA, 1992.
49. R. Krishnapuram, J. M. Keller, and Y. Ma. Quantitative Analysis of Properties and Spatial Relations of Fuzzy Image Regions. *IEEE Transactions on Fuzzy Systems*, 1(3):222–233, 1993.
50. S. Kullback. *Information Theory and Statistics*. Wiley, New York, 1959.

51. R. H. Lee and R. Leahy. Multi-Spectral Classification of MR Images Using Sensor Fusion Approaches. In *SPIE Medical Imaging IV: Image Processing*, volume 1233, pages 149–157, 1990.
52. T. Lee, J. A. Richards, and P. H. Swain. Probabilistic and Evidential Approaches for Multisource Data Analysis. *IEEE Transactions on Geoscience and Remote Sensing*, GE-25(3):283–293, 1987.
53. H. Leung. Neural Networks Data Association with Application to Multiple-Target Tracking. *Optical Engineering*, 35(3):693–700, March 1996.
54. J. D. Lowrance, T. M. Strat, L. P. Wesley, T. D. Garvey, E. H. Ruspini, and D. E. Wilkins. The Theory, Implementation and Practice of Evidential Reasoning. SRI project 5701 final report, SRI, Palo Alto, June 1991.
55. A. De Luca and S. Termini. A Definition of Non-Probabilistic Entropy in the Setting of Fuzzy Set Theory. *Information and Control*, 20:301–312, 1972.
56. H. Maître. Image Fusion and Decision in a Context of Multisource Images. In *9th Scandinavian Conference on Image Analysis*, volume 1, pages 139–153, Uppsala, Sweden, June 1995.
57. H. Maître. Entropy, Information and Image. In H. Maître and J. Zinn-Justin, editors, *Progress in Picture Processing, Les Houches Session LVIII*, pages 881–1115. Springer Verlag, 1996.
58. G. M. T. Man and J. C. H. Poon. A new Similarity Measurement Method for Fuzzy-Attribute Graph Matching and its Application to Handwritten Character Recognition. In *Int. Carnahan Conf. on Security Technology*, pages 46–49, Lexington, KY, October 1992.
59. J.-F. Mangin, J. Regis, I. Bloch, V. Frouin, Y. Samson, and J. Lopez-Krahe. A Markovian Random Field based Random Graph Modelling the Human Cortical Topography. In *CVRMed'95*, pages 177–183, Nice, France, April 1995.
60. L. Mascarilla. Rule Extraction based on Neural Networks for Satellite Image interpretation. In *SPIE Image and Signal Processing for Remote Sensing*, volume 2315, pages 657–668, Rome, Italy, 1994.
61. S. Mascle, I. Bloch, and D. Vidal-Madjar. Unsupervised Multisource Remote Sensing Classification using Dempster-Shafer Evidence Theory. In *SPIE/EUROPTO Conference on Image and Signal Processing for Remote Sensing*, volume 2579, Paris, France, September 1995.
62. D. McKeown, W. A. Harvey, and J. McDermott. Rule-Based Interpretation of Aerial Imagery. *IEEE Trans. on Pattern Analysis and Machine Intelligence*, 7(5):570–585, 1985.
63. M. Ménard, E. H. Zahzah, and A. Shahin. Mass Function Assessment: Case of Multiple Hypotheses for the Evidential Approach. In *Europto Conf. on Image and Signal Processing for Remote Sensing*, Taormina, Italy, September 1996.
64. H. Moissinac, H. Maître, and I. Bloch. Markov Random Fields and Graphs for Uncertainty Management and Symbolic Data Fusion in a Urban Scene Interpretation. In *SPIE/EUROPTO Conference on Image and Signal Processing for Remote Sensing*, Paris, France, September 1995.
65. R. E. Neapolitan. A Survey of Uncertain and Approximate Inference. In L. Zadeh and J. Kaprzyk, editors, *Fuzzy Logic for the Management of Uncertainty*, pages 55–82. J. Wiley, New York, 1992.
66. H. Ogawa. A Fuzzy Relaxation Technique for Partial Shape Matching. *Pattern Recognition Letters*, 15:349–355, 1994.
67. S. K. Pal. Fuzzy Set Theoretic Measures for Automatic Feature Evaluation. *Information Science*, 64:165–179, 1992.

68. J. Pearl. Fusion, Propagation, and Structuring in Belief Networks. *Artificial Intelligence*, 29:241–288, 1986.
69. W. Pedrycz. Fuzzy Sets in Pattern Recognition: Methodology and Methods. *Pattern Recognition*, 23(1/2):121–146, 1990.
70. E. Piat. Fusion de croyances dans le cadre combiné de la logique des propositions et de la théorie des probabilités, application à la reconstruction de scène en robotique mobile. PhD thesis, Université de Technologie de Compiègne, 1996.
71. A. Ralescu and R. Hartani. Modeling the Perception of Facial Expressions from Face Photographs. In *10th Fuzzy System Symposium*, pages 405–408, Ozaka, Japan, June 1994.
72. H. S. Ranganath and L. C. Chipman. Fuzzy Relaxation Approach for Inexact Scene Matching. *Image and Vision Computing*, 10(9):631–640, 1992.
73. N. S. V. Rao and S. S. Iyengar. Distributed Decision Fusion under Unknown Distributions. *Optical Engineering*, 35(3):617–624, March 1996.
74. J. B. Romine and E. W. Kamen. Modeling and Fusion of Radar and Imaging Sensor Data for Target Tracking. *Optical Engineering*, 35(3):659–673, March 1996.
75. A. Rosenfeld. Fuzzy Digital Topology. *Information and Control*, 40:76–87, 1979.
76. A. Rosenfeld. The Fuzzy Geometry of Image Subsets. *Pattern Recognition Letters*, 2:311–317, 1984.
77. A. Rosenfeld, R. Hummel, and S. Zucker. Scene Labeling by Relaxation Operations. *IEEE Transactions on Systems, Man and Cybernetics*, 6:420–433, 1976.
78. F. Russo and G. Ramponi. An Image Enhancement Technique based on the FIRE Operator. In *IEEE Int. Conf. on Image Processing*, volume I, pages 155–158, Washington DC, 1995.
79. F. Salzenstein and W. Pieczynski. Unsupervised Bayesian Segmentation using Hidden Fuzzy Markov Fields. In *IEEE Int. Conf. on Acoustics, Speech and Signal Procesing*, Detroit, Michigan, 1995.
80. G. Shafer. *A Mathematical Theory of Evidence*. Princeton University Press, 1976.
81. J. C. Simon. *From Pixels to Features*. V-X, North Holland, Amsterdam, 1989.
82. P. Smets. Medical Diagnosis: Fuzzy Sets and Degree of Belief. In *Colloque International sur la Théorie et les Applications des Sous-Ensembles Flous*, Marseille, September 1978.
83. P. Smets. The Combination of Evidence in the Transferable Belief Model. *IEEE Transactions on Pattern Analysis and Machine Intelligence*, PAMI-12(5):447–458, 1990.
84. P. Smets. Belief Functions: The Disjunctive Rule of Combination and the Generalized Bayesian Theorem. *International Journal of Approximate Reasoning*, 9:1–35, 1993.
85. P. Smets. The Transferable Belief Model for Uncertainty Representation. Technical Report TR/IRIDIA/95-23, IRIDIA, Université Libre de Bruxelles, Bruxelles, Belgium, 1995.
86. Léa Sombé. *Raisonnements sur des informations incomplètes en intelligence artificielle*. Teknea, Marseille, 1989.
87. H. E. Stephanou and S. Y. Lu. Measuring Consensus Effectiveness by a Generalized Entropy Criterion. In *First Conference on Artificial Intelligence Applications*, pages 518–523, Denver, December 1984.
88. T. M. Strat. Decision Analysis using Belief Functions. Technical Note 472, SRI, September 1989.

89. D. Y. Suh, R. M. Mersereau, R. L. Eisner, and R. I. Pettigrew. Automatic Boundary Detection on Cardiac Magnetic Resonance Image Sequences for Four Dimensional Visualization of the Left Ventricle. In *First Conference on Visualization in Biomedical Computing*, pages 149–156, Atlanta GE, 1990.
90. H. Tahani and J. M. Keller. Information Fusion in Computer Vision Using the Fuzzy Integral. *IEEE Transactions on System, Man and Cybernetics*, 20(3):733–741, 1990.
91. S. C. A. Thomopoulos. Sensor Integration and Data Fusion. *Journal of Robotics Systems*, 7(3):337–372, 1990.
92. F. Tupin, H. Maître, J-F. Mangin, J-M. Nicolas, and E. Pechersky. Linear Feature Detection on SAR Images: Application to the Road Network. Technical report, Ecole Nationale Supérieure des Télécommunications (96D006), May 1996.
93. J. van Cleynenbreugel, S. A. Osinga, F. Fierens, P. Suetens, and A. Oosterlinck. Road Extraction from Multi-temporal Satellite Images by an Evidential Reasoning Approach. *Pattern Recognition Letters*, 12:371–380, 1991.
94. A. Winter, H. Maître, N. Cambou, and E. Legrand. Object Detection Using a Multiscale Probability Model. In *IEEE Int. Conf. on Image Processing ICIP'96*, volume I, pages 269–272, Lausanne, September 1996.
95. R. R. Yager. Connectives and Quantifiers in Fuzzy Sets. *Fuzzy Sets and Systems*, 40:39–75, 1991.
96. B. Yan. Semiconormed Possibility Integrals and Multi-Feature Pattern Classification. *Pattern Recognition*, 26(12):1855–1862, 1993.
97. L. A. Zadeh. Fuzzy Sets. *Information and Control*, 8:338–353, 1965.
98. L. A. Zadeh. Fuzzy Sets as a Basis for a Theory of Possibility. *Fuzzy Sets and Systems*, 1:3–28, 1978.
99. E. Zahzah. *Contribution à la représentation des connaissances et à leur utilisation pour l'interprétation automatique des images satellites*. Thèse de doctorat, Université Paul Sabatier, Toulouse, 1992.
100. R. Zwick, E. Carlstein, and D. V. Budescu. Measures of Similarity Among Fuzzy Concepts: A Comparative Analysis. *International Journal of Approximate Reasoning*, 1:221–242, 1987.

Fuzzy Linguistic Methods for the Aggregation of Complementary Sensor Information

G. MAURIS, E. BENOIT, L. FOULLOY

LAMII/CESALP
Université de Savoie, 41 Avenue de la Plaine,
BP 806 F-74016 ANNECY FRANCE

ABSTRACT

The problem of the aggregation of complementary information is a crucial point in the monitoring of large intelligent systems. This paper deals with the cases in which there is no analytical mathematic model to derive new information from the basic measurements. Our artificial intelligence approach consists in using linguistic knowledge provided by experts. In the second section, we explain within the framework of the fuzzy subset theory how the numeric and linguistic representations could be used in the measurement aggregation problems. Next, the third section proposes an interpolation mechanism that creates a fuzzy partition of the numeric multi-dimensional space of the basic features. In section four, we present a formal symbolical representation of fuzzy If ... Then ... rules for the combination of basic features. The proposed methods are then applied in section five to the problem of the navigation of a mobile robot equipped with two ultrasonic range finding sensors, giving the proximity, the orientation, and the danger of an obstacle by the aggregation of the distance information provided by each one.

1 Introduction

In the last ten years, the interest in multisensor intelligent systems has increased rapidly [1][2]. The reason for this is that the conventional single sensor approach cannot be used in a particular environment due to their physical limitations (e.g., acoustic sensors in space), while others are limited due to either technical or economical factors. Adding more sensors, providing redundant and complementary information, offers the capability of resolving most complex situations, and will lead to a richer description of the world. The use of multisensor-systems is particularly necessary in applications where the requirement is that the system interacts with and operates in an unstructured environment without the full control of a human operator. Furthermore, existing multisensor systems are used in the following areas of applications [3]: industrial tasks like material handling, part manufacturing, inspection assembly, military command and control, mobile robot navigation and control, ...

If the benefit of multisensor systems seems promising (improved accuracy, increased range detection, enhanced reliability, features impossible to perform with individual sensors, shorter acquisition time, decreased costs, ...), the way to monitor such systems leads to many problems [2][4][5]:

- fusion of information : the sensors used in sensor systems are of vastly different characteristics due to their differing physical principles, their outputs can be of entirely different types, and they are geared towards a variety of applications. Thus, it is difficult to combine such different quantities in a systematic and standardised way.

- sensor monitoring : how to allow for sensory system configuration as a means to use a sensor only as needed, how to obtain a greater tolerance for sensing device failures; and how to facilitate future incorporation of additional devices, in a way that is optimal from the point of view of the task to be fulfilled?

- system issues : how can sensing activities be best organized? Sensors may operate independently and in parallel. The sensors may cooperate with each other and with knowledge residing in the system. How can communication between sensors be achieved?

In fact, all of these aspects constitute the multisensor integration problem, whose issues are largely linked to the type of information under consideration. In this paper, we will focus on the first problem, i.e. the fusion of information, because it conditions largely the other problems. Two fusion methodologies are generally considered [6][7] : the first based on statistical decision theory, the second based on artificial intelligence.

Most multisensor systems to date, have been designed around a method of fusion which includes the use of a statistical measurement model of each sensor [3]. Different more or less complex techniques are available : weighted average, Kalman filter, Bayesian approach, Dempster-Shafer evidential reasoning. They have shown their effectiveness in many applications, essentially in the case where the fusion consists in merging the outputs of many sensors providing redundant information of the same type. However they have drawbacks. The most important one is that the decision making is performed on low level information (numeric or stochastic characterization of sensor output) and under severe restrictions (keeping the consistency of the world model representation, stochastic model of the sensor, Gaussian hypothesis for the noise, ...). In fact, the statistical techniques are not adapted to the cases where the fusion involves complementary information of various nature, and requires prior knowledge coming from experienced users or other sensors to derive the requisite information [8].

For example, raw data in an image might represent intensity, colour, range, texture, surface shape, ... Different types of data will suggest different representations, and consequently different aggregation methods : intensities simply as numerical values of continuous quantities; texture might be best represented as a certain statistical distribution of intensity; colour might be represented as a vector of intensity values for the primary colours red, green and blue or in terms of hue, saturation and brightness; surface shapes might be represented by a set of symbolic names (square, rectangle, ...). Moreover, the knowledge used for understanding the information in the scene may include declarative statements such as "a red disc signifies danger".

Therefore to determine the operations leading to the acquisition of new information, we must know:

- first, what type of information is obtained from the elementary sensors,

- second, what type of information is wanted for the information resulting from the combination of the basic data.

- third, how the link between the basic information and the resulting information is expressed.

In fact, the process of aggregation is largely directed by the latter aspect, which determines the method to use in order to combine the elementary information, according to the level of knowledge at our disposal. If an analytical relationship between the input and output information (i.e. a high level knowledge) is known, conventional statistical methods could be used [5]. If the relation between the basic features is not precisely known, but must express some specificities of the application considered, such as compromise between the features, relative importance of the features, tolerance of the supervisor, ..., different fuzzy operators could be used [9][10][11].

When we have only sampled input-output numeric data pairs, neural methods are efficient to approximate the relation between them [12][13]. Our field of research is concerned with the situations where linguistic knowledge is involved in the process of aggregation. For these cases, artificial intelligence approaches based on fuzzy subset theory are available [14].

In the second section, the common and different aspects of the numeric and linguistic representation of an informational entity are explained. Then, we will present an interpolative aggregation method and a rule-based aggregation method, both based on a linguistic description of information by means of fuzzy subset theory. Our approach is then applied to the problem of the navigation of a mobile robot equipped with two ultrasonic range finding fuzzy sensors, giving the orientation of an obstacle by the aggregation of the distance information provided by each one, and giving a description of danger by aggregating the obstacle proximity and the orientation. The conclusion points out the interest of the proposed approach, discusses the benefits and drawbacks of the different underlying methods, and lists the remaining problems related to the aggregation of complementary information by linguistic methods.

2 The linguistic information aggregation problem

2.1 Symbolic and numeric representations

According to measurement theory [15] [16], the measurement process consists in representing observations by mathematical images (symbols or numbers) keeping the relations obtained from experiments. Generally, physical quantities are evaluated in a numerical way. The numerical representation presents many advantages. It is precise, not redundant, complete, and provides a lot of arithmetic relations. Methods are available to take into account the imprecision of information (error calculus, probability theory). Moreover, numerical measurements are based on an objective operational procedure. Nevertheless, numerical measurements are not always available, because of problems of acquisition or storage of information, because of the multi-dimensional nature of the analysed property, or because it is concerned with attributes describing human behaviour. In such cases, one is led to make a qualitative description of the observed phenomenon with words of the natural language. Natural language could be seen as a qualitative representation. In fact, it is a nominal measurement providing an equivalence relation between the words. At first sight, symbolic representation seems less interesting than numerical representation. It is less precise, redundant, and is subjective in the sense that it can not be experimentally verified by an objective procedure (i.e. independent from the observer). But the symbolic qualitative description is compact (it reduces the problem of the storage of information), easily understood by human beings (it allows reasoning), and has a richness that numerical measurements do not have, i.e. background knowledge upon the global purpose (i.e. a great pragmatic value).

A feature V could be associated both with a numerical set X or with a symbolic set $\mathcal{L}$, for example the temperature can be defined on the numerical Celsius's scale or on the set of terms $\mathcal{L}=\{$cold, middle, warm$\}$; or V could only be associated with a symbolic set $\mathcal{L}$, for example the danger can not be defined on a numeric scale but on a set of terms $\mathcal{L}$, e.g. $\mathcal{L}=\{$small, moderate, high$\}$. One means to overcome the threshold effects of the symbolic representation is to authorize the affectation of a degree between 0 and 1 for the terms of $\mathcal{L}$, i.e. to have a fuzzy symbolic characterization of V. This fuzzy symbolic characterization (e.g. $\{0.8/$cold, $0.2/$middle, $0/$warm$\}$) can be provided directly by expert, or deduced from the meanings of the terms L of $\mathcal{L}$, i.e. membership functions $\mu_{M(L)}$ defined on the set $\mathcal{F}(X)$ of the fuzzy subset of X. The grade of membership $\mu_{M(L)}(x)$ represents to which degree x belongs to the meaning of L.

Examples of membership functions of fuzzy meanings for many terms of distance are given in fig.1. Thus a fuzzy characterization could be deduced by a fuzzy description defined as a fuzzy mapping from the measurement set X in the set of the fuzzy subsets of symbols $\mathcal{F}(\mathcal{L})$. So the fuzzy description is characterized by its membership function noted as $\mu_{D(x)}$. The grade of membership $\mu_{D(x)}(L)$ represents to which degree the term L characterizes the numerical value x. There is a fundamental relation between the fuzzy description and the fuzzy meanings, that allows to deduce the linguistic description of a measurement from the fuzzy meanings of the symbolic set : $\mu_{D(x)}(L)=\mu_{M(L)}(x)$, e.g. $\mu_{D(150)}(small)=0$, $\mu_{D(150)}(medium)=0.8$, $\mu_{D(150)}(high)=0$. If we implement these meanings in the knowledge base of a sensor, the latter could then make a linguistic description of the observed phenomenon, and thus is called a fuzzy sensor [18].

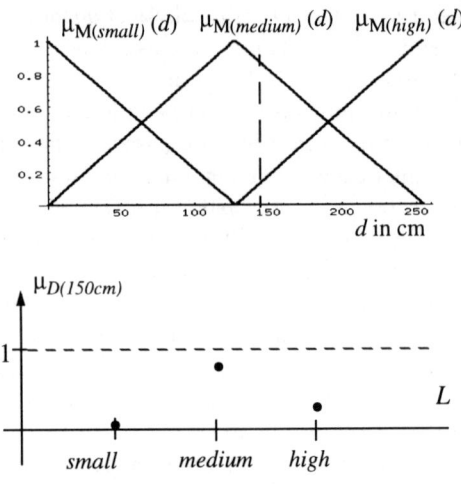

Fig. 1 Examples of fuzzy meanings and fuzzy descriptions.

Therefore, with the fuzzy subset approach, we introduce quantitative aspects in the qualitative representation, thus having the benefits of the two representations using a linguistic variable.

2.2 Linguistic aggregation methods

In many situations, the information sought after is not available from a single entity, but from many different complementary entities. In measurement theory, this case corresponds to the so-called conjoint measurement. Here, we will call it the aggregation of complementary information. The problem considered consists in obtaining a linguistic description of a high level entity from the knowledge of the basic features involved in the definition of this high level entity.

In the following, each basic features V_i can be characterized by a numeric variable x_i belonging to X_i. Each basic feature can also be characterized by a variable L_i whose values are linguistic terms taken in the set $\mathcal{L}_i$. The aggregated entity will be denoted as V. Let X be the set of the associated numerical measurement vectors v's and $\mathcal{L}$ be the set of the linguistic terms. In terms of the formalism developed in the preceding paragraph, and according to the preceding notations, aggregating linguistic information means that we want to obtain the fuzzy description of every basic measurement vector $v=(x_1, x_2, ..., x_n)$. In a coherent way with the link between the fuzzy meaning and the fuzzy description, we have for every L belonging to $\mathcal{L}$: $\mu_{D(v)}(L) = \mu_{M(L)}(v)$, two strategies could be considered.

- First, building indirectly $\mu_{D(v)}(L)$ by computing first $\mu_{M(L)}(v)$. Here we propose to compute this multi-dimensional membership function by parts and by an affine interpolation between a few characteristic points given by an expert. In fact, with this method we work on the numerical cartesian product $X_1 \times X_2 \times ... \times X_n$.

- Secondly, building directly $\mu_{D(v)}(L)$ from the descriptions of the elementary measurements x_i's, that is $\mu_{D(x_i)}(L_i)$, e.g. provided by basic fuzzy sensors. The link between $\mu_{D(v)}(L)$ and the $\mu_{D(x_i)}(L_i)$ could be obtained, for example, from a set of rules relating the L's to the L_i's of $\mathcal{L}_i$, as we will describe later. In fact, with this method we work on the linguistic cartesian product $\mathcal{L}_1 \times \mathcal{L}_2 \times ... \times \mathcal{L}_n$, and we assume that the aggregated entity V is decomposable under the basic variables V_i's.

3 Interpolative aggregation method

This method consists in defining first the fuzzy meaning of each term L on the numerical multi-dimensional space of the basic features. Then, the fuzzy description of a measurement vector $v=(x_1, x_2, ..., x_n)$ is deduced from the link between the fuzzy meaning and the fuzzy description. The second stage of this policy being clearly evident, we will focus our statement on the first stage, i.e. the building of the fuzzy meaning of each term.

The universe of discourse of the terms of the aggregated entity being multi-dimensional, a global definition of the membership function by an expert is generally difficult to obtain. In return, the expert could easily define the meaning of a term for a particular characteristic measurement vector denoted here as v_i. Generally the points v_i's correspond entirely to the meaning of a particular term, i.e. the degree of the membership function for these characteristic points is equal to 1 for a particular term, to 0 for the other terms. The proposed method for the building of the membership function of every term consists in a multi-linear interpolation between these characteristic points. In order to detail the chosen strategy to make this interpolation, we will consider first their crisp membership function and then their fuzzy extension.

3.1 Crisp membership functions

The proposed approach consists in using characteristic points v_i, typical of the different terms considered, to build a partition of the universe of discourse. This partition is realized by considering for each characteristic point v_i, the set $F(v_i)$ composed of the measurement points, which are closer to v_i than to the other characteristic points v_j, i.e. :

$F(v_i) = \{v \text{ / for every } j \neq i, d(v_i, v) < d(v, v_j)\}$,

where d is a distance defined on the numerical multi-dimensional space $X_1 \times X_2 \times ... \times X_n$. This tessellation of the universe of discourse in polyhedrons is called a Voronoi's diagram [19]. Then we define the meaning of the term L corresponding to v_i by the following characteristic function :

$\mu_{M(L)}(v) = \mu_{F(v_i)}(v) = 1$ if $v \in F(v_i)$, 0 otherwise.

If a term is defined by several characteristic points, then the meaning is defined by :

$\mu_{M(L)}(v) = \mu_{\cup_i F(v_i)}(v) = 1$ if $v \in \cup_i F(v_i)$, 0 otherwise.

An example of such a partition in the two-dimensional case. can be seen in fig. 2.

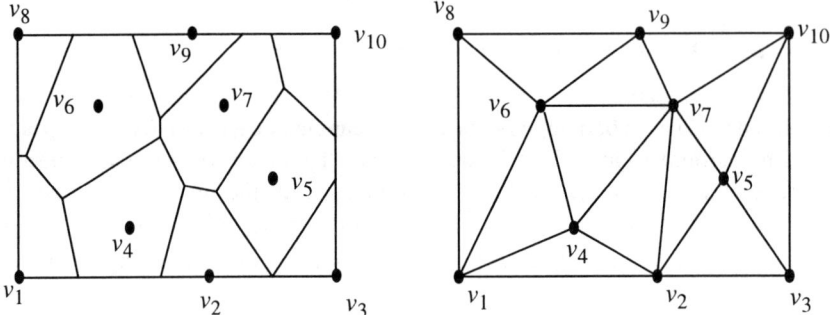

Fig. 2 Voronoi's diagram and Delaunay's diagram in the two-dimensional case.

3.2 Fuzzy membership functions

The preceding method presents many disadvantages. First, there are ambiguous measurements, for which $d(v_i, v) = d(v, v_j)$. This situation imposes an arbitrary choice. Secondly, we do not take the graduality of the terms into account. The aim of the fuzzy membership function is to overcome these difficulties in relating the membership grade to the Euclidian distance between the measurement point and the characteristic point. This objective is a bit too ambitious, because it leads to a high-order equation system. So we will restrict it, by imposing the following conditions. The membership function of the associated term is equal to 1 for a considered characteristic point v_i, 0 for the other terms. For measurement points far from v_i, the membership function will be equal to 0, and for measurement points close to v_i, the membership function will be related to the distance to v_i.

To implement this idea properly, we begin by joining the characteristic points, whose Voronoi's polyhedrons have a common face (see figure 2). So we obtain in the two-dimensional case a partition in triangles of the universe of discourse called a Delaunay's triangulation [20]. Now, the membership function of the meaning of a term L associated to the particular characteristic point v_i will be defined by parts on each triangle :

- if the considered triangle does not contain the characteristic point v_i, then the membership function is equal to 0 for all the points belonging to this triangle,

- if the considered triangle contains the characteristic point, then we take $\mu_{M(L)}(v_i)=1$ and $\mu_{M(L)}(v_j)=0$ for the other vertices of the triangle. For all the other points $v=(x_1, x_2)$ belonging to the triangle, we define the membership function by an affine interpolation by taking $\mu_{M(L)}(x_1, x_2)=\alpha_1 x_1 +\alpha_2 x_2 +\alpha_3$; the coefficients α_1, α_2 and α_3 are determined by the preceding conditions on the three vertices of the triangle (system of three equations with three unknowns).

This process is performed on each triangle and for each term. Hereafter an example is shown.

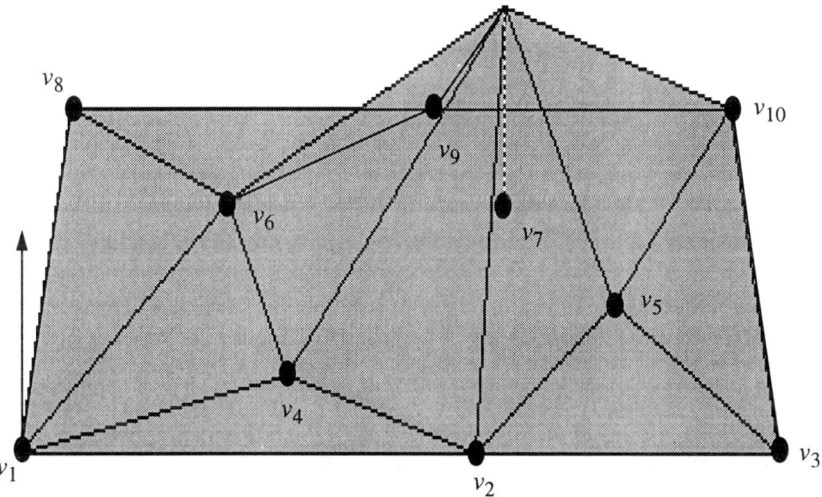

Fig. 3 Membership function associated to the characteristic point v_7.

The above explanations entail several remarks.

- This method may then be extended to the n-dimensional case. The triangles are transformed in so called n-simplexes. We have to determine now n coefficients α_i's. It is always possible to find these coefficients because of the independent linearity of the $n+1$ vertices of a n-simplex. In this way, the terms constitute a fuzzy partition, i.e. $\forall v \in X_1 \times X_2 \times ... \times X_n, \Sigma_{L \in \mathcal{L}}\ \mu_{M(L)}(v)=1$. But, if we want to consider all the points of $X_1 \times X_2 \times ... \times X_n$ the vertices of this multi-dimensional universe must have been defined as characteristic points.

- The area constituted by the points such as $\mu_{M(L)}(v) > 0.5$ is not equal to the Voronoi's area of the point v_i.

- Several distances could be considered on the multi-dimensional space $X_1 \times X_2 \times ... \times X_n$. But, if we consider the Euclidian distance, the membership function is equivalent for every point on a line $[v_i,v_j]$ to the ratio of the distance of v to these two characteristic points, i.e. $\mu_{M(L)}(v) = d(v_j,v) / d(v_i,v_j)$. Moreover, with this well-known distance, efficient algorithms already exist to perform the partitioning of the multidimensional space in simplexes [21].

- The meanings of the terms for the characteristic points could be taken not only in the set $\{0,1\}$, but also in the interval $[0,1]$. But in this case, to guarantee that the terms forms a fuzzy partition, we must have for every v_i belonging to $X_1 \times X_2 \times ... \times X_n, \Sigma_{L \in \mathcal{L}}\ \mu_{M(L)}(v_i)=1$.

- If a term L is defined by a set of a few characteristic points, then the resulting fuzzy meaning can be considered as the union of the fuzzy meaning associated with each point taken in the set of characteristic points. The union operator $\bot$, i.e. a triangular conorm, must satisfy $x \bot y = x+y$ if $x+y \leq 1$, for example $x \bot y = min(x+y,1)$, in order to always have a fuzzy partition for the meanings of the terms [22].

With this method, the knowledge needed to configure the sensor is very compact. It can be acquired during a supervised learning phase by a communication with a human or an expert system. During the configuration phase, the supervisor and the sensor analyse the same phenomenon and the supervisor gives his description to the sensor. The sensor increases its knowledge with its measurements associated with the supervisor's descriptions. Therefore, we have a method to calibrate linguistically the fuzzy sensor, which is moreover freed from systematic numeric measurement errors.

Hereafter, we will apply this method to a two-dimensional case, i.e. to the description of orientation from two distance measurements (see section 5.2). An example of a three-dimensional case concerning a colour problem could be found in [23].

4 Rule-based method

The purpose of this approach is to replace the numerical algorithm used to aggregate the numerical values provided by conventional sensors, by a rule-based formalism working with a linguistic description of the measurement provided by fuzzy sensors.

4.1 Representation of the rules

For the sake of writting simplicity, we will consider only rules with two premises linked by an "**and**", that is:

$$\textit{If } V_1 \textit{ is } L_1^i \textbf{ and } V_2 \textit{ is } L_2^i \textbf{ then } V \textit{ is } L^k \text{ with } (i, j, k) \in K, \qquad (1)$$

Here, we will consider that the rules represent a link between the different linguitic terms of the antecedent and the conclusion. Therefore, the set of rules could be viewed as a fuzzy relation $\mathcal{R}$ between the symbols L_1's of $\mathcal{L}_1$ (set of terms characterizing V_1), the symbols L_2's of $\mathcal{L}_2$ (set of terms characterizing V_2) and L's of $\mathcal{L}$ (set of terms characterizing V). Let E be a fuzzy subset of $\mathcal{L}_1 \times \mathcal{L}_2$. The image of E by the fuzzy relation $\mathcal{R}$ is given by Zadeh's compositional rule of inference:

$$\forall L \in \mathcal{L}, \mu_F(L) = \sup\nolimits_{(L_1, L_2) \in \mathcal{L}_1 \times \mathcal{L}_2} T\ [\mu_E(L_1, L_2), \mu_{\mathcal{R}}(L_1, L_2, L)] \qquad (2)$$

Let us assume that the fuzzy subet E is separable, that is:

$$\mu_E(L_1, L_2) = T'[\mu_{E_1}(L_1), \mu_{E_2}(L_2)]. \qquad (3)$$

Let us also assume that $\mathcal{L}_1 \times \mathcal{L}_2$ is finite. Thus, the supremum can be replaced by the maximum operator. By extending the maximum to any t-conorm, Eq. (2) becomes :

$$\forall L \in \mathcal{L}, \mu_F(L) = \perp_{(L_1, L_2) \in \mathcal{L}_1 \times \mathcal{L}_2} T\ \{T'[\mu_{E_1}(L_1), \mu_{E_2}(L_2)], \mu_{\mathcal{R}}(L_1, L_2, L)\} \qquad (4)$$

4.2 Discussion

In the preceding sections, we have described a symbolical representation of If ... Then rules, but a numerical representation which involves the numerical fuzzy subsets aoosciated to each terms could be also considered. In the symbolic view, the output of the rule inference is a fuzzy linguistic characterization. To derive the output, only a linguistic characterization of the inputs is required, and the validity of the rule could also be taken into account. Thus, this approach is also able to aggregate fuzzy symbolic variables not related to a numeric space. Moreover, in the case where this characterization is a description obtained from the meanings of the input terms, the meanings of the output terms are built indirectly by inference on the numerical sets of the inputs. This way a symbolic entity defined on a multi-dimensional numerical space is built. This view is therefore more efficient than the numerical one, because it could be applied even in the case where the inputs and the output are not defined on a numerical universe. In addition, it clearly separates the numerical linguistic conversion (made by a fuzzy sensor), and the knowledge inference (made by a fuzzy aggregator), which leads at the hardware level to a distribution in different fuzzy components, thus allowing a simple cascading of information. This provides a means to get around the decomposability problem in the case where a hierarchization of the involved information can be made. This aspect will be illustrated in section 5.3 by a description of danger from two pieces of sub-information themselves defined from elementary measurements. Finally, if we want to obtain a numerical output using a symbolic view of the rule, it is enough to make a symbolic defuzzification of the fuzzy linguistic output. A linguistic variable defined on a mono-dimensional numerical space is built by this means. A good choice of this symbolic defuzzification even leads to the same value as the one obtained with a numerical view of the rule [24].

So, as a summary of this discussion, we could say that even though the numerical view of the rule is widely used in control problems, we recommend a symbolic representation of the rules for the aggregation problems.

5 Applications to the perception of environment by robot sensors

5.1 Introduction

Fuzzy logic methods have proven efficient, when the process is difficult to model and when there is a significant heuristic knowledge from human operators. This is the case in the robotics field, where the environment is often unknown or unstructured, leading to a difficulty of an analytical modelling. But the robots are often developed to achieve tasks that human beings usually do. These reasons explain the amount of research involving the fuzzy subset theory in robotics applications [25]. In the field of mobile robots, much attention has been devoted to navigation problems, which could be classified into two types [26] : those in which the world model is precisely known, leading to the use of analytical methods (configuration space, potential field approach, ...) and those in which the environmental information is extracted from sensors and the world model is not known precisely. In the latter case, the robot must monitor the environment and determine its moving from sensory data. In this approach, generally the global navigation system uses a globally defined symbolic task (e.g. reach the target) from the high level reasoning system, and produces a sequence of locally defined symbolic tasks (e.g. avoid the obstacle, follow the wall,...). These qualitatively defined local tasks must then in turn be transformed into a quantitative guidance for the low level controller of the robot, using for instance the fuzzy subset theory.

The fuzzy approach consists in representing relations between inputs and outputs in an If ... Then ... manner and constructing a knowledge base from human experience [27] [28]. The difficulties are of two types : expressing the heuristic knowledge from operators and extracting linguistic information from sensor measurements. In order to execute the subtasks, and to make the switching between them, aggregated information of different levels is required. In this section we will consider the aggregation of two basic distance pieces of information, often provided by ultrasonic sensors [29] [30] [31] in order to obtain location information [32]. In order to illustrate the two ways (interpolation and rule) of fusing information, we will consider the following configuration, that represents a robot equipped with two range-finding fuzzy sensors $c1$ and $c2$ [33].

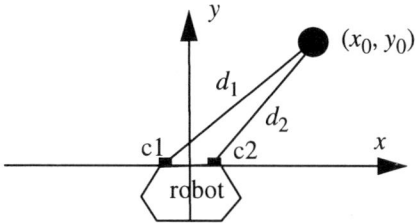

Fig. 4 Configuration of the problem considered.

The following problem will be studied: how to obtain a linguistic description of the proximity and the orientation of an obstacle with the terms *left, right, front,* from the linguistic descriptions of the distances with the terms *small, medium, high*. Therefore only the interpolative and the symbolic rule-based methods are available in this case. We will also consider the combination of orientation and proximity to define danger information aimed at activating safety procedures.

5.2 Interpolative method

To apply this method, we have chosen four characteristic points on the plane (d1,d2).

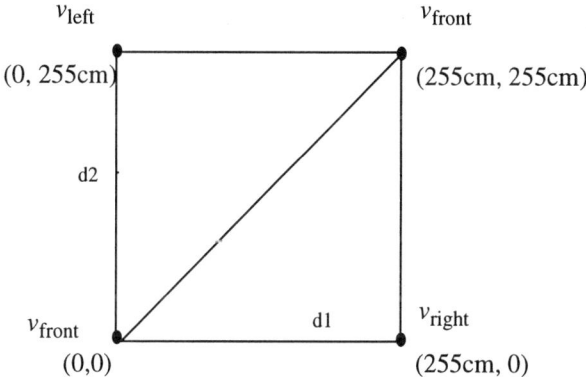

Fig. 5 Triangulation of the (d_1,d_2) space.

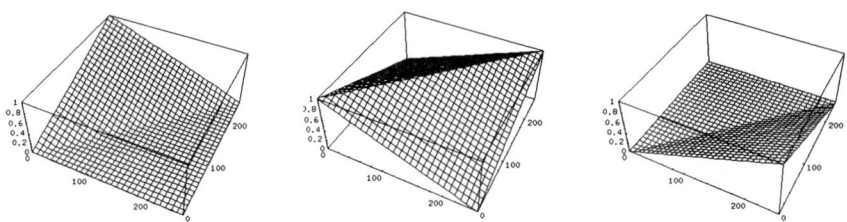

Fig. 6 Meaning of the terms "front, left, right" from examples.

In fig. 6, the meanings of the terms *front, left, right* obtained by the linear interpolation described in section 3 are plotted. The general form of these meanings is quite natural. This is due to the good choice made by the expert of the characteristic points.

5.3 Rule-based method

Orientation and proximity descriptions

According to the position of the two sensors, we propose to take the following rules for the aggregation.

Rule base for the proximity					Rule base for the orientation				
		d1					d1		
	Distance	small	Medium	high		Distance	small	Medium	high
d2	small	Close	Quite_Close	Quite_Far	d2	small	Front	Right	Right
	Medium	Quite_Close	Quite_Far	Far		Medium	Left	Front	Right
	high	Quite_Far	Far	Far		high	Left	Left	Front

Fig. 7 Sets of rules for the definition of proximity and orientation.

We could interpret these rules within the frame of a symbolic view of the rules. Through the sets of rules of fig. 7 and the meanings of the terms of fig.1, the meanings of the proximity and the orientation (plotted fig.8) are built by using formula (14).

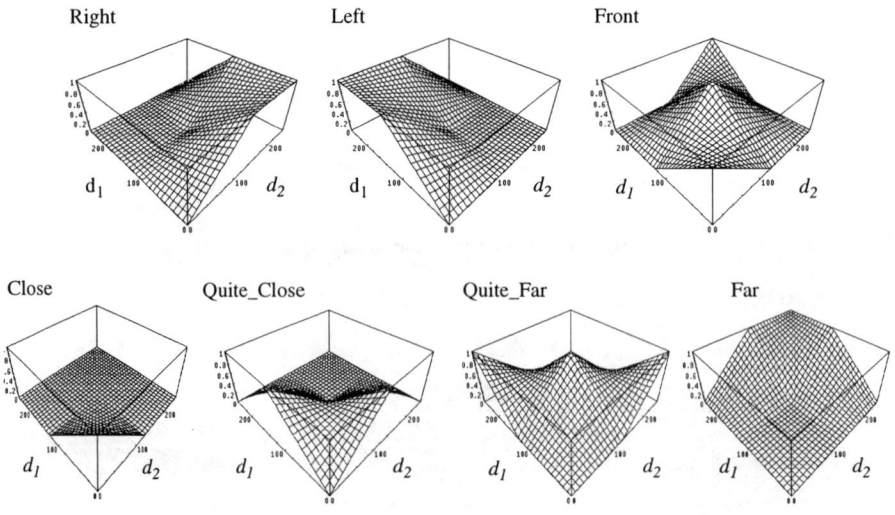

Fig. 8 Meanings of the aggregated terms.

Note that the meanings of the term *front* has not a regular shape in the sense that for measurements such as $d_1=d_2$, the membership is not always equal to one as expected. This is due to the rectangular partitioning of the space made by the rules. So for the orientation description, the interpolative method proposes better results (see Fig. 6).

Danger description

In order to illustrate the cascading of linguistic information, we will consider here a description of danger by the following rules.

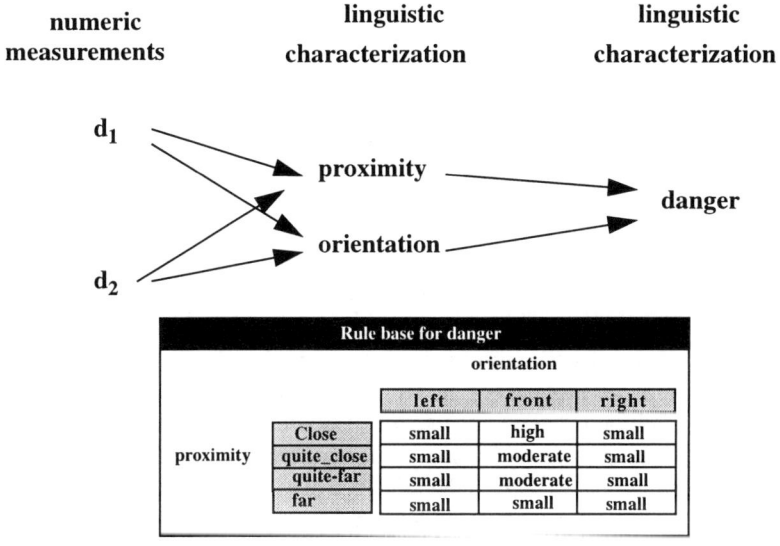

Fig. 9 Definition of danger.

The linguistic characterization of danger is obtained in a hierarchical way from the basic numeric measurements d_1 and d_2. Only the meanings of the terms of distance small, medium, high are required. Then the inference is made using the symbolic view of the rules. For example, let us assume that d_1=90cm and d_2=230cm. According to the definition of the distance meanings (cf. fig.1), their respective symbolic fuzzifications are:

$\phi(d_1)$={0.3/small, 0.7/medium, 0/high} and $\phi(d_2)$={0/small, 0.2/ medium, 0.8/high}.

Using the set of rules of fig. 9, the meanings of proximity and orientation are deduced by the compositional rule of inference using the set of rules of fig. 7:

proximity={0/close, 0.06/quite_close, 0.38/quite_far, 0.56/far}

orientation={0.86/left, 0.14/front, 0/right}.

Finally, entering these characterizations in the set of rules of fig. 7, we obtain: danger={0.94/small, 0.06/moderate, 0/high}.

The definition of danger by hierarchization is a means to avoid defining danger directly from the linguistic description of d_1 and d_2, which would not be easy. Nevertheless, we could have used typical examples on the numeric space (d_1, d_2) and the interpolative method to define danger, but in this case we the information of proximity and orientation would heve been lost.

6 Conclusion

This paper has been concerned with the aggregation of complementary information using linguistic information. The cornerstone of the aggregation problem is how the link between the basic features and the resulting information is expressed. When this link cannot be represented by a numerical model, we have proposed to use a set of characteristic examples or to use a rule-based linguistic model of the aggregation. With the interpolative method, the computation load is high, because it works on a multi-dimensional numerical space. But it provides an easy way to calibrate the output information linguistically according to the user, when only typical examples are known. With the fuzzy rule-based methods, we have seen the superiority of symbolic semantics, which leads to a better distribution of the operations involved in the aggregation process. In addition to the explanatory model provided by the rules, the symbolic view easily allows the addition of new features, and also a hierarchised cascading of basic information, thus leading to the definition of abstract entities. An important problem, which has not been considered here, is the influence of measurement errors in the basic features. Some research is currently in progress in order to take into account such errors in the proposed aggregation methods.

7 References

[1] Ren C. Luo, M.H. Lin, R.S. Scherp, *"Dynamic multisensor data fusion system for intelligent robots"*, IEEE J. of Robotics and Automation, Vol. 4, No 4, Aug. 1988, pp. 386-396.

[2] J.K. Aggarwal, Y.F. Wang, *"Sensor fusion in robotics systems"*, Control and Dynamic Systems, Vol. 39, 1991, pp. 435-462.

[3] Ren C. Luo, Michael G. Kay, *"Multisensor integration and fusion in intelligent systems"*, IEEE trans. on Systems, Man, and Cybernetics, Vol. 19, No 5, Oct. 1989, pp. 901-931.

[4] T.C. Henderson, E. Shilcrat, *"Logical sensor systems"*, J. of Robotic Systems, 1984, pp. 169-193.

[5] J. Manyika, H. Durrant-White, *Data fusion and sensor management*, Ellis Horwood Ed., New York, 1994.

[6] S.S. Iyengar, R.L. Kayshyap, R.N. Madan, *"Distributed sensor networks - introduction to the special section"*, IEEE/SMC, Vol. 21, No 5, Sept-Oct. 1991, pp. 1027-1031.

[7] Lauber A., *"Intelligent multisensor fusion"*, Int. Conf. on Fault Disgnosis, TOOLDIAG 93, Toulouse, France, April 5-7 1993, pp. 140-144.

[8] T.D. Garvey, *"A survey of AI approaches to the the integration of information"*, Proc. of SPIE, Vol. 782, 1987, pp. 68-82.

[9] D. Dubois, H. Prade, *"A review of fuzzy set aggregation connectives"*, Information Science, 36, 1985, pp. 85-121.

[10] R.R. Yager, *"Connectives and quantifiers in fuzzy sets"*, Fuzzy Sets and Systems, 40, 1991, pp. 39-75.

[11] M. Grabisch, *"Fuzzy integral in multi-criteria decision making"*, Fuzzy Sets ans Systems, 69, 1995, pp. 279-298.

[12] T. Fukuda, K. Shimojima, F. Arai, H. Matsuura, *"Multi-sensor integration system based on fuzzy inference and neural network for industrial application"*, IEEE Int. Conf. on fuzzy systems, San Diego, USA, March 1992, pp. 907-914

[13] D.A. Dornfeld, *"Sensor fusion"*, In *handbook of Intelligent Sensors for Industrial Automation*, Nello Zuech Ed., Addison-Wesley, 1991, pp. 419-508..

[14] L.A. Zadeh, *"Quantitative fuzzy semantics"*, Information Sciences, Vol. 3, 1971, pp. 159-176.

[15] Krantz D.H., Luce R.D., Suppes P., Tversky A., *Foundations of measurement*, Academic press, New York, Vol. 1, 1971.

[16] Finkelstein L., *"Measurement : fundamentals principles"*, in : Ed. Finkelstein L. and Grattan K.T.V, *Concise Encyclopedia of Measurement and Instrumentation*, 1994, pp. 201-205.

[17] Zadeh L. A., *"The concept of a linguistic variable and its application to approximate reasoning"*, Information Sciences, part 1:Vol 8, No 3, pp. 199-249, part 2: Vol 8,pp. 301-357, part 3:Vol 9, pp.43-80, 1975.

[18] G. Mauris, E. Benoit, L. Foulloy, *"Fuzzy symbolic sensors: from concepts to applications"*, Measurement, Vol. 12, No 4, 1994, pp. 357-384.

[19] Preparata F.P., Shamos M.I., *Computational geometry : an introduction*, Ed. Springer-Verlag, 1985.

[20] Bowyer A., *"Computing Dirichlet tessellations"*, The Computer Journal, Vol. 24, No 2, 1991, pp. 162-166.

[21] George P.L., Hermeline F., *"Maillage de Delaunay d'un polyèdre convexe en dimension d. Extension à un polyèdre quelconque"*, Research report INRIA, N^O 969, Feb.1989, 43 pages.

[22] G. Mauris, E. Benoit, Foulloy L., *"Fuzzy sensors for the fusion of information"*, Proc. of the IMEKO XIII, Turin, Italy, Sept. 1994, pp. 1009-1014.

[23] Benoit E., Mauris G., Foulloy L., *"A fuzzy colour sensor"*, Proc. of the IMEKO XIII World Congress, Turin, Italy, Sept. 1994, pp. 1015-1020.

[24] Foulloy L., Galichet S., *"Typology of fuzzy controllers"* in *Theoretical Aspects of Fuzzy Control*, (H.T. Nguyen, M. Sugeno, R. Tang, R. Yager Eds), Wiley , 1995, pp. 65-90.

[25] A. Ollero, A. Garcia-Cerezo, *"Design of fuzzy logic control systems. Applications to robotics "*, Proc. of the European Workshop on Industrial Fuzzy Control and Applications (IFCA 93), Terrassa, Spain, April 93.

[26] S. Ishikawa, *"A method of indoor mobile robot navigation by using fuzzy control "*, IEEE Int. Workshop on Intelligent Robots and Systems, Osaka, Nov. 1991, pp. 10/3-10/9.

[27] H. Graham, *"A fuzzy logic approach for safety and collision avoidance in robotic system"*, Proc. of Int. Conf. on human aspects of advanced manufacturing and hybrid automation, Amsterdam, Aug. 1992, pp. 493-498

[28] Yen, N. Pfluger, B. Lea, M. Murphy, Y. Jani, *" Employing fuzzy logic for navigation and control in an autonomous mobile system"*, Proc. of the American Control Conference, San francisco, June 1993, pp. 1850-1854.

[29] K.T. Song, J.C. Tai, *" Fuzzy navigation of a mobile robot"*, IEEE/RJS Int. Conf. on Intelligent Robots and Systems, Raleigh, July 1992, pp. 621-627.

[30] R. Kuc, *" Three dimensional docking using qualitative sonar"*, IEEE/RJS Int. Conf. on Intelligent Robots and Systems, July 1993, pp. 480-488.

[31] Mauris G., Benoit E., Foulloy L., *"Ultrasonic smart sensors : the importance of the measurement principle"*, Proc. of the IEEE/SMC, Le Touquet, France, October 1993, Vol. III pp. 55-60.

[32] A. Bagchi, H. Hatwal, *"Fuzzy logic based techniques for motion planning of a robot manipulator amongst unknown moving obstacle"*, Robotica, Vol. 10, 1992, pp. 563-573.

[33] G.Mauris, Benoit E., Foulloy L., *"An intelligent ultrasonic range finding sensor for robotics"*, World Congress IFAC 96, San francisco, USA, July 1996, Vol A, pp. 487-492.

Uncertain Data Aggregation in Classification and Tracking Processes

Alain Appriou

ONERA
BP 72, 92322 Châtillon Cedex
France

Abstract. To identify or localize a target, multisensor analysis has to be able to recognize one situation out of a set of possibilities. To do so, it uses measurements of more or less doubtful origin and prior knowledge that is understood to be often poorly defined, and whose validity is moreover difficult to evaluate under real observation conditions. The present synthesis proposes a generic modeling of this type of information, in the form of mass sets of the theory of evidence, with closer attention being paid to the most common case where the data originates from statistical processes. On the one hand robust target classification procedures can be achieved by applying appropriate decision criteria to these mass sets, on the other hand they can be integrated rigorously into a target tracking process, to reflect the origin of the localization measurements better. In all cases, the solutions found are placed in relation to those of the main competitive approaches currently used.

1 Problem Formulation

Ordinarily, when analyzing a situation, the available sensors have to be used under unfavorable conditions, inducing uncertainties at different levels :

- measurements that are imprecise, erroneous, incomplete, or ill-suited to the problem,

- ambiguous observations (*e.g.* a position or velocity measurement not necessarily related to the object in question),

- knowledge (generated by learning, models, and so forth) that is, in theoretical terms, incomplete, poorly defined, and especially more or less representative of reality, in particular in light of the varying context.

Moreover, the disparity of the data delivered by the various sensors, which is intended to remedy the individual insufficiencies of each, requires a detailed evaluation of each of them, based on any exogenous information that might characterize their pertinence to the problem at hand and the context investigated, while such information is itself often very subjective and imprecise.

Theories of uncertainty offer an attractive federative framework in this context. But they run up against a certain number of difficulties in practice : intepretation and modeling of the available information in appropriate theoretical frameworks, choice of an association architecture and combination rules, decision principles to be adopted, constraints concerning the speed and volume of the necessary computations.

To provide solutions to these questions, we will first consider a generic problem in which we attempt to characterize the likelihood of I hypotheses H_i theoretically listed in an exhaustive and exclusive set E. These hypotheses may typically concern the presence of entities, target or navigation landmark identities, vector or target localization, or the status of a system or of a situation.

Such a likelihood function may then be integrated either into :

- a choice strategy, to declare the most likely hypothesis (target identification, intelligence, and so on),

- a filtering process (such as target tracking or navigation updating),

- a decision aid process for implementing means of analysis, electronic warfare, or intervention.

The likelihood functions we want have to be developed from the data provided by J sensors S_j. Each of them is assumed to be associated with processes that extract a measurement or a set of measurements s_j, pertinent to the targeted discrimination function, from the raw signals or images it generates.

In the framework of the generic problem we will be considering first, we assume that each measurement s_j can be used to generate I criteria C_{ij}, on the basis of any *a priori* knowledge, having values in [0, 1] capable of characterizing the likelihood of each hypothesis H_i. A quality factor q_{ij} with values in [0, 1] is also associated with each likelihood C_{ij}. Its purpose is to express the aptitude of the criteion C_{ij} to discriminate the hypothesis H_i under the given observation conditions, on the basis of a dedicated learning process or exogenous knowledge. This factor includes mainly the confidence that can be accorded to the validity of the *a priori* knowledge used for generating C_{ij}. As concerns, for example, the representativity of a learning process in a varying context, it will typically depend on the quality, volume, and exactness of the available preliminary data, and on any marring of the corresponding measurements.

The developments presented are conducted in the theory of evidence framework [4], which happens to be the broadest and best-suited to the interpretation of the data considered, and also the most federative in terms of synergy with the related processes (especially as concerns the stochastic filtering). The generic problem considered thus leads to an axiomatic search for the mass sets on the set of hypotheses H_i capable of ensuring the synthesis of the set of (C_{ij}, q_{ij}).

A similar approach is also followed in the most commonly encountered concrete cases in which we have statistical learning $p(s_j/H_i)$ of each measurement s_j under the different hypotheses H_i. Although they are obtained using axioms specific to this latter situation, the corresponding models can be considered strictly as the special cases of the models found for the generic problem, as long as we have the appropriate expression for the criteria C_{ij} as a function of the $p(s_j/H_i)$.

The models developed in the two cases are then used for an application to target classification. A decisional procedure is then defined to guide the choices as regards the mass sets delivered. The very simple solutions that result from this can then be placed in relation to competitive approaches.

Lastly, the same models are used for a tracking application involving multiple, varied, moving targets in a dense environment, on the basis of observations output by a set of disparate and possibly delocalized sensors. The hypotheses considered are then the joint identity and localization hypotheses. Thanks to a set approach to the localization problem, the discernment frameworks can be conveniently managed to allow all the available data to be merged and, at the same time, generate an exact set of information that can be injected into a Bayesian filter of usual form. In addition to a richer and more appropriate exploitation of the data, the concept proposed integrates the classification function into the tracking function, which cannot be done formally with the usual probabilistic approaches, and intrinsically performs the matching of disparate multi-sensor data.

2 Generic Model

The very general problem of discrimination introduced in section 1 will be considered here in the practical case of interest when the criteria C_{ij} are generated by separate information channels, for which reason they are differentiated according to their pertinence by the factors q_{ij}. We further assume that we are in the most frequently encountered context where the criteria C_{ij} taken separately are always at least of refutation value, in the sense that, when zero, this guarantees that the associated hypothesis H_i is not verified.

This leads to a formal construction of the problem on the basis of two axioms :

Axiom 2.1 : Each of the I*J pairs [C_{ij}, q_{ij}] constitutes a distinct source of information having the focal elements H_i, ¬H_i, and E, in which the frame of discernment E represents the set of the I hypotheses.

Axiom 2.2 : When $C_{ij} = 0$ is valid ($q_{ij} = 1$), we can assert that H_i is not verified.

Axiom 2.1 requires that I*J mass sets $m_{ij}(.)$ be generated from the I*J respective pairs [C_{ij}, q_{ij}]. For each, the mass of focal elements H_i, ¬H_i, and E is at first defined by the value of the corresponding criterion C_{ij}, which can be interpreted only in terms of credibility or plausibility of H_i. Axiom 2.2 then limits the number of allowable interpretations to two. The first interpretation leads to :

$$Cr_{ij}(H_i) = 0 \quad et \quad Pl_{ij}(H_i) = C_{ij} \qquad (2.1)$$

and the second to :

$$Cr_{ij}(H_i) = Pl_{ij}(H_i) = C_{ij} \qquad (2.2)$$

Then, including the confidence factor q_{ij} for C_{ij} by disccounting at the rate $(1-q_{ij})$ provides the desired mass set $m_{ij}(.)$. This leads to the two possible models :

Model 1 :

$$m_{ij}(H_i) = 0 \qquad (2.3)$$
$$m_{ij}(\neg H_i) = q_{ij}*(1-C_{ij}) \qquad (2.4)$$
$$m_{ij}(E) = 1-q_{ij}*(1-C_{ij}) \qquad (2.5)$$

Model 2 :

$$m_{ij}(Hi) = q_{ij}*C_{ij} \qquad (2.6)$$
$$m_{ij}(\neg Hi) = q_{ij}*(1-C_{ij}) \qquad (2.7)$$
$$m_{ij}(E) = 1-q_{ij} \qquad (2.8)$$

A mass set $m(.)$ synthesizing all the evaluations is then obtained by computing the orthogonal sum of the different mass sets $m_{ij}(.)$ in the framework of each model :

$$m(.) = \bigoplus_{i,j} m_{ij}(.) \qquad (2.9)$$

It should be noted that Model 1 is consonant, and therefore lends itself (but it alone) to interpretation in the framework of possibility theory :

$$N_{ij}(H_i) = 0 \qquad (2.10)$$
$$\Pi_{ij}(H_i) = 1-q_{ij}*(1-C_{ij}) \qquad (2.11)$$

Different conjunction operators are then possible for combining the I*J sources thus formalized [7]. We can, for example, use as reference the idempotent triangular norm, for its fundamental properties, which differentiate it the most from the orthogonal sum used in the evidence theory approach, for its ease in calculation, and for its widespread use when motivated choices are lacking :

$$N(H_i) = 0 \qquad (2.12)$$

$$\Pi(H_i) = \min_j \{ \Pi_{ij}(H_i) \} / \max_i \{ \min_j \{ \Pi_{ij}(H_i) \} \} \qquad (2.13)$$

The practical determination of the C_{ij} and q_{ij} terms is still, in all cases, a problem specific to the type of application at hand. The tie-in with probabilistic approaches can nonetheless be established only for a special expression of the data compatible with this theory. The following section covers this type of situation, which is a common one in practice.

3 Model With Statistical Learning

The problem dealt with now assumes that each of the measurements s_j has first been subjected to a learning of the *a priori* probability distributions $p(s_j/H_i)$, under the various hypotheses H_i. Most systems do in fact allow a certain number of preliminary measurements in different real or simulated situations, from which histograms can be generated to get a numerical or analytical model of the distributions $p(s_j/H_i)$. The I*J values of probability density $p(s_j/H_i)$ associated respectively with the J local measurements s_j constitute the inputs for the processes discussed hereafter.

If we consider the most common case, where the measurements s_j can be assumed to be statistically independent, since the sensors are generally chosen for the complementary nature of the data they generate, the likelihood of each hypothesis H_i can be established immediately by the Bayesian approach, which typically calls for an evaluation of the *a posteriori* probability $P(H_i/s_1,...,s_J)$ of each hypothesis H_i using :

$$P(H_i/s_1,...,s_J) = \{[\prod_j p(s_j/H_i)]*P(H_i)\} / \sum_k \{[\prod_j p(s_j/H_k)]*P(H_k)\} \qquad (3.1)$$

in which $P(H_i)$ designates its *a priori* probability.

However, this kind of approach quickly runs into difficulty when the real observation conditions differ from the available learning conditions, or when the measurement bank is not sufficient for a suitable learning process. The lack of control that can be seen at this level in most applications does in effect lead us to use distribution models that turn out to be more or less representative of the data actually encountered. In addition, it is often difficult to find a set of *a priori* probabilities $P(H_i)$ capable of reflecting the real situation with fidelity.

3.1 Specification of the Approach

What we want to do here is to find a modeling based solely on the knowledge of $p(s_j/H_i)$ and capable of integrating any information concerning the reliability of the various distributions, whether this come from a more or less partial knowledge of the observations conditions or from a qualification of a data bank.

According to the generic approach introduced in section 1, any available qualitative information is assumed to be synhesized in the form of I*J coefficients $q_{ij} \in [0,1]$, each being representative of a degree of confidence in the knowledge of each of the I*J distributions $p(s_j/H_i)$.

Dealing with this problem in the terms of evidence theory requires finding, for each source S_j, a model of its I *a priori* probabilities $p(s_j/H_i)$ and their I respective confidence factors q_{ij} in the form of a mass set $m_j(.)$, characterized by a credibility function $Cr_j(.)$, and by a plausibility function $Pl_j(.)$. Since the sources S_j are distinct, a global evaluation $m(.)$ can then be obtained by computation of the orthogonal sum of the $m_j(.)$. The appropriate frame of discernment is of course the set of the I *a priori* listed hypotheses H_i.

To do this, we propose to conduct an exhausive and exact search of all the models that might satisfy three fundamental axioms in the context considered. These three axioms are chosen beforehand on the basis of their legitimacy in most of the applications concerned. They are :

Axiom 3.1 : Consistency with the Bayesian approach in the case where the learned distributions $p(s_j/H_i)$ are perfectly representative of the densities actually encountered ($q_{ij}=1, \forall i,j$) and where the *a priori* probabilities $P(H_i)$ are known.

Axiom 3.2 : Separability of the evaluation of the hypotheses H_i ; that is, each probability must be considered as a distinct source of information generating a particular mass set $m_{ij}(.)$, mainly capable of integrating the confidence factor q_{ij} specific to it. We thus require that each mass set $m_j(.)$ be the orthogonal sum of the I mass sets $m_{ij}(.)$ considered for $i \in [1,I]$. Also, considering the way the $p(s_j/H_i)$ probabilities are generated, the focal elements of the mass set $m_{ij}(.)$ can be only H_i, $\neg H_i$, or E, where the frame of discernment E is the set of hypotheses H_i.

Axiom 3.3 : Consistency with the probabilitic association of the sources ; for independent sources Sj and densities $p(s_j/H_i)$ perfectly representative of reality, the modeling procedures retained must lead to the same result if we compute the orthogonal sum of the $m_j(.)$ modeled from the $p(s_j/H_i)$ or if we model directly the joint probabilities $p(s_1,\ldots,s_J/H_i)$ given by :

$$p(s_1,\ldots,s_J/H_i) = \prod_j p(s_j/H_i) \tag{3.2}$$

The search for models satisfying these three axioms will be conducted hereafter by progressively restricting the set of possible models, taking the axioms into account in the order stated.

3.2 Axiom 3.1 : Consistency With the Bayesian Approach

Development

Let $m_0(.)$ be the mass set representative of information source S_0 consisting of the *a priori* probabilities $P(H_i)$. $m_o(.)$ is then a Bayesian mass set defined by :

$$m_0(H_i) = P(H_i) , \quad \forall\ i \in [1,I] \tag{3.3}$$
$$m_0(A) = 0 , \quad \forall\ A \neq H_i , \ i \in [1,I] \tag{3.4}$$

The desired consistency requires that the orthogonal sum of the mass sets $m_j(.)$ and $m_0(.)$ produces a Bayesian mass set $m_b(.)$ in conformity with the Bayesian inference (3.1) whenever the distributions $p(s_j/H_i)$ are perfectly representative of the densities actually encountered, and thus whenever $q_{ij}=1$ for any i and j. This axiom should, in particular, remain true for any subset of combined sources S_j delimited by $j \in J' \subset [1,J]$. Concretely :

$$m_b(.) = \{ \bigoplus_{j \in J'} m_j(.) \} \oplus m_0(.) \tag{3.5}$$

should under these conditions therefore verify :

$$m_b(H_i) = \{ [\prod_j p(s_j/H_i)]*P(H_i) \} / \sum_k \{ [\prod_j p(s_j/H_k)]*P(H_k) \} , \forall\ H_i \in E \tag{3.6}$$

Moreover, equations (3.3), (3.4), and (3.5) lead to :

$$m_b(H_i) = \{ [\prod_j Pl_j(H_i)]*P(H_i) \} / \sum_k \{ [\prod_j Pl_j(H_k)]*P(H_k) \} , \forall\ H_i \in E \tag{3.7}$$

By satisfying (3.6) and (3.7) jointly for any $J' \subset [1,J]$ we lastly define each $m_j(.)$ by its plausibility function using the I equations :

$$Pl_j(H_i) = K_j * p(s_j/H_i) , \quad i \in [1,I] \tag{3.8}$$

in which K_j is a unique parameter for the I equations, defined simply by :

$$K_j \in [\{\sum_i p(s_j/H_i)\}^{-1}, \{\max_i[p(s_j/H_i)]\}^{-1}] \tag{3.9}$$

These bounds on K_j are required only by the intrinsic nature of the idea of plausibility, which has to remain less than unity, while the sum of the values it takes for events constituting a partition of E (the H_i themselves, here) must be greater than unity.

Comments

The conclusion thus drawn from Axiom 3.1 calls for a few comments. Firstly, in the general case where I>2, for each value of K_j other than the minimum required by (3.9), there exists an infinite number of possible mass sets, defined by a system of I+1 equations (I equations (3.8) and the sum of the masses equal to 1) with 2^I-1 unknowns.

For the minimum value of K_j, the result obtained always amounts to a unique, and moreover Bayesian, mass set :

$$m_j(H_i) = p(s_j/H_i) / \sum_k p(s_j/H_k), \quad \forall i \in [1,I] \tag{3.10}$$

$$m_j(A) = 0, \quad \forall A \neq H_i, \quad i \in [1,I] \tag{3.11}$$

Of the various solutions obtained for the maximum value of K_j, there exists a consonant solution, unique on the set of solutions found, that corresponds to the model proposed by G. SHAFER on the basis of this characteristic alone, for a context similar to that of the present Axiom 3.1 [4]. To give a practical expression to this solution, let us suppose that the $p(s_j/H_i)$ are arranged such that $p(s_j/H_1) \geq p(s_j/H_2) \geq \ldots \geq p(s_j/H_I)$. The focal elements are the I subsets of E :

$$A_i = \bigcup_{k \leq i} H_k, \quad i \in [1,I] \tag{3.12}$$

and the corresponding masses are given by :

$$m_j(A_I) = K_j * p(s_j/H_I) \tag{3.13}$$
$$m_j(A_i) = K_j * \{p(s_j/H_i) - p(s_j/H_{i+1})\}, \quad \text{pour } 1 \leq i \leq I-1 \tag{3.14}$$

It should nonetheless be pointed out that this last solution does not satisfy Axioms 3.2 and 3.3, and that it therefore cannot be retained in the following.

Let us lastly say that, in the ideal case where the distributions $p(s_j/H_i)$ are perfectly representative of the densities acually encountered, a maximum likelihood procedure requires retaining the hypothesis H_i that will maximize $p(s_1,\ldots,s_I/H_i)$. Yet since the hypotheses H_i are singletons of the frame of discernment E, and $p(s_1,\ldots,s_I/H_i)$ is the product of the $p(s_j/H_i)$ provided by the I independent sources S_j, the plausibility $Pl(H_i)$ obtained after associating the sources S_j is expressed, using (3.8), by :

$$Pl(H_i) = K_f * p(s_1,\ldots,s_I/H_i) \, , \, \forall \, i \in [1,I] \tag{3.15}$$

in which the coefficient K_f, independent of H_i, integrates the K_j terms and the inconsistency of the combination. To remain consistent with this particular case, any decision procedure to designate the most realistic hypothesis must, for our problem, exclusively maximize a monotonic increasing function of the plausibility $Pl(H_i)$ alone, obtained after combining the sources S_j.

3.3 Axiom 3.2 : Separability of Hypothesis Evaluations

This axiom consists in considering that each mass set $m_j(.)$ sought is itself the result of a combination between I mass sets $m_{ij}(.)$ ($i \in [1,I]$) :

$$m_j(.) = \underset{i}{\oplus} \, m_{ij}(.) \tag{3.16}$$

A mass set $m_{ij}(.)$ also has three focal elements (H_i, $\neg H_i$, and E), whose masses depend only on the value $p(s_j/H_i)$ and the corresponding factor q_{ij}.

Since the hypotheses H_i are the singletons of the frame of discernment E, the plausibility $Pl_j(H_i)$ is proportional to the product by k of the $Pl_{kj}(H_i)$ associated with the $m_{kj}(.)$. After normalization in relation to the product by k of the $Pl_{kj}(\neg H_k)$, it is finally expressed :

$$Pl_j(H_i) = Kf_j * \{m_{ij}(H_i) + m_{ij}(E)\} / \{m_{ij}(\neg H_i) + m_{ij}(E)\}, \, i \in [1,I] \tag{3.17}$$

in which the factor Kf_j is independent of the hypothesis H_i concerned.

Holding to constraint (3.8) as required by Axiom 3.1 will then permit the probability $p(s_j/H_i)$ alone to be associated with mass set $m_{ij}(.)$ alone, for $q_{ij}=1$, only if :

$$\{m_{ij}(H_i) + m_{ij}(E)\} / \{m_{ij}(\neg H_i) + m_{ij}(E)\} = R_j * p(s_j/H_i) \tag{3.18}$$

in which R_j is a normalization constant independent of H_i, whose possible values depend only on the distributions $p(s_j/H_i)$ actually taken into account, as we shall see in the following. In pracitice, this constant allows us to consider the general framework where the $p(s_j/H_i)$ are known only relatively, *i.e.* to within a normalization gain.

Expressed parametrically as a funcion of the level of uncertainty $m_{ij}(E)$, (3.18) procures the desired mass set :

$$m_{ij}(H_i) = \{R_j*p(s_j/H_i)-m_{ij}(E)\}/\{1+R_j*p(s_j/H_i)\} \qquad (3.19)$$
$$m_{ij}(\neg H_i) = \{1-R_j*p(s_j/H_i)*m_{ij}(E)\}/\{1+R_j*p(s_j/H_i)\} \qquad (3.20)$$
$$m_{ij}(E) = f[R_j*p(s_j/H_i)] \in [0, R_j*p(s_j/H_i)] \qquad (3.21)$$

in which f is any function verifying simply (3.21).

This condition (3.21) is required by the mass idea (included between 0 and 1), which also limits the possible values of R_j as a function of the distributions $p(s_j/H_i)$ used, and does so independently of the measures s_j actually observed :

$$R_j \in [0, (\max_{s_j,i}\{p(s_j/H_i)\})^{-1}] \qquad (3.22)$$

It is furthermore possible to show that these conditions are sufficient in order for the coefficient K_j in expression (3.8), calculated for the combination (3.16), to verify the constraint (3.9). This can be done simply by showing that the expression for K_j is then an increasing monotonic function of each $m_{ij}(E)$, whose extreme values make it possible to satisfy the interval (3.9).

If we introduce the factor q_{ij} into the expressions (3.19), (3.20), and (3.21) in terms of discounting, the $m_{ij}(.)$ are finally given by :

$$m_{ij}(H_i) = q_{ij}*\{R_j*p(s_j/H_i)-A_i\}/\{1+R_j*p(s_j/H_i)\} \qquad (3.23)$$
$$m_{ij}(\neg H_i) = q_{ij}*\{1-R_j*p(s_j/H_i)*A_i\}/\{1+R_j*p(s_j/H_i)\} \qquad (3.24)$$
$$m_{ij}(E) = 1-q_{ij}+q_{ij}*A_i \qquad (3.25)$$

in which R_j is still defined by (3.22), and A_i by :

$$A_i = f[R_j*p(s_j/H_i)] \in [0, R_j*p(s_j/H_i)] \qquad (3.26)$$

The general expression of the models $m_j(.)$ that satisfy Axioms 3.1 and 3.2 is thus found by (3.16) applied to (3.23), (3.24), and (3.25). An infinite number of solutions thus still fit our problem.

3.4 Axiom 3.3 : Consistency With the Probabilistic Association of the Sources

Considering the special structure (3.16) of the mass sets $m_j(.)$ complying with Axioms 3.1 and 3.2, and the associativity of the orthogonal sum, Axiom 3.3 will be satisfied for models such that, if the q_{ij} are equal to 1, the mass set $m_i(.)$ defined by :

$$m_i(.) = \bigoplus_j m_{ij}(.) \qquad (3.27)$$

$$m_{ij}(.) = F[R_j*p(s_j/H_i)] \qquad (3.28)$$

is identical to the mass set $m'_i(.)$ obtained by direct modeling, using the same function $F(.)$:

$$m'_i(.) = F[\prod_j \{R_j*p(s_j/H_i)\}] \qquad (3.29)$$

The $m_{ij}(.)$ verifying (3.19), (3.20), (3.21), and (3.22), in the combination (3.27), yield :

$$m_i(H_i) = (V*X-Y*W)/(V*X+X-Y*W) \qquad (3.30)$$
$$m_i(\neg H_i) = (X-Y*W)/(V*X+X-Y*W) \qquad (3.31)$$
$$m_i(E) = Y*W/(V*X+X-Y*W) \qquad (3.32)$$

with the definitions :

$$V = \prod_j \{R_j*p(s_j/H_i)\} \qquad (3.33)$$

$$W = \prod_j \{1+R_j*p(s_j/H_i)\} \qquad (3.34)$$

$$X = \prod_j \{1+m_{ij}(E)\} \qquad (3.35)$$

$$Y = \prod_j m_{ij}(E) \qquad (3.36)$$

and the constraints :

$$m_{ij}(E) = f[R_j*p(s_j/H_i)] \in [0, R_j*p(s_j/H_i)] \qquad (3.37)$$
$$R_j \in [0, (\max_{s_j, i}\{p(s_j/H_i)\})^{-1}] \qquad (3.38)$$

At the same time, the mass set $m'_i(.)$ is written :

$$m'_i(H_i) = \{V-m'_i(E)\}/\{1+V\} \qquad (3.39)$$
$$m'_i(\neg H_i) = \{1-V*m'_i(E)\}/\{1+V\} \qquad (3.40)$$

$$m'_i(E) = f[\prod_j \{R_j^* p(s_j/H_i)\}] \in [0, \prod_j \{R_j^* p(s_j/H_i)\}] \qquad (3.41)$$

in which V is still given by (3.33), and the R_j are also constrained by (3.38).

We can go about comparing the mass sets $m_i(.)$ and $m'_i(.)$ by letting $m_i(E)=m'_i(E)$ in (3.32). Then expressions (3.39) and (3.40) are equivalent to expressions (3.30) and (3.31), respectively, which means that under all circumstances $m_i(H_i)=m'_i(H_i)$ and $m_i(\neg H_i)=m'_i(\neg H_i)$. On the other hand, (3.37) and (3.41) will be equivalent for the same function f, through (3.32) still under the constraint $m_i(E)=m'_i(E)$, only for the following two functions f :

$$f(x) = 0, \forall x \qquad (3.42)$$
$$f(x) = x \qquad (3.43)$$

After examination of Axiom 3.3, only two models are left that simultaneously satisfy the three axioms. Both are defined by (3.23), (3.24), (3.25). They differ by the fact that $A_i=0$ for one while $A_i=R_j^* p(s_j/H_i)$ for the other, the R_j being constrained by (3.22) in both cases.

3.5 Summary of Models Obtained

There are finally only two models, then, that jointly satisfy the three desired axioms. Both meet the decomposition :

$$m_j(.) = \bigoplus_i m_{ij}(.) \qquad (3.44)$$

Model 1 is particularized by :

$$m_{ij}(H_i) = 0 \qquad (3.45)$$
$$m_{ij}(\neg H_i) = q_{ij}^* \{1-R_j^* p(s_j/H_i)\} \qquad (3.46)$$
$$m_{ij}(E) = 1-q_{ij}+q_{ij}^* R_j^* p(s_j/H_i) \qquad (3.47)$$

and *Model 2* by :

$$m_{ij}(H_i) = q_{ij}^* R_j^* p(s_j/H_i)/\{1+R_j^* p(s_j/H_i)\} \qquad (3.48)$$
$$m_{ij}(\neg H_i) = q_{ij}/\{1+R_j^* p(s_j/H_i)\} \qquad (3.49)$$
$$m_{ij}(E) = 1-q_{ij} \qquad (3.50)$$

In both cases, the normalization factor R_j is simply constrained by :

$$R_j \in [0, (\max_{s_j, i}\{p(s_j/H_i)\})^{-1}] \tag{3.51}$$

3.6 Tie-in With the Generic Problem

The problem of discriminating on the basis of a statistical learning process, as described in section 3, is in fact a special case of the general problem discussed in section 2. Both models provided by (2.3) to (2.5) and (2.6) to (2.8) in section 2 are in fact strictly equivalent to the two models found here in (3.45) to (3.47) and in (3.48) to (3.50), if we adopt the following respective definitions for the C_{ij}:

for model 1 : $\qquad C_{ij} = R_j * p(s_j/H_i)$ \hfill (3.52)

for model 2 : $\qquad C_{ij} = R_j * p(s_j/H_i)/[1+R_j * p(s_j/H_i)]$ \hfill (3.53)

in which R_j is still, of course, the normalization gain constrained by (3.51).

This outcome is in fact legitimate if we note that Axiom 2.1 is expressed directly by Axiom 3.2, and that the solutions required by Axioms 3.1 and 3.3 automatically verify Axiom 2.2. Axioms 3.1 and 3.3 simply make it possible to specify the inclusion of the particular information $p(s_j/H_i)$ in the expression for the criterion, C_{ij}.

It should nonetheless be pointed out that the probabilistic nature of the constraints that allow us to define more precisely the problem discussed here (consistency with Bayes) prevents any formal development in the framework of possibility theory, as the ideas of Bayesian and consonant credibility are incompatible.

Furthermore, whether for the generic problem or for the problem with statistical learning, the absence of C_{ij} or $p(s_j/H_i)$ data, characterized by $q_{ij}=0$, does amount to ignoring the corresponding mass set $m_{ij}(.)$, since in this case the mass set is trivial ($m_{ij}(E)=1$) and is therefore a neutral element of the orthogonal sum. So the models proposed always have the faculty of handling incomplete data sets. Moreover, the elementary models $m_{ij}(.)$ are simply defined on $\{H_i, \neg H_i\}$ with no requirement concerning the content of $\neg H_i$. In particular, they allow the association of sources defined in different frames of reference, either by intrinsic inclusion of the referential differences, or by integration into more general methods in terms of the considered mass sets, as the models remain entirely compatible with these methods [8, 9].

Lastly, when the data s_j are discrete values (local identity declarations, for example), the generalized Bayes theorem defined by P. SMETS in the framework of evidence theory can be applied, for the case of statistical learning, to the cartesian product of the set of data and set of hypotheses. It then strictly yields Model 1 developed here [10].

4 Target Classification

The target classification function consists in recognizing the type of target observed, or even identifying a target, on the basis of the different discriminating features s_j delivered by the sensors S_j used. So the question is to designate the most likely hypothesis H_i^* in light of the information generated. Such a decision, which is immediate when a probability can be associated *a posteriori* with each hypothesis, becomes quite delicate when the evaluations are presented in terms of the mass sets of evidence theory. The whole difficulty revolves around the non-exclusivity of the evaluations, which raises the practical problem of interpretation and relative inclusion of the masses attached to those focal elements of cardinal 2 or greater, in the designation of a unique singleton. This problem, which is general to evidence theory and unavoidable in the present context, has, as of today, been addressed only by more or less satisfactory intuitive solutions.

So below, we propose three different global approaches to the problem of choosing the most likely hypothesis H_i^*, considering an arbitrary mass set m(.) on the frame of discernment E={H_1,..., H_I}, when no other *a priori* basis for discriminating among the Hi is retained.

A synthesis of the resulting procedures provides a decision law suited to the classification problem. When applied to the modeled mass sets, either in the framework of the generic problem or in that of statistical learning, this law supplies classification methods of noteworthy interest.

4.1 "Focalized" Global Approach

This approach consists in defining I certain mass sets $m_i(.)$, each of them being respectively focused on each of the I hypotheses H_i of the frame of discernment E ($m_i(H_i)$=1). The inconsistency K_i, provided by the orthogonal sum of the mass set $m_i(.)$ and the available mass set m(.), reflects their disagreement, and so represents the conflict between the assessment m(.) and the fact that hypothesis H_i is actually true. According to this, we have to choose the hypothesis H_i^* that ensures a minimal inensistency K_i. As K_i can be written :

$$Ki = 1 - Pl(Hi) \tag{4.1}$$

in which Pl(.) is the plausibility function associated with m(.), we have to choose the hypothesis that provides a maximal plausibility.

The interest in this inconsistency criterion is confirmed by the idea of entropy that is connected with it [5].

4.2 "Bayesian" Global Approach

The idea here is to consider a given set of "equiprobable" Bayesian masses $m_0(.)$ on the frame of discernment E ($m_0(H_i)=1/I$, $\forall\, i\in[1,I]$). Endowing this mass set $m_0(.)$ with a role similar to that of equiprobable *a priori* probabilities in the Bayesian inference, a mass set $m_c(.)$ can be determined by orthogonal sum of the mass set $m_0(.)$ and the available mass set $m(.)$. $m_c(.)$ is then a Bayesian mass set defined by:

$$m_c(H_i) = Pl(H_i) / \{ \sum_{k\in[1,I]} Pl(H_k) \}, \quad i\in[1,I] \tag{4.2}$$

$$m_c(A) = 0, \quad \forall\, A\neq H_i, \quad i\in[1,I] \tag{4.3}$$

in which $Pl(.)$ is the plausibility function associated with $m(.)$. By reference to the maximum *a posteriori* probability, the decision procedure obviously consists in retaining the hypothesis H_i^* that has the maximum mass, and thus the maximum plausibility $Pl(H_i^*)$, here again.

Conceptually, the principle of this approach consists in substituting an equal confidence between the singletons of the frame of discernment in place of the total *a priori* uncertainty, so as to force the discrimination among these elements alone.

4.3 General Approach of the Decision

Here we look for a solution with reference to a more general decisional context summarized, for example, in [6]. The purpose is to choose one of a number of possible actions to take, a_h, on the basis of the evaluation provided by the mass set $m(.)$ on the frame of discernment E.

This choice can be made by maximizing a cost function $C(a_h)$ on the set of possible actions, knowing the weight $G(a_h/B_k)$ assigned to each potential action a_h when the event B_k, a subset of E, occurs:

$$C(a_h) = \sum_{B_k \subset E} \{G(a_h/B_k) * m(B_k)\} \tag{4.4}$$

The whole difficulty of using such a procedure in practice, and hence its credibility, resides in the evaluation of the weights, $G(a_h/B_k)$, which is usually very subjective. While we may in general consider that the weights relative to the singletons H_i of E are given by the system or user, those relative to the subsets B_k of cardinal 2 or higher must, on the other hand, be determined intuitively, possibly in accordance with a preferred "attitude" [6].

Yet in our case, this subjective character can be greatly attenuated by the one-to-one correspondence we have to establish between the set of actions and the frame of discernment E, as each action a_h consists in declaring a hypothesis H_i to be true. So, with no specific requirements, the weights are legitimately given by :

$$G(a_i/B_k) = 1 \quad \text{si} \quad H_i \in B_k \qquad (4.5)$$
$$G(a_i/B_k) = 0 \quad \text{si} \quad H_i \notin B_k \qquad (4.6)$$

so as to conform with the associated idea of mass $m(B_k)$ as introduced by evidence theory, *i.e.* as an evaluation of one of the elements of B_k, though we cannot specify of which element of B_k it is.

Under these conditions, (4.4) also leads to the designation of hypothesis H_i^* of maximum plausibility $Pl(H_i^*)$ as the most likely.

4.4 Synthesis

The three global approaches presented all converge to the same decisional procedure, which consists in retaining the most likely hypothesis H_i^* such that :

$$Pl(H_i^*) = \max_{i \in [1,I]} \{ Pl(H_i) \} \qquad (4.7)$$

It should also be noted that this decision law is the one that satisfies the constraint developed at the end of section 3.2.

For all the models discussed in sections 2 and 3, as the hypotheses H_i are singletons of the frame of discernment E, the plausibility $Pl(H_i)$ is proportional to the j and k product of the $Pl_{kj}(H_i)$ associated with the $m_{kj}(.)$. After normalization by the j and k product of the $Pl_{kj}(\neg H_k)$, we come to designate H_i^* by the criterion :

$$Pl(H_i^*) = \max_{i \in [1,I]} \left\{ \prod_j \left([m_{ij}(H_i)+m_{ij}(E)] / [m_{ij}(\neg H_i)+m_{ij}(E)] \right) \right\} \qquad (4.8)$$

4.5 Generic Problem

The criterion established above can be applied to the two models provided for the generic problem, leading to the two respective solutions :

Solution 1: max $\{ \prod_i \prod_j [1-q_{ij}*(1-C_{ij})] \}$ (4.9)

Solution 2: max $\{ \prod_i \prod_j [1-q_{ij}*(1-C_{ij})]/[1-q_{ij}*C_{ij}] \}$ (4.10)

It should be noted that Solution 1 also meets a maximum credibility criterion.

Concerning the reference possibilistic approach, a maximum possibility criterion is unavoidable, as $N(H_i)$ is zero. It generates the corresponding solution:

max $\{ \min_{i,j} [1-q_{ij}*(1-C_{ij})] \}$ (4.11)

The behaviors of solutions 1 and 2 have been compared by various simulations to the possibilistic solution, and also to "reference" solutions generated by ordinary techniques, when individual confusion matrices are available for each sensor [1,3]. The best performance is always obtained with the solutions proposed, essentially due to the fineness of the combination operator used, to centralized association under all circumstances, and to the inclusion of all the available information in an appropriate framework.

The simplicity of the calculations and ease of use of these solutions is also worth noting.

4.6 Problem With Statistical Learning

Criterion (4.8), applied to the two models obtained for dealing with statistical learning, generate the following two solutions, respectively:

Solution 1: max $\{ \prod_i \prod_j [1-q_{ij}+q_{ij}*R_j*p(s_j/H_i)] \}$ (4.12)

Solution 2: max $\{ \prod_i \prod_j ([1-q_{ij}+R_j*p(s_j/H_i)]/[1+(1-q_{ij})*R_j*p(s_j/H_i)]) \}$ (4.13)

in which R_j is still constrained by (3.51).

If we try to compare these two procedures considering the optimized expressions as two functions of the variables q_{ij} and $[R_j*P(m_j/H_i)]$, it appears that they are numerically very close. They satisfy the same limiting values and the same monotonies, with a maximum difference of a few percent for certain median values. The values taken by the function relative to Model 2 are also always greater than those of the function established on Model 1. In practice, these two formulations will therefore yield generally equivalent performance, which has in effect been observed in practice [2]. In such a situation, it is clear that our preference should go to the simpler formulation, *i.e.* the one associated with Model 1.

Let us also note that, when all the q_{ij} are 1, *i.e.* when the distributions $p(s_j/H_i)$ are perfectly representative of reality, the two approaches do in fact reduce to a maximum likelihood procedure.

The simulations [2] that have been conducted moreover reveal the robustness imparted to the solutions proposed, as well as the aptitude of the q_{ij} factors to integrate linguistic or subjective information, considering the low sensitivity of the results to the choice of a given value for these coefficients. Implementation of Solution 1 using neuro-fuzzy techniques further brings out an automatic learning of the factors q_{ij} for complex situations [11].

Lastly, the remark of section 4.5 concerning simplicity holds well here, too.

5 Target Tracking

The problem dealt with here is that of tracking a moving target of any possible nature, in a dense environment, using observations delivered by a set of disparate and possibly delocalized sensors. One of the main purposes is to overcome the problem of spurious sources present in the vicinity of the target. These sources may be due to intelligent countermeasures, artifacts, or vehicles that are untracked, for operational or technical reasons. A situation of major practical interest appears when the tracking is initialized on objects that are very close together or even at the same point, such as when a fighter plane enters an airspace hidden behind or close to an airliner.

Simultaneous tracking of multiple targets may also be suitably handled with the proposed approach, as is shown by the extension proposed at the end of the present discussion.

Ordinarily, systems operating in this field make sequential use of detection, filtering, and, if any, classification functions. It is then possible to match measurements from different sensors before or after filtering, by associating plots or tracks.

The essential distinguishing feature of the concept proposed resides in the fact that it performs a filtering directly on the signals available in the different resolution cells of each sensor, rather than on plots provided by a detection procedure. The signal processing consists in extracting discriminatory features concerning the target identiy, and generating an identity likelihood function for each resolution cell concerned. These likelihoods are combined among themselves and then combined with the likelihood stemming from the prediction, to update the filters adapted to various identities.

Although it constitutes no particular limitation on the concept proposed, the discussion here presumes two simplifying assumptions :

- The target tracked is the only one of its particular identity in the space being processed. Any number of other targets may exist, but with different identities.

- The targets are resolved, which means that a given resolution cell contains at most one target of any given identity.

5.1 General Principle

The technique used for the filtering aspects is inspired directly from the Probabilistic Data Association Filter (PDAF) family of methods developed by Y. BAR SHALOM from the ordinary KALMAN filter, to handle multiple detections [13]. These methods differ essentially from the KALMAN filter by the estimate updating phase, in which they proceed in two steps :

- First the statistical gating selects the detected plots located in a given vicinity of the predicted position. The vicinity is determined so as to contain the target with an *a priori* probability greater than a given threshold.

- Then the estimate and its covariance are updated on the basis of an innovation determined by linear combination of the innovations individually due to each plot retained as potential successor of the processed track. The weighting coefficients are the *a priori* probabilities for each of these plots to actually be due to the target, considering the detection and false alarm probabilities of the detector used, the predicted position and its covariance, and the statistical gating threshold.

In a first approximation, the method proposed here can be interpreted as a PDAF whose detection would operate at minimum threshold, with Detection Probability = False Alarm Probability = 1. At the level of the statistical gating, then, this is equivalent to retaining and processing one plot per resolution cell located within the vicinity defined around the predicted position.

The "*a priori*" probability that weights the innovation due to each of these plots in updating the estimate is, on the other hand, modified to reflect the likelihood of the identity present in the corresponding cell — information generated from the recognition of identity features extracted from the signal isolated by the spatial resolution of the sensors.

The expression for the filtering, prediction, and statistical gating modules specifically for this modified version of the PDAF is given in the appendix. The following discussion concerns the special development of the weightings then necessary for the innovation.

Two indispensable ideas supporting these developments must nonetheless be defined :

- The sensors are said to be "aligned" if they break the validation gate down into the same resolution cells. For convenience here, the sensors are assumed to be classed in groups of sensors that are aligned among themselves, while two sensors of two different groups are necessarily unaligned. Each sensor will thus be denoted S_j^l, where l designates to which of the L groups of aligned sensors the sensor in question belongs, and j is its sequence number within the group of J sensors.

- If, for a group l of aligned sensors, x^{ln} designates the nth of N resolution cells of nonzero intersection with the validation gate, then the sensors in question "resolve" the gate if the gate entirely includes each x^{ln}.

The extraction of features in each resolution cell x^{ln} by each sensor S_j^l is also assumed to provide information of the type considered by the generic model (section 2), or more specifically by the model with statistical learning (section 3), such that these models are applicable. We therefore have I*J mass sets $m_{ij}(.)$ per resolution cell x^{ln}, with each cell being defined either by (2.3) to (2.5), or (2.6) to (2.8), or (3.45) to (3.47), or (3.48) to (3.50). So they are, from now on, denoted $m_{ij}^{ln}(.)$, by reference to the resolution cell x^{ln} to which it relates, and their respective frames of discernment are denoted $E_i^{ln} = \{H_i^{ln}, \neg H_i^{ln}\}$. It will be noted that the use of the models established above is advantageous in light of their suitability to the problems generally encountered, but that this is not indispensable : the discussion here starts with any given mass sets $m_{ij}^{ln}(.)$, which can be obtained by any other means.

The object of the procedure to be developed is therefore to combine the various sources $m_{ij}^{ln}(.)$, each being specific to a sensor S_j^l, a resolution cell x^{ln}, and a particular evaluated identity H_i. The combination is performed in such a way as to provide the likelihood of each possible distribution of identity hypotheses (including target absence) on the M resolution cells x^m of the validation gate. The x^m cells are the intersections of the x^{ln} cells of the various groups l of sensors, so that the combination processes applied offer the best spatial resolution at the end of the process.

Thanks to a special property of evidence theory, the combination of the resulting likelihoods with the *a priori* localization probabilities (α^0, α^m) of the tracked target delivered by the filter prediction, directly generate the *a posteriori* probabilities (β^0, β^m) of the target in question. The probabilities α^m and β^m are relative to the presence of the target in the cell x^m, while the probabilities α^0 and β^0 concern its absence in the gate. The probabilities β^m and β^0 are used to weight the innovation due to each of the cells x^m in the estimate update, as was introduced above.

This concept also naturally provides track validation criteria, typically based on the likelihood of the actual presence of a target of the desired identity in the validation gate, in consideration of the various features observed.

It should also be noted that such a filter is by nature suited to a given identity, the purpose of the proposed concept being to reject as effectively as possible those signals that might be due to neighboring targets of different identity. In track initialization phase, a battery of different filters suitable for different identities should therefore be used. The filter whose identity is most likely can be chosen progressively using track validation criteria. This organization also makes it possible to adopt the most appropriate dynamic model for the identity processed, for each filter.

5.2 Association Procedures and Filter Expression

For convenience of explanation, we will assume hereafter that the identity of the tracked target is designated by H_I.

Considering the nature of the problem, the required associations must be performed by orthogonal sum of all of the sources, to obtain their conjunction. This must be done in the finest common frame of discernment, which is the set E^F of the possible identity distributions on the various cells x^m of the validation gate. As the orthogonal sum is commutative, the association order of the various sources is theoretically arbitrary. To simplify the calculations, however, the approach chosen consists in associating the sources by order of decreasing similarity of their frame of discernment, whereas applying appropriate refinements at each step. Figure 1 shows the resulting logic of operations.

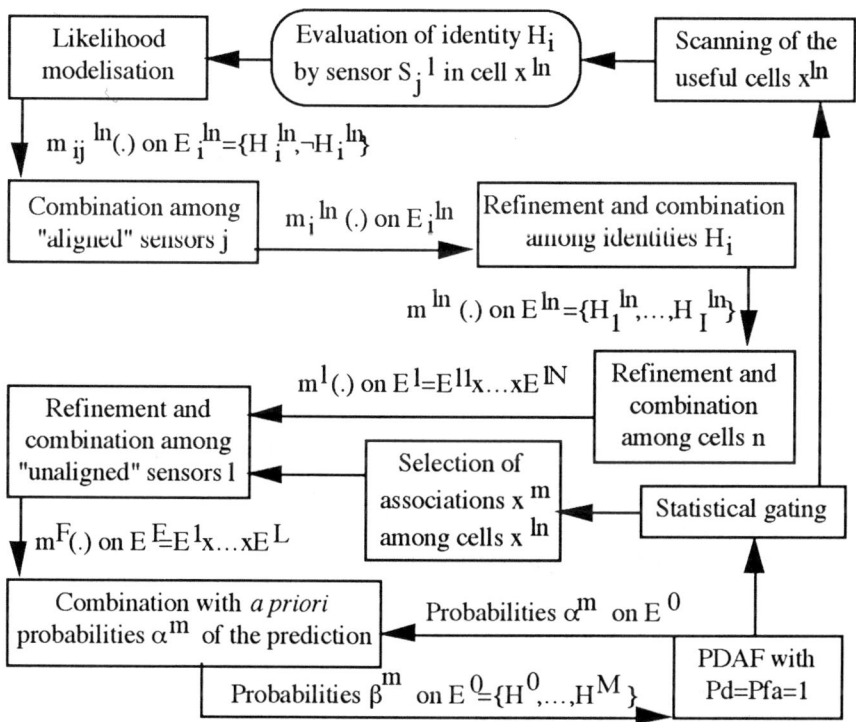

Fig. 1. Combination processing.

Procedure at the Level of Each Resolution Cell

The first step consists of associating the mass sets $m_{ij}^{ln}(.)$ defined on the same frame of discernment $E_i^{ln} = \{H_i^{ln}, \neg H_i^{ln}\}$, among the various sensors j of a given alignment group l. For each E_i^{ln}, their orthogonal sum directly yields the mass set $m_i^{ln}(.)$ defined by:

$$m_i^{ln}(H_i^{ln}) = \{ \prod_{j=1}^{J} [m_{ij}^{ln}(H_i^{ln}) + m_{ij}^{ln}(E_i^{ln})] - \prod_{j=1}^{J} m_{ij}^{ln}(E_i^{ln}) \} / (1 - K_i^{ln}) \quad (5.1)$$

$$m_i^{ln}(\neg H_i^{ln}) = \{ \prod_{j=1}^{J} [m_{ij}^{ln}(\neg H_i^{ln}) + m_{ij}^{ln}(E_i^{ln})] - \prod_{j=1}^{J} m_{ij}^{ln}(E_i^{ln}) \} / (1 - K_i^{ln}) \quad (5.2)$$

$$m_i^{ln}(E_i^{ln}) = \{ \prod_{j=1}^{J} m_{ij}^{ln}(E_i^{ln}) \} / (1 - K_i^{ln}) \quad (5.3)$$

in which K_i^{ln} represents the inconsistency of the combination, which does not call for an explanation for the purposes of the remaining discussion.

The refinement of E_i^{ln} in the set $E^{ln} = \{H_1^{ln}, \ldots, H_I^{ln}\}$ of the identities attached to the cell x^{ln} allows the minimum extension, in the common frame of discernment E^{ln}, of the $m_i^{ln}(.)$ relative to these different identities. The orthogonal sum of the mass sets obtained leads to the mass set $m^{ln}(.)$. Then all we have to do is express the plausibilities of H_1^{ln} and of $\neg H_1^{ln}$, which are all that is needed for the rest of the discussion.

$$Pl^{ln}(H_1^{ln}) = (1-K^{ln})^{-1} * [m_1^{ln}(H_1^{ln}) + m_1^{ln}(E_1^{ln})] * \prod_{i=1}^{I-1} [m_i^{ln}(\neg H_i^{ln}) + m_i^{ln}(E_i^{ln})] \quad (5.4)$$

$$Pl^{ln}(\neg H_1^{ln}) = (1-K^{ln})^{-1} * [m_1^{ln}(\neg H_1^{ln}) + m_1^{ln}(E_1^{ln})] * [\prod_{i=1}^{I-1} [m_i^{ln}(\neg H_i^{ln}) + m_i^{ln}(E_i^{ln})]$$

$$- \prod_{i=1}^{I-1} m_i^{ln}(\neg H_i^{ln}) + \sum_{i=1}^{I-1} \{ m_i^{ln}(H_i^{ln}) * \prod_{\substack{i'=1 \\ i' \neq i}}^{I-1} [m_{i'}^{ln}(\neg H_{i'}^{ln}) + m_{i'}^{ln}(E_{i'}^{ln})] \}] \quad (5.5)$$

in which K^{ln} is the combination inconsistency, the expression of which is not useful for the following.

Summary at the Level of the Validation Gate

So two cases should be distinguished, depending on whether the x^{ln} resolve the validation gate or not (see section 5.1). If they do, then the E^{ln} relative to each x^{ln} cell need only be refined in the set $E^l = E^{l1} x \ldots x E^{lN}$ of possible identity distributions on the cells in question, and the orthogonal sum of the resulting minimum extensions can be performed. The very special nature of the associated focal elements, each being specific to a distinct component of the cartesian product, allows a relatively simple expression for the only plausibilities we now have to evaluate on the basis of (5.4) and (5.5). These plausibilities concern the N hypotheses H^{ln} of the presence of identity H_I, respectively, in the cell x^{ln}, to the exclusion of any other cell, and the hypothesis H^0 of the absence of identity H_I in the gate. These hypotheses are in fact specific subsets of E^l, as there exists one and only one target of identity H_I in the gate, according to the axiom adopted to start with (section 5.1). This leads to :

$$Pl^l(H^0) = \prod_{n=1}^{N} Pl^{ln}(\neg H_I^{ln}) \qquad (5.6)$$

$$Pl^l(H^{ln}) = Pl^{ln}(H_I^{ln}) * \prod_{\substack{n'=1 \\ n' \neq n}}^{N} Pl^{ln'}(\neg H_I^{ln'}) \qquad (5.7)$$

If the x^{ln} do not resolve the validation gate, an additional prior refinement should be performed on each E^{ln} in a set $E'^{ln} = \{H_1^{ln0}, H_1^{ln1}, \ldots, H_I^{ln0}, H_I^{ln1}\}$ to separate each hypothesis H_i^{ln} between, on the one hand, a similar hypothesis H_i^{ln1} simply relative to the part of x^{ln} covering the gate, and, on the other hand, an additional hypothesis H_i^{ln0} relative to the part of x^{ln} outside the gate. The operations conducted in the case where the gate is resolved are then conducted on the modified set $E^l = E'^{l1} x \ldots x E'^{lN}$. However, hypothesis H^{ln} is now reduced to the presence of the identity H_I just in that part of cell x^{ln} covering the gate, and excluding any other cell defined in the gate. Hypothesis H^0, though its definition remains unchanged, also corresponds to a different subset of E^l. The result is the modified expressions :

$$Pl^l(H^0) = 1 \qquad (5.8)$$

$$Pl^l(H^{ln}) = Pl^{ln}(H_I^{ln}) \qquad (5.9)$$

The groups 1 of unaligned sensors are then combined in both cases by refining the E^l in the common set $E^F = E^1 x \ldots x E^L$, and performing the orthogonal sum of the associated minimal extensions. This step reflects the fact that we are interested in the intersections x^m of the cells x^{ln} (see section 5.1). As each of the associated focal elements here is still specific to a distinct component of the cartesian product, the plausibilities of H^0 and of the hypotheses H^m of presence of identity H_I in cell x^m, to the exclusion of any other cell, are expressed for the resulting mass set $m^F(.)$ by :

$$Pl^F(H^0) = \prod_{l=1}^{L} Pl^l(H^0) \qquad (5.10)$$

$$Pl^F(H^m) = \prod_{\substack{l=1 \\ x^m \subset x^{ln}}}^{L} Pl^l(H^{ln}) \qquad (5.11)$$

Determination of Weightings β^m and Practical Expression for the Filter

The last step consists in combining this result with the *a priori* probabilities α^m, which constitute a Bayesian mass set on $E^0 = \{H^0, H^1, \ldots, H^M\}$. As E^0 is a partition of a subset of E^F, conditioning and coarsening $m^F(.)$ of E^F on E^0 makes it possible to take the orthogonal sum with the set of α^m. The resulting mass set, which is Bayesian over E^0, is directly the set of probabilities β^m we are looking for :

$$\beta^0 = \alpha^0 * Pl^F(H^0) / \{\alpha^0 * Pl^F(H^0) + \sum_{m=1}^{M} \alpha^m * Pl^F(H^m)\} \qquad (5.12)$$

$$\beta^m = \alpha^m * Pl^F(H^m) / \{\alpha^0 * Pl^F(H^0) + \sum_{m'=1}^{M} \alpha^{m'} * Pl^F(H^{m'})\} \qquad (5.13)$$

Expressions (5.1) to (5.13) can then be summarized by :

$$\beta^0 = \alpha^0 / \{\alpha^0 + \sum_{m=1}^{M} \alpha^m * Q^m\} \qquad (5.14)$$

$$\beta^m = \alpha^m * Q^m / \{\alpha^0 + \sum_{m'=1}^{M} \alpha^{m'} * Q^{m'}\} \qquad (5.15)$$

in which : $Q^m = Pl^F(H^m) / Pl^F(H^0) = \prod_{\substack{l=1 \\ x^m \subset x^{ln}}}^{L} Q^{ln} \qquad (5.16)$

with, for the sensor groups l that resolve the gate :

$$Q^{ln} = \prod_{j=1}^{J} A_{Ij}^{ln} / \{1 - \prod_{i=1}^{I-1} (1 - \prod_{j=1}^{J} B_{ij}^{ln}) + \sum_{i=1}^{I-1} (\prod_{j=1}^{J} A_{ij}^{ln} - \prod_{j=1}^{J} B_{ij}^{ln})\} \quad (5.17)$$

and, for the sensor groups l that do not resolve the gate :

$$Q^{ln} = \prod_{j=1}^{J} A_{Ij}^{ln} / \{1 - \prod_{i=1}^{I} (1 - \prod_{j=1}^{J} B_{ij}^{ln}) + \sum_{i=1}^{I} (\prod_{j=1}^{J} A_{ij}^{ln} - \prod_{j=1}^{J} B_{ij}^{ln})\} \quad (5.18)$$

In both cases, the coefficients A_{ij}^{ln} and B_{ij}^{ln} represent, respectively, the expressions :

$$A_{ij}^{ln} = \{m_{ij}^{ln}(H_i^{ln}) + m_{ij}^{ln}(E_i^{ln})\} / \{m_{ij}^{ln}(\neg H_i^{ln}) + m_{ij}^{ln}(E_i^{ln})\} \quad (5.19)$$

$$B_{ij}^{ln} = m_{ij}^{ln}(E_i^{ln}) / \{m_{ij}^{ln}(\neg H_i^{ln}) + m_{ij}^{ln}(E_i^{ln})\} \quad (5.20)$$

Formulas (5.14) to (5.20) therefore express the probabilities β^m the filter requires, starting with the probabilities α^m it generates, while (5.19) and (5.20) are fed by any of the models developed in sections 2 or 3. The full expression for the appropriate filter is given in the appendix. This filter will hereafter be designated the Multiple Signal Filter (MSF).

Let us note that, if we consider only two identity hypotheses (*i.e.* absence or presence of a target in each cell in question), if the available information is of the probabilistic type used by the models in section 3, and if the distributions are perfectly representative of reality (all $q_{ij}^{ln}=1$), then there exists a Bayesian solution to the problem. It is easy to verify that this solution actually is the special case of the filter proposed for the stated conditions. Such a solution was, for example, used in the PDAFAI to include the amplitude of the observed signal [14].

On the other hand, as soon as the number of identity hypotheses exceeds 2 (absence/presence of the tracked target, in the cell considered), no formal probabilistic approach is possible any more, since the prediction can no longer provide the *a priori* probabilities of the different identities needed for the Bayesian inference of the update. One of the advantages of the approach proposed is therefore to obtain an exact solution for these situations, which are especially of concern here (see introduction to section 5). The method described also makes it possible to manage the uncertainty on the models and to include data that is not necessarily probabilistic.

5.3 Joint Tracking of Multiple Targets

We now propose to extend the single-target concept above to the joint tracking of multiple targets whose validation gates overlap. The purpose is therefore to develop a new Joint Multiple Signal Filter (JMSF) from the Joint Data Association Filter (JDAF) of

Y. BAR SHALOM [13], using the aproach that allowed us to establish the MSF from the PDAF. Let P be the number of tracks concerned. All the notation used up till now is conserved, with an added subscript p to indicate the track to which the notation refers.

The formulation sought can be obtained by refining the mass sets $mF_p(.)$, defined at the level of each track by (5.10) and (5.11), of $E^F{}_p$ in the cartesian product $E^X = E^F{}_1 \times ... \times E^F{}_P$, and performing their orthogonal sum in this new reference system. The result should then be conditioned and coarsened on the cartesian product of the $E^0{}_p = \{H^0{}_p, ..., H^{M_P}{}_p\}$, minus the different target position combinations in which more than one target is located in the same resolution cell. The mass set obtained can then be combined with the weighting coefficients $\alpha^{m_P}{}_p$ that would be used in a JPDAF operating at Detection Probability = 1, similar to those found in (A1) and (A2) for the PDAF. These coefficients are actually identical to *a priori* probabilities in the reference system considered. This leads us to the *a posteriori* weighting coefficients $\beta^{m_P}{}_p$ needed for updating the filters associated, respectively, with each track p :

$$\beta^{m_P}{}_p = D^{-1} * \sum_{\substack{m_{p'} \in [0, M_{p'}] \\ p' \in [1,P]-\{p\} \\ x^{m_1} \neq ... \neq x^{m_P}}} \{\alpha^{m_P}{}_p * Pl_p(H^{m_P}{}_p) * \prod_{p' \in [1,P]-\{p\}} [\alpha^{m_{p'}}{}_{p'} * Pl_{p'}(H^{m_{p'}}{}_{p'})] \} \quad (5.21)$$

where, by convention, $m_p = 0$ corresponds to a position of target p outside the gate, and where D is the normalization factor that guarantees :

$$\sum_{m_p=0}^{M_p} \beta^{m_P}{}_p = 1 \quad (5.22)$$

In practice, (5.21) is therefore expressed :

$$\beta^{m_P}{}_p = D^{-1} * \alpha^{m_P}{}_p * Q^{m_P}{}_p * \sum_{\substack{m_{p'} \in [0, M_{p'}] \\ p' \in [1,P]-\{p\} \\ x^{m_1} \neq ... \neq x^{m_P}}} \prod_{p' \in [1,P]-\{p\}} \alpha^{m_{p'}}{}_{p'} * Q^{m_{p'}}{}_{p'} \quad (5.23)$$

with, for each track p, $Q^0{}_p = 1$ and $Q^{m_P}{}_p$ given by (5.16) to (5.20).

5.4 Advantage of the Approach

Various operational advantages are illustrated in the framework of simulations conducted elsewhere [12]. These stem directly from the intrinsic features of the method proposed, which it was possible to develop in the framework of the theory of evidence by adopting a set approach to localization problem :

- Firstly, it is the likelihood functions that are combined at all levels and filtered, and not the local detection decisions used by traditional procedures. All the accessible data is therefore used as finely as possible, which avoids any compression of information that could only be harmful to the quality of the result, either in its precision, robustness against noise, or track management aspects.

- The classification features are also integrated at the level farthest upstream of the tracking filter, which the traditional Bayesian formalism does not allow in the presence of multiple targets of different identities, since the prediction function cannot produce all the necessary *a priori* probabilities. The sharpening of the resulting discrimination power allows the use of filters adapted to different target identities, and thus leads to a reciprocal improvement of the tracking and classification functions. Integrating the proposed procedure into methods of the Interacting Multiple Models (IMM) type using multiple models of motion could also make it possible to make use of models that are not only representative of a target motion dynamic, but also of an identity compatible with this dynamic. Such a synergy could be provided by an approach of the type used in [14] for the PDAFAI.

- When ordinary PDAF filtering is used, an optimum detection threshold has to be determined, and the resulting detection probability must be known. These requirements soon become critical in most practical applications, especially at the performance limits, because of the uncertainties in the expected signatures. The proposed technique offers better robustness against these uncertainties thanks to the suitable management of learning representativity defects possible with the generic models used at the outset.

- The ambiguities of spatial matching of data coming from disparate sensors are intrinsically processed at the level farthest upstream of the filter by including the likelihood of all possible associations of resolution cells, as developed on the basis of the discriminating features. Beyond the management and automatic processing of ghosts, false alarms, non-detections, and so forth, this approach offers better localization precision and better classification of targets separated this way, while avoiding the use of any unstable nonlinear filters.

- Tracks can be conveniently managed (initialization, validation, and so forth) on the basis of a track validity criterion reflecting the likelihood for a target of right identity to be effectively inside the tracking gate. Typically, a criterion like this has to guarantee a sufficiently low value of $Pl^F(H^0)$ provided by (5.10), and a sufficiently high maximum value between the various $Pl^F(H^m)$ calculated by (5.11).

It should lastly be noted that the combining procedures described here are not specific to a PDAF type filter, and could, for example, be used in synergy with methods based on hypothesis tests, such as the Multiple Hypothesis Filter (MHF) developed by D.B. REID [15]. The likelihood $m^F(.)$ of the spatial distributions of identities in the gate can in fact be included directly for generating likelihood criteria relative to the various association hypotheses, which then integrate the identity.

6 Conclusion

Simple, robust procedures have been developed in the framework of the theory of evidence for target tracking and classification functions in multiple-sensor analysis. They combine a generic modeling, for the purpose of integrating all the accessible information suitably, with original classification and tracking concepts, offering a centralized, global approach to the problems starting with arbitrary discriminating mass sets. A formal integration of the data matching, combining, classification, and tracking functions is possible with the proposed approach, with all the advantages this implies (see section 5.4).

References

1. A. Appriou : Formulation et traitement de l'incertain en analyse multi-senseurs. Conférence invitée, 14ème Colloque GRETSI, Juan-les-Pins, 13-16 septembre 1993.
2. A. Appriou : Probabilités et incertitude en fusion de données multi-senseurs. Revue Scientifique et Technique de la Défense, n°11, 1991-1, pp 27-40.
3. A. Appriou : Intérêt des théories de l'incertain en fusion de données. Conférence invitée, Colloque International sur le Radar, Paris, 24-28 avril 1989.
4. G. Shafer : A mathematical theory of evidence. Princeton University Press, Princeton, New Jersey, 1976.
5. R.R. Yager : Entropy and specificity in a mathematical theory of evidence. International Journal General Systems, Vol. 9, 1983, pp 249-260.
6. R.R. Yager : A general approach to decision making with evidential knowledge. Uncertainty in Artificial Intelligence, L.N. Kanal & J.F. Lemmer éd., Elsevier Science Publishers, B.V. North-Holland, 1986.
7. D. Dubois, H. Prade : Théorie des possibilités : application à la représentation des connaissances en informatique", Masson, Paris, 1988.
8. F. Janez, A. Appriou : Théorie de l'évidence et cadres de discernement non exhaustifs. Revue Traitement du Signal, Vol. 13, n° 2, 1996.
9. F. Janez, A. Appriou : Fusion of sources defined on different non-exhaustive frames. IPMU' 96, Granada, Spain, July 1-5 1996.
10. V. Nimier, A. Appriou : Utilisation de la théorie de Dempster-Shafer pour la fusion d'informations. GRETSI, 15ème Colloque sur le Traitement du Signal et des Images, Juan-les Pins, 18-22 septembre 1995.
11. M.C. Perron-Gitton : Apport d'une approche neuro-floue dans un contexte de fusion de données basé sur la théorie de l'évidence. IPMU' 94, Paris, 4-8 juillet 1994.
12. A. Appriou : Multiple signal tracking processes. Aerospace Science and Technology, n° 2, February 1997.
13. Y. Bar Shalom, T.E. Fortmann : Tracking and data association. Academic Press, New York, 1988.
14. D. Lerro, Y. Bar Shalom : Interacting multiple model tracking with target amplitude feature. IEEE Transactions on Aerospace and Electronic Systems, Vol. 29, n° 2, april 1993.

15. J. Dezert : Vers un nouveau concept de navigation autonome d'engin. Un lien entre le filtrage à association probabiliste de données et la théorie de l'évidence. Thèse de doctorat, Université Paris-XI, 27 septembre 1990, ONERA.

Appendix : Filter Expression

The filtering, prediction, and statistical gating modules are those of a PDAF that would operate at minimum threshold with Pd=Pfa=1.

Filtering

The α^m coefficients are given by :

$$\alpha^0 = M * (2*\prod/\gamma)^{r/2} * (1-Pg) / C_r \qquad (A1)$$

$$\alpha^m = \exp[-0.5 * (x^m - x_k)^T V_k^{-1} (x^m - x_k)] \qquad (A2)$$

in which : $C_r = \prod^{r/2} / \Gamma(1+r/2)$ \qquad (A3)

x_k and V_k designate the predicted position and its covariance, at time k. r is the common dimension of x_k and x^m. Pg represents the *a priori* probability that the target is in the validation gate, considering the choice of statistical gating threshold γ.

The β^m coefficients are determined from the α^m coefficients by (5.14) to (5.20). The estimated state $X_{k/k}$ and its covariance $P_{k/k}$, which are outputs of the procedure, are then updated at time k by :

$$X_{k/k} = X_{k/k-1} + G_k * z_k \qquad (A4)$$

$$P_{k/k} = \beta^0 * P_{k/k-1} + (1 - \beta^0) * (I - G_k * H) * P_{k/k-1} + P_k \qquad (A5)$$

in which : $P_k = G_k * [(\sum_{m \neq 0} \beta^m * z_k^m * z_k^{mT}) - z_k * z_k^T] * G_k^T$ \qquad (A6)

$$z_k^m = x^m - x_k \qquad (A7)$$

$$z_k = \sum_{m \neq 0} (\beta^m * z_k^m) \qquad (A8)$$

$$G_k = P_{k/k-1} * H^T * V_k^{-1} \qquad (A9)$$

H is the position observation matrix.

Prediction

The predicted state, $X_{k/k-1}$, and its covariance, $P_{k/k-1}$, used above for updating the filter, are calculated from the state $X_{k-1/k-1}$ and its covariance $P_{k-1/k-1}$ estimated at the time of previous observation k-1 by the filtering module :

$$X_{k/k-1} = F * X_{k-1/k-1} \tag{A10}$$
$$P_{k/k-1} = F * P_{k-1/k-1} * F^T + Q \tag{A11}$$

in which F is the state transition matrix from one observation time to the next, and Q the noise covariance matrix on the state.

The predicted position measurement x_k and its covariance V_k, used by the filtering and gating modules, are then determined by :

$$x_k = H * X_{k/k-1} \tag{A12}$$
$$V_k = H * P_{k/k-1} * H^T + R \tag{A13}$$

with R designating the noise covariance matrix on the position measurement.

Statistical Gating

The cells x^{ln} and x^m to be processed (figure 1) are selected by the tests :

$$(x^{ln} - x_k)^T * V_k^{-1} * (x^{ln} - x_k) \geq \gamma \tag{A14}$$
$$(x^m - x_k)^T * V_k^{-1} * (x^m - x_k) \geq \gamma \tag{A15}$$

A Generalized Fuzzy Clustering Model Based on Aggregation Operators and its Applications

Mika Sato[1] and Yoshiharu Sato[2]

[1] Hokkaido Musashi Women's Junior College,
Kita 22, Nishi 13, Kita-ku, Sapporo 001, Japan,
e-mail : mika@huie.hokudai.ac.jp
[2] Hokkaido University,
Kita 13, Nishi 8, Kita-ku, Sapporo 060, Japan,
e-mail : ysato@huie.hokudai.ac.jp

Abstract. This paper proposes a fuzzy clustering model which defines a generalized structural model of similarity between a pair of objects.
The structure of an observed similarity is usually unknown and complicated, and so various fuzzy clustering models are required to identify the latent structure of the similarity data. Therefore, we define the general class of fuzzy clustering models, so as to represent many different structures of a similarity data. In order to define the generalized fuzzy clustering model, we use aggregation operators for representing the degree of simultaneous belongingness of a pair of objects to a cluster, and define some required conditions for the operators. T-norms are examples to satisfy these conditions.
Moreover, asymmetric aggregation operators are proposed to apply asymmetric similarity data. The asymmetric operators are defined by using generator function of T-norms. The validity of this model is shown by investigating the characteristic feature of the model and numerical applications.

1 Introduction

The main purpose of unsupervised classification (clustering) of a set of objects is to detect *natural subgroups* (clusters) based on the dissimilarity (or similarity) between a pair of objects. Depending on the definition or interpretation of natural subgroups, many different algorithms have been proposed.

In usual clustering, namely a partition of a set of n objects into K mutually exclusive clusters, the state of clustering is expressed by an $n \times K$ matrix $U = (u_{ik})$, where $u_{ik} = 1$ if object i belongs to the cluster k, otherwise $u_{ik} = 0$. To ensure that the clusters are disjointed and non-empty, u_{ik} must satisfy the following conditions.

$$\sum_{k=1}^{K} u_{ik} = 1 \quad i = 1, \cdots, n, \tag{1}$$

$$u_{ik} \in \{0, 1\} \quad i = 1, \cdots, n; k = 1, \cdots, K. \tag{2}$$

Condition (1) means that for arbitrary i, u_{ik} must satisfy $\sum_{k=1}^{K} u_{ik} = 1$. However, in practical situations, there are many cases in which the exclusive clusters are not suitable for natural subgroups. Therefore, the notion of a fuzzy cluster has been proposed. The essential part of this idea is to replace condition (2) with

$$u_{ik} \in [0, 1] \quad i = 1, \cdots, n; k = 1, \cdots, K.$$

This means that the cluster (natural subgroup) is considered to be a fuzzy subset on a set of objects. That is, u_{ik} represents "the degree of belonging" of the i-th object to the k-th cluster. A pioneering work to apply the concept of fuzzy sets to a cluster analysis was made by E. Ruspini in 1969 [6]. Since the fuzzy k-means clustering algorithm was proposed by J.C. Bezdek and J.C. Dunn [1], [2], several methods of fuzzy clustering have developed rapidly and many applications have been suggested.

On the other hand, the clustering model in hard cluster analysis has been proposed in [9], which is intended to find the structure of the similarity between the pair of objects. In the case of the clustering model, a cluster is defined as the group whose elements share common properties. By introducing the concept of the fuzzy cluster into the clustering model, we can construct a natural fuzzy clustering model which is possible to interpret as the structure of similarity. A fuzzy grade shows the degree to which the object shares the common properties of each cluster. The essential merits of the fuzzy clustering models are 1) the amount of computations for the identification of the models are much fewer than a hard clustering model and 2) fewer number of clusters are needed to get a suitable fitness compared with the hard clustering model [7], [8].

2 Additive Clustering Model (ADCLUS)

The additive clustering model in hard cluster analysis is intended to find the structure of the similarity between a pair of objects. The model is denoted by the following:

$$s_{ij} \approx \hat{s}_{ij} = \sum_{k=1}^{K} w_k p_{ik} p_{jk},$$

where s_{ij} $(i, j = 1, 2, \cdots, n)$ is the similarity between objects i and j, and $\hat{s}_{ij}$ is an estimate of s_{ij}. K is the number of clusters, and w_k is a weight representing the salience of the property corresponding to cluster k. If object i has the property of cluster k, then $p_{ik} = 1$; if object i does not have the property of cluster k, then $p_{ik} = 0$. Notice that the product $p_{ik}p_{jk}$ is unity if and only if both objects i and j belong to cluster k and the similarity between the pair of objects is defined to be a common property of the objects. Moreover, if the pair of objects shares some common properties, the grade which the pair of objects contributes to the similarities is assumed to be mutually independent.

The algorithm to find the solution of ADCLUS is regarded as a combinatorial optimization problem. Therefore, it is difficult to obtain the global optimal solution of ADCLUS.

3 Generalized Fuzzy Clustering Model

Suppose that there exist K fuzzy clusters on a set of n objects, that is, the partition matrix $U = (u_{ik})$ is given. s_{ij} is a similarity between objects i and j, and ε_{ij} is the additive measurement error between the observed similarity s_{ij} and the model. Let $\rho(u_{ik}, u_{jl})$ be a function which denotes a degree of simultaneous belongingness of a pair of objects i and j to clusters k and l, namely, a degree of sharing common properties. Then a general model for the similarity s_{ij} is defined as follows:

$$s_{ij} = \varphi(\rho_{ij}) + \varepsilon_{ij}, \tag{3}$$

$$\rho_{ij} = (\rho(u_{i1}, u_{j1}), \cdots, \rho(u_{i1}, u_{jK}), \cdots\cdots, \rho(u_{iK}, u_{j1}), \cdots, \rho(u_{iK}, u_{jK})).$$

Where, φ is an arbitrary real valued function of feasible solutions u_{ik} and w_{kl}. We assume that if all of $\rho(u_{ik} u_{jl})$ is multiplied by α, then the similarity is also multiplied by α. Therefore, the function φ itself must satisfy the condition of "positively homogeneous of degree 1 in the ρ", that is,

$$\alpha\varphi(\rho_{ij}) = \varphi(\alpha\rho_{ij}), \quad \alpha > 0.$$

We consider the following functions as typical functions of φ:

$$s_{ij} = \{\sum_{k=1}^{K}\sum_{l=1}^{K} w_{kl}\rho^r(u_{ik}, u_{jl})\}^{\frac{1}{r}} + \varepsilon_{ij}, \quad 0 < r. \tag{4}$$

$$s_{ij} = \{\prod_{k=1}^{K}\prod_{l=1}^{K} w_{kl}\rho(u_{ik}, u_{jl})\}^{\frac{1}{K^2}} + \varepsilon_{ij}. \tag{5}$$

The function (5) is essentially the same as (4), because (5) is transformed as follows:

$$\log s_{ij} = \frac{1}{K^2}\sum_{k=1}^{K}\sum_{l=1}^{K} \log w_{kl}\rho(u_{ik}, u_{jl}) + \varepsilon_{ij}.$$

Therefore, we will deal with only (4) ($r = 1$) hereafter.

The degree ρ is assumed to satisfy the following conditions:

1. **Boundary conditions**

$$0 \leq \rho(u_{ik}, u_{jl}) \leq 1, \quad \rho(u_{ik}, 0) = 0, \quad \rho(u_{ik}, 1) = u_{ik}. \tag{6}$$

2. **Monotonicity**

$$\rho(u_{ik}, u_{jl}) \leq \rho(u_{sk}, u_{tl}), \text{ whenever } u_{ik} \leq u_{sk}, u_{jl} \leq u_{tl}. \tag{7}$$

3. **Symmetry**

$$\rho(u_{ik}, u_{jl}) = \rho(u_{jl}, u_{ik}). \tag{8}$$

The first condition means that if one object j belongs completely to cluster k, then the degree of simultaneous belongingness to cluster k equals u_{ik}, and if one object j does not belong to cluster k, then ρ is 0. The second condition shows that the greater the degree of belongingness of objects s and t to clusters k and l, the greater the degree of simultaneous belongingness of the pair of objects s and t. The third condition is that the degree of simultaneous belongingness of objects i and j is equivalent to the degree of objects j and i. In particular, both T-norms [11] and Yager's operator [12] satisfy the above conditions.

In the forgoing sections 3.1, 3.2, and 3.3, we will describe the examples of model (4) using representative aggregation operators so as to present this model move clearly. Moreover, the model corresponding to each aggregation needs individual algorithms.

3.1 Algebraic Product

In the case of $\rho(u_{ik}, u_{jl}) = u_{ik} u_{jl}$, (4) is represented as follows:

$$s_{ij} = \sum_{k=1}^{K} \sum_{l=1}^{K} w_{kl} u_{ik} u_{jl} + \varepsilon_{ij}, \qquad (9)$$

where u_{ik} is a fuzzy grade which represents the degree of belongingness of object i to cluster k, and satisfies the following condition:

$$u_{ik} \geq 0, \quad \sum_{k=1}^{K} u_{ik} = 1. \qquad (10)$$

In this case, we introduce the following transformation, to avoid the constrained optimization problem:

$$\begin{pmatrix} u_{i1} \\ u_{i2} \\ u_{i3} \\ \vdots \\ u_{iK-1} \\ u_{iK} \end{pmatrix} = \begin{pmatrix} \cos^2 \theta_{i1} \\ \sin^2 \theta_{i1} \cos^2 \theta_{i2} \\ \sin^2 \theta_{i1} \sin^2 \theta_{i2} \cos^2 \theta_{i3} \\ \vdots \\ \sin^2 \theta_{i1} \sin^2 \theta_{i2} \cdots \cos^2 \theta_{iK-1} \\ \sin^2 \theta_{i1} \sin^2 \theta_{i2} \cdots \sin^2 \theta_{iK-1} \end{pmatrix}.$$

From this, we know that condition (10) is satisfied for any $\theta_{ik} \in [0, 2\pi]$, that is,

$$\sum_{k=1}^{K} u_{ik}(\theta_{i1}, \theta_{i2}, \cdots, \theta_{iK-1}) = 1, \quad u_{ik}(\theta_{i1}, \theta_{i2}, \cdots, \theta_{iK-1}) \geq 0.$$

In (9), the weight w_{kl} is considered to be a quantity which shows the asymmetric similarity between the pair of clusters. That is, we assume that the asymmetry of the similarity between the objects is caused by the asymmetry of the similarity between the clusters. If $k = l$, $u_{ik}, u_{jl} = 1$ and $w_{kl} > 1$, then the right hand side of (9) clearly exceeds 1.0. Therefore, we need at least the following condition

$$0 \leq w_{kl} \leq 1, \qquad (11)$$

because $0 \leq s_{ij} \leq 1$. We avoid the constrained problem by the following transformation

$$w_{kl} = \frac{1}{1 + e^{-b_{kl}}},$$

where $-\infty \leq b_{kl} \leq \infty$, and notice that w_{kl} does not attain 0 or 1, but only approximately.

The method of the fuzzy clustering based on this model is to find partition matrix $U = (u_{ik})$ as well as $W = (w_{kl})$, which respectively satisfy conditions (10) and (11), and which have the best fitness for model (9). Then we find U and W, which minimize the following sum of square error η^2 under conditions (10) and (11),

$$\eta^2 = \frac{\sum_{i \neq j=1}^{n}(s_{ij} - \sum_{k=1}^{K}\sum_{l=1}^{K}u_{ik}u_{jl})^2}{\sum_{i \neq j=1}^{n}(s_{ij} - \bar{s})^2}, \quad \bar{s} = \sum_{i \neq j=1}^{n}\frac{s_{ij}}{n(n-1)}.$$

The details of this model will be found in the literatures by Sato, *et al.* [7], [8].

3.2 Minimum Operator

Let $\rho(u_{ik}, u_{jl})$ be $\min(u_{ik}, u_{jl})$, then (4) is

$$s_{ij} = \sum_{k=1}^{K}\sum_{l=1}^{K} w_{kl} \min(u_{ik}, u_{jl}) + \varepsilon_{ij}.$$

The purpose is to find u_{ik} and w_{kl} which minimize the following sum of the squared errors $J_{\min}$:

$$J_{\min} = \sum_{i \neq j=1}^{n}(s_{ij} - \sum_{k=1}^{K}\sum_{l=1}^{K}\min(u_{ik}, u_{jl}))^2 = \sum_{i \neq j=1}^{n}\varepsilon_{ij}^2. \quad (12)$$

If we denote

$$F_{ij} = \sum_{k=1}^{K}\sum_{l=1}^{K} w_{kl} \min(u_{ik}, u_{jl})$$
$$= \sum_{k=1}^{K}\sum_{l=1}^{K} w_{kl} \frac{1}{2}\{(u_{ik} + u_{jl}) - |u_{ik} - u_{jl}|\},$$

then (12) is $J_{\min} = \sum_{i \neq j=1}^{n}(s_{ij} - F_{ij})^2 = \sum_{i \neq j=1}^{n}\varepsilon_{ij}^2.$

In this minimization problem, the descent vector is determined as follows:

$$\frac{\partial J_{\min}}{\partial u_{ar}} = -\sum_{i \neq a=1}^{n}\sum_{k=1}^{K}(s_{ia} - \frac{1}{2}w_{kr}\{(u_{ik} + u_{ar})$$
$$- |u_{ik} - u_{ar}|\})(w_{kr} - \text{sign}^+|u_{ik} - u_{ar}|)$$
$$- \sum_{j \neq a=1}^{n}\sum_{l=1}^{K}(s_{aj} - \frac{1}{2}w_{rl}\{(u_{ar} + u_{jl})$$
$$- |u_{ar} - u_{jl}|\})(w_{rl} - \text{sign}^+|u_{ar} - u_{jl}|),$$
$$a = 1, \cdots, n, \quad r = 1, \cdots, K,$$

where

$$\text{sign}^+(u_{ar} - u_{jr}) = \begin{cases} 1 & (u_{ar} - u_{jr} \geq 0) \\ -1 & (u_{ar} - u_{jr} < 0) \end{cases}.$$

$$\frac{\partial J_{\min}}{\partial w_{st}}$$
$$= 2\sum_{i\neq j=1}^{n}\{s_{ij} - \sum_{k,l=1}^{K} w_{kl}\min(u_{ik},u_{jl})\}\min(u_{is},u_{jt})$$
$$= \sum_{i\neq j=1}^{n} s_{ij}\min(u_{is},u_{jt}) \quad (13)$$
$$- \sum_{k,l=1}^{K}\{\sum_{i\neq j=1}^{n} s_{ij}\min(u_{ik},u_{jl})\min(u_{is},u_{jt})\}w_{kl}$$
$$= 0.$$

In order to get matrix representation of (13), we put

$$c_{st} = \sum_{i\neq j=1}^{n} s_{ij}\min(u_{is},u_{jt}),$$

$$g_{(st)(kl)} = \sum_{k,l=1}^{K}\{\sum_{i\neq j=1}^{n} s_{ij}\min(u_{ik},u_{jl})\min(u_{is},u_{jt})\},$$

$$C = (c_{st}), \quad G_{(st)(kl)} = (g_{(st)(kl)}), \quad W = (w_{kl}),$$

where C is $K \times K$-matrix, G is $K^2 \times K^2$-matrix and W is $K \times K$-matrix. Then (13) is

$$G \times \text{vec}(W) = \text{vec}(C),$$

where vec($\cdot$) denotes vec-operator. Therefore, we get W as the following solution:

$$\text{vec}(W) = G^+ \times \text{vec}(C),$$

where G^+ is the the Moore-Penrose generalized inverse matrix. Gradient method is used to find the optimal solutions.

3.3 Hamacher Product and Einstein Product

The operator which is defined by

$$\rho(u_{ik}, u_{jl}) = (u_{ik}u_{jl})/(p + (1-p)(u_{ik} + u_{jl} - u_{ik}u_{jl})) \quad (14)$$

is called the Hamacher Product [3]. Using this operator, (4) is represented by

$$s_{ij} = \sum_{k,l=1}^{K} w_{kl}(u_{ik}u_{jl})/(p + (1-p)(u_{ik} + u_{jl} - u_{ik}u_{jl})) + \varepsilon_{ij}.$$

In (14), if $p = 1.0$, then the usual algebraic product is obtained. When $\rho(u_{ik}, u_{jl}) = (u_{ik}u_{jl})/\{2 - (u_{ik} + u_{jl} - u_{ik}u_{jl})\}$, this operator is called the Einstein product [10], and (4) is denoted as

$$s_{ij} = \sum_{k=1}^{K}\sum_{l=1}^{K} w_{kl}(u_{ik}u_{jl})/\{2 - (u_{ik} + u_{jl} - u_{ik}u_{jl})\} + \varepsilon_{ij}.$$

Gradient method is used to find the optimal solutions. The algorithm for finding the solutions u_{ik} and w_{kl} is almost the same as the case in section 3.1, which will be found in literatures [7], [8].

4 The Feature of Aggregation Operators

If aggregation operators are monotonic to each other, the results of the proposed model are almost the same. In this case, there would be no reason to use the other monotonic aggregation operator. Therefore, we will discuss the monotonicity of the aggregation operator.

Suppose Π is the direct product space of $[0, 1] \times [0, 1]$, and Φ is a function family that satisfies conditions (6), (7), and (8). For $\forall (u_a, u_b), \forall (u_c, u_d) \in \Pi, \forall \rho_1, \forall \rho_2 \in \Phi$, if

$$\rho_1(u_a, u_b) > \rho_1(u_c, u_d) \Rightarrow \rho_2(u_a, u_b) > \rho_2(u_c, u_d), \tag{15}$$

then we define the aggregation operator ρ which satisfies conditions (6), (7), and (8) as monotone. Figures 1~4 are surfaces of $\rho(x, y)$ with respect to bivariate x, y. In these figures, surfaces are convex with respect to plane $x = y$, and concave with respect to plane $x + y = 1$, but there is a difference of degree of unevenness. Figures 5~8 show values of ρ_1 and ρ_2 with respect to 100 points (x, y), where

$$\{(x, y) |\ (0.0, 0.0), (0.0, 0.1), \cdots, (0.0, 1.0),$$
$$(0.1, 0.0), (0.1, 0.1), \cdots, (0.1, 1.0),$$
$$\cdots \cdots$$
$$(1.0, 0.0), (1.0, 0.1), \cdots, (1.0, 1.0)\}.$$

Figure 5 shows (ρ_1, ρ_2), if ρ_1 is Minimum and ρ_2 is each Algebraic product, Einstein product, and Hamacher product ($p = 0.1$). Figures 6~8 show the results when we assume that ρ_1 is each Algebraic product, Einstein product, Hamacher product. From Figures 5~8, ρ does not satisfy the condition (15), and clearly it is not monotone.

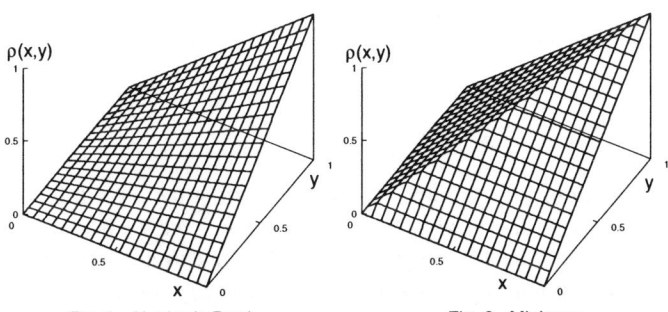

Fig. 1. Algebraic Prod.   Fig. 2. Minimum

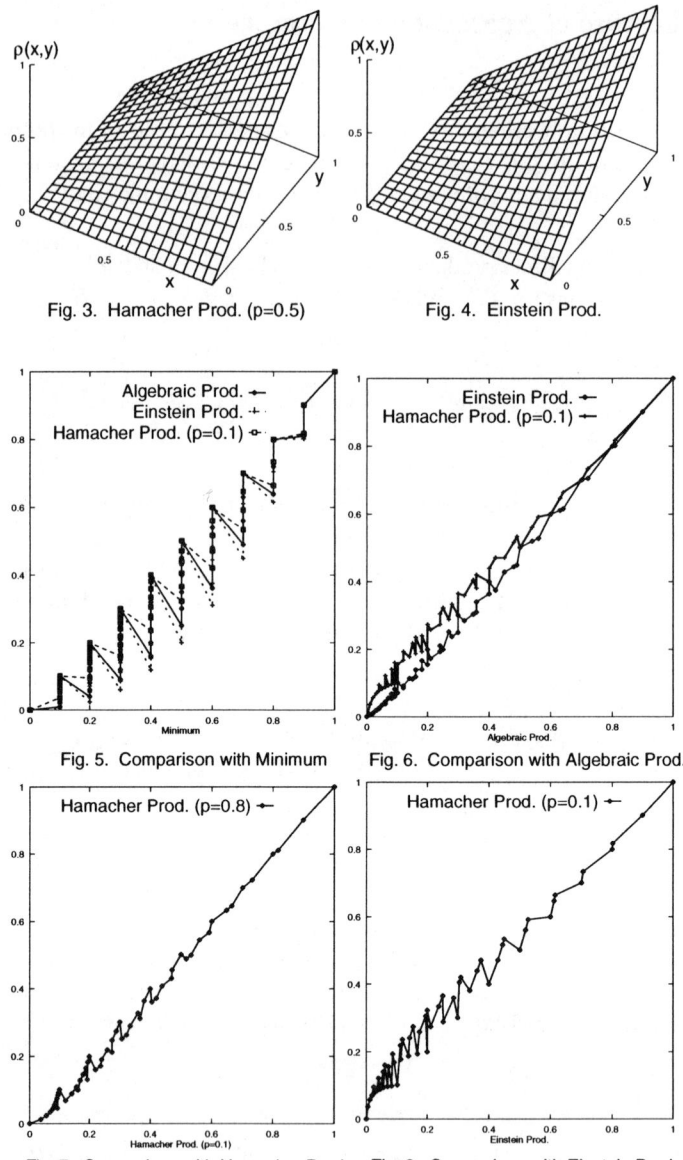

Fig. 3. Hamacher Prod. (p=0.5)

Fig. 4. Einstein Prod.

Fig. 5. Comparison with Minimum

Fig. 6. Comparison with Algebraic Prod.

Fig. 7. Comparison with Hamacher Prod.

Fig. 8. Comparison with Einstein Prod.

5 Robustness

In a model fitting problem, it is important whether or not the algorithm of the fitting is able to reproduce the structure when the observed data have the actual structure of the model. And whether the small change of the data has an extreme influence on the result or not is also an important problem. This is the robustness of the model. By the Monte Carlo simulation, we shall investigate this problem. The method of simulation for the robustness is shown as follows:

(step 1) Using uniform pseudorandom numbers for the interval $[0, \frac{\pi}{2}]$, $\theta_{ik}, (i = 1, \cdots, n, k = 1, \cdots, K-1)$ is generated. Calculate $u_{ik} = u_{ik}(\theta_{i1}, \theta_{i2}, \cdots, \theta_{iK-1})$ and $s_{ij} = \sum_{k=1}^{K} \rho(u_{ik}, u_{jk})$.

(step 2) Generate the error Δs_{ij} using pseudorandom numbers, and calculate the following:

$$s_{ij}^* = s_{ij} + \Delta s_{ij}.$$

(step 3) Regarding s_{ij}^* as a given similarity data, estimate the u_{ik}^* of each fuzzy clustering model using the initial matrix $u_{ik}^{(0)} = u_{ik}(\theta_{i1} + \Delta\theta_{i1}, \theta_{i2} + \Delta\theta_{i2}, \cdots, \theta_{iK-1} + \Delta\theta_{iK-1})$, where $\Delta\theta_{ik}$ is generated by uniform pseudorandom numbers for the interval $[0, \frac{\pi}{2}]$.

(step 4) Using s_{ij}, s_{ij}^* and u_{ik}, u_{ik}^*, calculate,

$$\varepsilon_s^2 \equiv \frac{\sum_{i \neq j=1}^{n}(s_{ij} - s_{ij}^*)^2}{\sum_{i \neq j=1}^{n}(s_{ij} - \bar{s})^2}, \quad \bar{s} = \frac{1}{n(n-1)}\sum_{i \neq j=1}^{n} s_{ij}. \quad (16)$$

$$\varepsilon_u^2 \equiv \frac{\sum_{i=1}^{n}\sum_{k=1}^{K}(u_{ik} - u_{ik}^*)^2}{\sum_{i=1}^{n}\sum_{k=1}^{K}(u_{ij} - \bar{u})^2}, \quad \bar{u} = \frac{1}{nK}\sum_{i=1}^{n}\sum_{k=1}^{K} u_{ik}. \quad (17)$$

The above simulation is repeated 60 times as the number of observations $n = 15$. The results are shown by Fig. 9 ~ Fig. 14. In these figures, the abscissa shows the normalized error of similarity (16) and the ordinate shows the normalized error of grade (17). From the results of these simulations, we can find the robustness of the models, because when the similarity has moderate error, the errors of the results are within the error of the similarity data. In particular, the best result is obtained by using a min-operator in this model. We can theorize that the reason depends on the region of the solution, because the region of min-operator is larger than that of other operators. Furthermore, increasing the value of p makes the normalized error of grade larger. Therefore, we find that the Hamacher product is better than the usual algebraic product in this case.

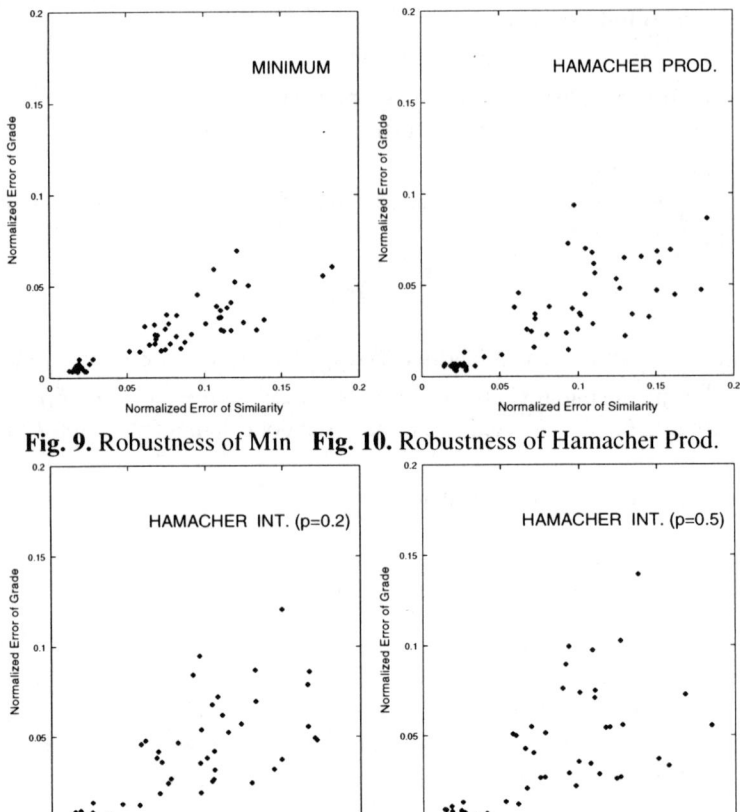

Fig. 9. Robustness of Min **Fig. 10.** Robustness of Hamacher Prod.

Fig. 11. Robustness of Hamacher Int. **Fig. 12.** Robustness of Hamacher Int.
(p=0.2) (p=0.5)

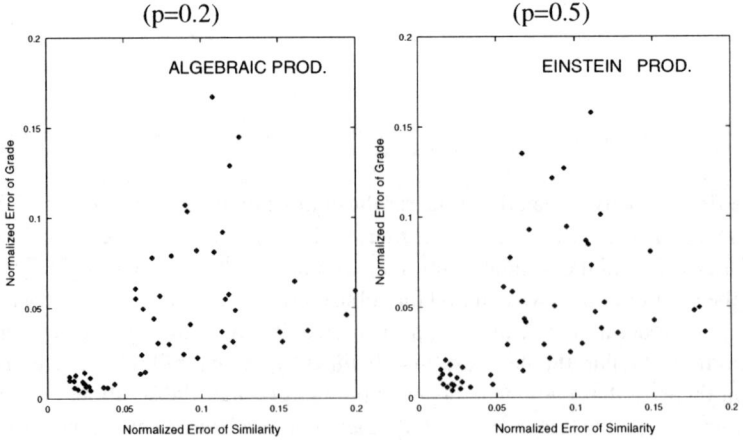

Fig. 13. Robustness of Algebraic Prod. **Fig. 14.** Robustness of Einstein Prod.

6 Asymmetric Aggregation Operator

We will propose a new concept of aggregation operators, which we call asymmetric aggregation operators. In this case, we have to create new aggregation operators which satisfy the following conditions: Boundary conditions, Monotonicity, and Asymmetry.

Suppose $f(x)$ is a generator function of T-norms, which is a continuous monotone decreasing function under these conditions:

$$f : [0,1] \to [0, \infty], \quad f(1) = 0.$$

$\phi(x)$ is also a continuous monotone decreasing function satisfying

$$\phi : [0,1] \to [0, \infty], \quad \phi(1) = 1.$$

Then we define the following $\gamma(x, y)$ as the asymmetric aggregation operators:

$$\gamma(x, y) = f^{[-1]}(f(x) + \phi(x)f(y)).$$

Where

$$f^{[-1]}(z) = \begin{cases} f^{-1}(z), & z \in [0, f(0)) \\ 0, & z \in [f(0), \infty] \end{cases}$$

$f^{[-1]}$ is the pseudo inverse of f, and f^{-1} is the usual inverse of f. From the definition of T-norms,

$$t(x, y) = f^{[-1]}(f(x) + f(y)),$$

We can easily find that T-norms satisfy the condition of symmetry, $t(x, y) = t(y, x)$.

Using the generator function of the Hamacher product, i.e. $f(x) = \dfrac{1-x}{x}$ and the monotone decreasing function $\phi(x) = \dfrac{1}{x^2}$, the asymmetric aggregation operator is defined as

$$\gamma(x, y) = \frac{x^2 y}{1 - y + xy},$$

which is shown in Fig. 15. In Fig. 16, the dotted curve shows the intersecting curve of the surface and the plane $x = y$, and the solid curve is the intersection with $x + y = 1$. From the solid curve, we find the asymmetry of the proposed aggregation operator. Figure 17 shows the asymmetric aggregation operator defined as

$$\gamma(x, y) = \frac{xy}{y + x(2 - x)(1 - y)},$$

where the generator function is the function of the Hamacher product, i.e. $f(x) = \dfrac{1-x}{x}$, and the monotone decreasing function is $\phi(x) = 2 - x$. Figure 18 shows intersecting curves with $x = y$ and $x + y = 1$. In the case of the generator function of the Algebraic product, i.e. $f(x) = -\log x$ and the monotone decreasing function $\phi(x) = 2 - x$ (shown in Fig. 19), the asymmetric aggregation operator is defined as

$$\gamma(x, y) = xy^{(2-x)}.$$

Intersecting curves are shown in Fig. 20.

Using the asymmetric aggregation operators, we define the model for asymmetric similarity data as follows:

$$s_{ij} = \sum_{k=1}^{K} \gamma(u_{ik}, u_{jl}) + \varepsilon_{ij}. \tag{18}$$

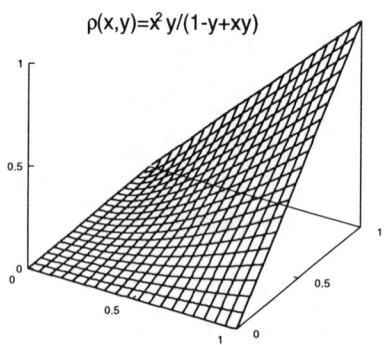

Fig. 15. Asymmetric Aggregation Operator

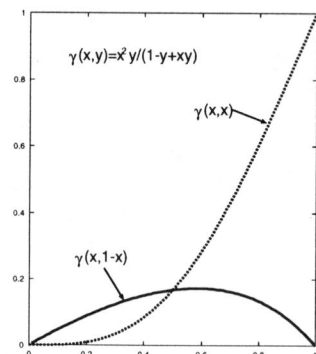

Fig. 16. Intersecting Curves

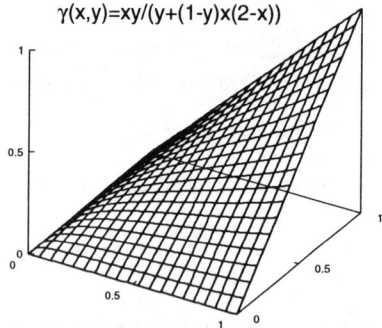

Fig. 17. Asymmetric Aggregation Operator

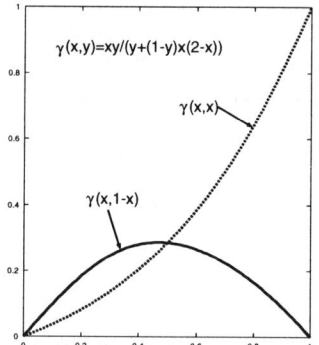

Fig. 18. Intersecting Curves

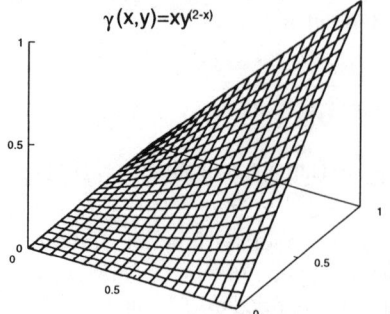

Fig. 19. Asymmetric Aggregation Operator

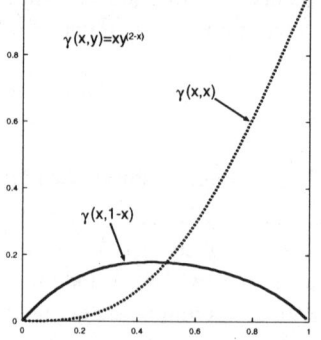

Fig. 20. Intersecting Curves

7 Example

To demonstrate the applications of models (9) and (18), we will use data which shows a number of high school students who moved from one prefecture to another for matriculation [13]. By a suitable normalization, standardizing by the total number of students, we can regard this as a similarity data. In the optimization algorithm used in this example, 20 sets of initial values are given by using uniform pseudorandom numbers in the interval $[0, \frac{\pi}{2}]$, and finally we select the best result. The number of clusters is determined based on the value of fitness. By increasing the number of clusters, the value of fitness decreases, but even if the number of clusters is greater than 4, the decrease of the fitness is not severe. From the principle of parsimony, it should be considered that the number of clusters is determined to be 4. The results of (9) using the minimum operator are shown in Fig. 21, Table 1, and Table 2. In Table 1, C_1, C_2, C_3, C_4 represent the four clusters, and the value shows the similarity between a pair of clusters. η^2 shows the value of fitness. Table 2 shows the degree of belongingness of the prefectures to clusters. In Fig. 21, monotone gradation shows the degree of belonging of a prefecture to a cluster. A darker tone means a larger grade. With the results, we find that geographical distance is closely connected with the moving students. This is shown by clusters C_1, C_2, and C_3. C_1 is the southern part, C_2 is located in the middle part, and C_3 is the northern part in Japan. These clusters seem to be related to cultural features. Cluster C_4 represents the prefectures which have large cities, and they have a large capacity for university students. Therefore, these cities become an influx point of the students. In particular, Tokyo is an area where students move to from all over Japan, so Tokyo is an isolated point in this data.

Moreover, Fig. 22 and Table 3 are results by (18) using the asymmetric aggregation operator defined in Fig. 15 and Fig. 16 ($\gamma(x, y) = x^2 y / 1 - y + xy$). In this case, the fitness is 0.014. From these results, we can find that clusters show small areas in which prefectures maintain communication with each other. In Fig. 22, clusters C_1, C_2 and C_3 demonstrate remarkable asymmetry between neighboring prefectures and remote prefectures. On the other hand, Cluster C_4 is the group in which Tokyo is the center, and the asymmetry is not so noticeable. This feature is also shown in Table 1.

Table 1 Asymmetric matrix

cluster	C_1	C_2	C_3	C_4
C_1	1.00	0.01	0.00	0.99
C_2	0.00	1.00	0.00	0.99
C_3	0.00	0.16	1.00	0.99
C_4	0.33	0.22	0.11	1.00

$$\eta^2 = 0.1866$$

Table 2 Fuzzy Grade (symmetric operator)

No. of Prefecture	C_1	C_2	C_3	C_4
Hokkaido	.67	.13	.08	.12
Aomori	.86	.06	.08	.00
Iwate	.86	.04	.10	.00
Miyagi	.79	.00	.08	.13
Akita	.88	.06	.07	.00
Yamagata	.83	.08	.09	.00
Fukushima	.81	.01	.11	.06
Ibaragi	.61	.00	.28	.11
Tochigi	.69	.03	.24	.04
Gunma	.68	.14	.17	.02
Saitama	.56	.00	.12	.31
Chiba	.57	.00	.12	.30
Tokyo	.25	.25	.25	.25
Kanagawa	.45	.00	.12	.43
Niigata	.73	.22	.03	.02
Toyama	.56	.44	.00	.00
Ishikawa	.50	.47	.01	.03
Fukui	.38	.62	.00	.00
Yamanashi	.60	.18	.20	.02
Nagano	.59	.36	.04	.01
Gifu	.35	.61	.03	.00
Shizuoka	.48	.30	.19	.03
Aichi	.28	.48	.01	.23
Miye	.28	.69	.03	.00
Shiga	.12	.78	.09	.00
Kyoto	.05	.55	.00	.40
Osaka	.00	.58	.00	.42
Hyogo	.00	.67	.08	.24
Nara	.01	.81	.07	.11
Wakayama	.16	.84	.00	.00
Tottori	.02	.66	.31	.00
Shimane	.07	.59	.34	.00
Okayama	.00	.53	.39	.08
Hiroshima	.00	.45	.46	.09
Yamaguchi	.00	.43	.54	.03
Tokushima	.02	.55	.43	.00
Kagawa	.01	.55	.44	.00
Ehime	.04	.46	.50	.00
Kochi	.01	.50	.49	.00
Fukuoka	.00	.22	.51	.27
Saga	.03	.25	.72	.00
Nagasaki	.11	.26	.63	.00
Kumamoto	.05	.22	.67	.06
Oita	.06	.28	.62	.05
Miyazaki	.15	.22	.63	.00
Kagoshima	.15	.19	.61	.06
Okinawa	.22	.14	.64	.00

Table 3 Fuzzy Grade (asymmetric operator)

No. of Prefecture	C_1	C_2	C_3	C_4
Hokkaido	.45	.18	.10	.26
Aomori	.56	.10	.04	.30
Iwate	.59	.09	.03	.29
Miyagi	.77	.01	.00	.22
Akita	.49	.13	.06	.32
Yamagata	.51	.11	.05	.33
Fukushima	.42	.10	.05	.43
Ibaragi	.20	.15	.09	.56
Tochigi	.24	.16	.09	.51
Gunma	.20	.16	.10	.54
Saitama	.03	.07	.04	.86
Chiba	.09	.10	.06	.75
Tokyo	.00	.00	.00	1.0
Kanagawa	.00	.01	.01	.98
Niigata	.25	.19	.08	.48
Toyama	.19	.37	.13	.31
Ishikawa	.15	.40	.17	.27
Fukui	.16	.38	.23	.22
Yamanashi	.16	.19	.11	.54
Nagano	.13	.26	.10	.51
Gifu	.07	.72	.07	.14
Shizuoka	.13	.33	.11	.44
Aichi	.00	.98	.00	.02
Miye	.08	.59	.17	.15
Shiga	.14	.28	.47	.11
Kyoto	.02	.17	.75	.06
Osaka	.00	.00	1.0	.00
Hyogo	.10	.16	.64	.10
Nara	.10	.21	.57	.12
Wakayama	.16	.19	.49	.16
Tottori	.20	.22	.34	.25
Shimane	.21	.22	.34	.23
Okayama	.20	.20	.42	.18
Hiroshima	.18	.18	.46	.18
Yamaguchi	.17	.17	.45	.20
Tokushima	.22	.22	.36	.19
Kagawa	.19	.22	.39	.20
Ehime	.24	.21	.36	.20
Kochi	.18	.21	.35	.26
Fukuoka	.00	.00	1.0	.00
Saga	.20	.14	.50	.16
Nagasaki	.22	.16	.43	.19
Kumamoto	.23	.17	.39	.20
Oita	.20	.16	.44	.20
Miyazaki	.23	.18	.36	.23
Kagoshima	.22	.19	.35	.24
Okinawa	.30	.25	.23	.21

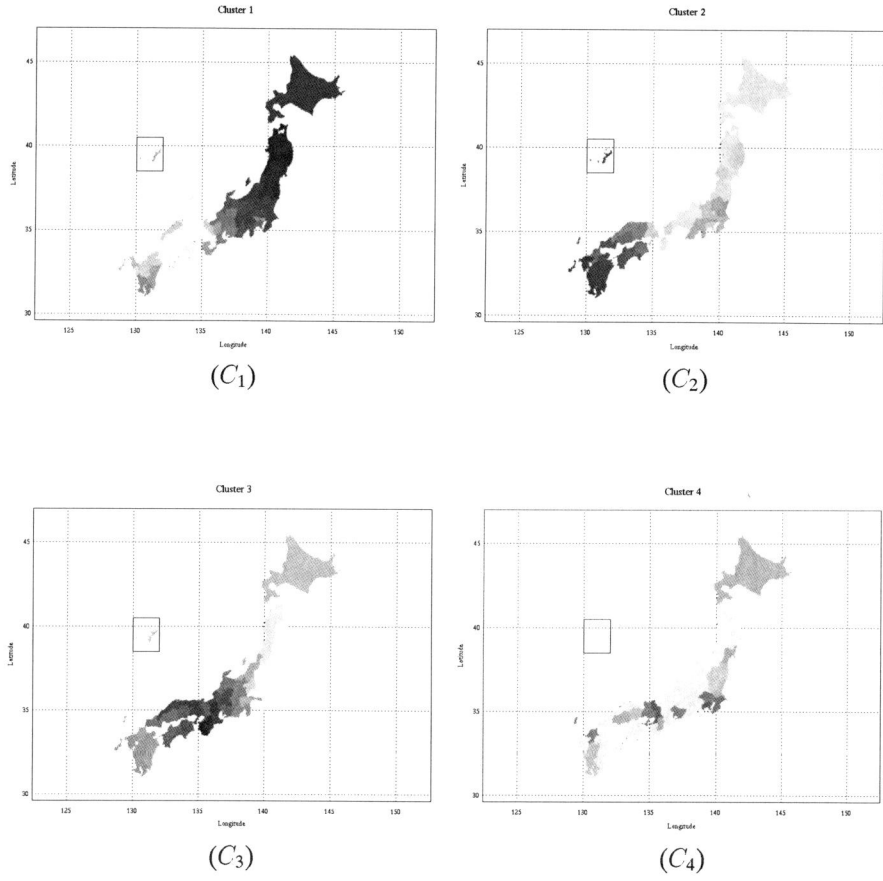

Fig. 21. Fuzzy Clustering

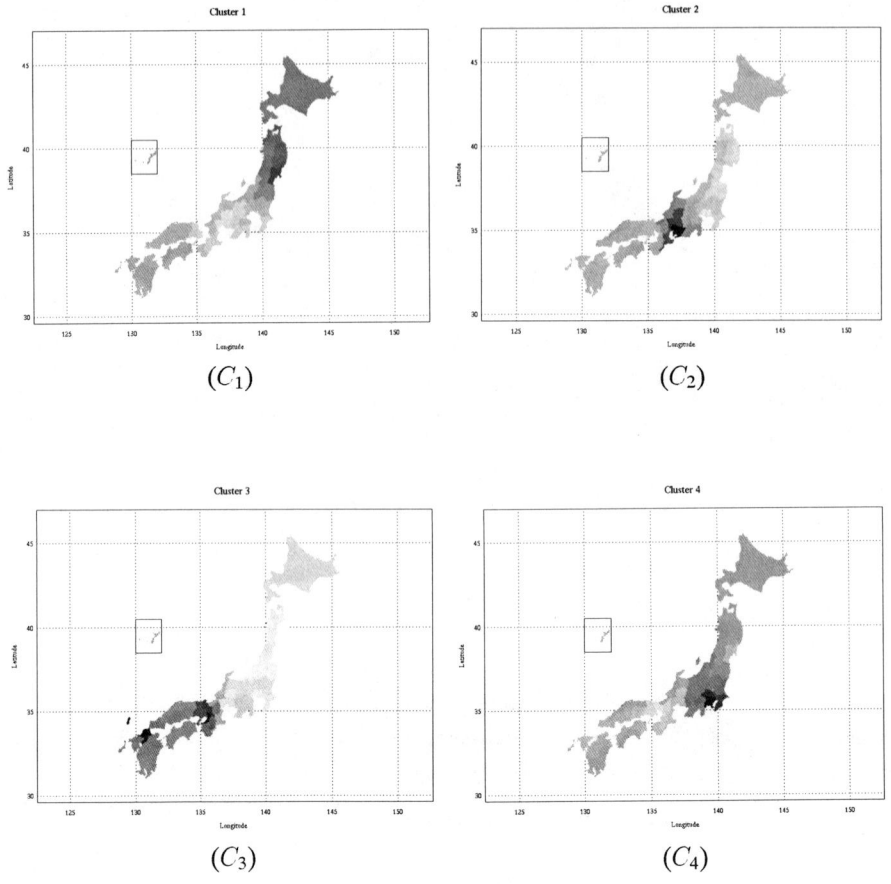

Fig. 22. Fuzzy Clustering

8 Conclusion

In the clustering model, how we define the degree of simultaneous belongingness of a pair of objects to a cluster, which means the degree of the sharing of common properties is an important problem. In order to represent the degree of belongingness, we use aggregation operators and define some required conditions for the operators. T-norm is an example to satisfy the conditions, but there is a difference of the presence of associativity. That is, the operators proposed in the model do not need to assume a condition of associativity, because the degree of belongingness is related to a pair of objects and a relationship among three objects is no consideration in the proposed model.

Moreover, clustering techniques based on asymmetric similarity have recently become common interests in several fields. In conventional clustering methods, Gordon [4] proposed a method using only the symmetric part of the data, and this method was based on the idea in which the asymmetry of the given similarity data can be regarded as errors of the symmetric similarity data, that is,

$$\tilde{S} = \frac{1}{2}(S + S'),$$

where S is an similarity matrix and S' is the transposed matrix of S. As for the data, $\tilde{S}$ was used.

Hubert [5] proposed the method to select a maximum element in the corresponding elements, that is,

$$\tilde{s}_{ij} = \max(s_{ij}, s_{ji}), \quad \tilde{s}_{ij} = \min(s_{ij}, s_{ji}).$$

However, there exist needs to analyze asymmetric data holding the asymmetry.

This paper proposes two methods for asymmetric similarity data, 1) the clustering method using asymmetric similarity between clusters, and 2) the clustering method using asymmetric aggregation operators which is defined by generator function of T-norms. As for the result of 1), we can obtain clusters in which objects are similar with each other, and we can find the asymmetry of the similarity data by the similarity between clusters. On the other hand, the result of 2) shows clusters in which objects are similar including the asymmetric structure, since the aggregation operator is asymmetric.

Moreover, the robustness of this model is confirmed by simulating, and the best robustness is obtained by using a minimum operator in the model. Some numerical applications show the validity of this model.

Acknowledgments

This research was supported in Grand-in-Aid for Encouragement of Young Scientists by the Japanese Ministry of Education.

References

1 Bezdek, J.C.: *Pattern Recognition with Fuzzy Objective Function Algorithms*. Plenum Press. (1987)
2 Dunn, J.C.: A Fuzzy Relative of the ISODATA Process and Its Use in Detecting Compact Well-Separated Clusters. *J. Cybernetics*. **3** No.3 (1973) 32-57
3 Fullér, R.: On Hamacher-sum of Triangular Fuzzy Numbers. *Fuzzy Sets and Systems*. **42** (1991) 205-212
4 Gordon, A.D.: A Review of Hierarchical Classification. *Journal of the Royal Statistical Society, Series A, 150*. (1987) 119-137
5 Hubert, L.: Min and Max Hierarchical Clustering Using Asymmetric Similarity Measures. *Psychometrika*. **38** (1973) 63-72
6 Ruspini, E.H.: A New Approach to Clustering. *Inform. Control.*. **15** No.1(1969) 22-32
7 Sato, M., Sato, Y.: An Additive Fuzzy Clustering Model. *Journal of Japan Society for Fuzzy Theory and Systems*. **6** (1994) 319-332
8 Sato, M., Sato, Y.: Extended Fuzzy Clustering Models For Asymmetric Similarity. Fuzzy Logic and Soft Computing. World Scientific. (1995) 228-237
9 Shepard, R.N., Arabie, P.: Additive Clustering: Representation of Similarities as Combinations of Discrete Overlapping Properties. *Psychological Review*. **86** (1979) 87-123
10 Smith, M. H.: Evaluation of Performance and Robustness of a Parallel Dynamic Switching Fuzzy System. *2nd Inter. Workshop on Industrial Fuzzy Control and Intelligent Systems*. (1992) 163-172
11 Weber, S.: A General Concept of Fuzzy Connectives, Negations and Implications Based on T-Norms and T-Conorms. *Fuzzy Sets and Systems*. **11** (1983) 115-134
12 Yager, R. R.: On A General Class of Fuzzy Connectives, *Fuzzy Sets and Systems*. **4** (1980) 235-242.
13 School Basic Survey. Ministry of Education in Japan. (1992)

Druck: betz-druck, D-64291 Darmstadt
Bindung: Großbuchbinderei J. Schäffer, D-67269 Grünstadt